ESO ASTROPHYSICS SYMPOSIA
European Southern Observatory

Series Editor: Bruno Leibundgut

Springer

Berlin
Heidelberg
New York
Hong Kong
London
Milan
Paris
Tokyo

Physics and Astronomy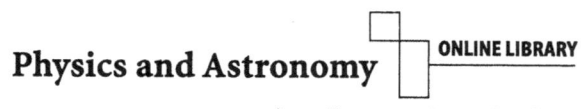

http://www.springer.de/phys/

ESO ASTROPHYSICS SYMPOSIA
European Southern Observatory

Series Editor: Bruno Leibundgut

Series homepage – http://www.springer.de/phys/books/eso/

Ralf Bender Alvio Renzini (Eds.)

The Mass
of Galaxies at Low
and High Redshift

Proceedings of the European Southern
Observatory and Universitäts-Sternwarte
München Workshop Held in Venice, Italy,
24-26 October 2001

Springer

Volume Editors

Ralf Bender
Universitäts-Sternwarte
Ludwig-Maximilians-Universität
Scheinerstrasse 1
81679 München, Germany

Alvio Renzini
European Southern Observatory
Karl-Schwarzschild-Strasse 2
85748 Garching, Germany

Series Editor

Bruno Leibundgut
European Southern Observatory
Karl-Schwarzschild-Strasse 2
85748 Garching, Germany

Cataloging-in-Publication Data applied for

A catalog record for this book is available from the Library of Congress.

Bibliographic information published by Die Deutsche Bibliothek
Die Deutsche Bibliothek lists this publication in the Deutsche Nationalbibliografie;
detailed bibliographic data is available in the Internet at http://dnb.ddb.de

ISBN 3-540-00205-7 Springer-Verlag Berlin Heidelberg New York

Springer-Verlag Berlin Heidelberg New York
a member of BertelsmannSpringer Science+Business Media GmbH

http://www.springer.de

© Springer-Verlag Berlin Heidelberg 2003
Printed in Germany

Typesetting: Camera-ready by the authors/editors
Cover design: Erich Kirchner, Heidelberg

Printed on acid-free paper   55/3141/du - 5 4 3 2 1 0

Preface

Measuring the masses of galaxies as a function of redshift is perhaps one of the most challenging open issues in current astronomical research. The evolution of the baryonic and dark matter components of galaxies is not only a critical test of the hierarchical formation paradigm, but ultimately also gives us new clues on the complex interplay between star formation, the cooling and heating of gas and galaxy merging processes.

To discuss this hot topic of modern cosmology, 117 scientists from 13 countries met for a 3-day workshop hosted by the Venice International University (VIU), from October 24 to 26, 2001. VIU is singularly located on the S. Servolo Island in the Venetian Lagoon, offering a peculiar point of view to Venice itself. The island, formerly a monastery, then an hospital for some 150 years, was deserted for more than 20 years until its many buildings and park have been fully restored to host the VIU. VIU is an association of six universities, namely the Venetian universities of Cà Foscari and Architecture, the Ludwig-Maximilians-Universität of Munich, the Universidad Autonoma de Barcelona, the Duke University at Durham, North Carolina, and the Tel Aviv University.

The program started with a review of current techniques to measure the baryonic (stellar) and dark masses of nearby galaxies, and then focussed on ongoing attempts at measuring these same quantities in galaxies at higher and higher redshifts. The last part of the meeting was devoted to future perspectives, with special emphasis on new survey projects and satellite missions. Substantial time was devoted to discussions, which continued during breaks taking advantage of the peaceful environment on the island, as well as during the morning and evening boat trips to and from the island. Sunny weather beautifully cooperated to the success of the meeting, along with the excellent Venetian dishes catered to the island at lunch time.

The workshop was organized jointly by the European Southern Observatory and the Observatory of the Ludwig-Maximilians-Universität. The Scientific Organizing Committee included R. Bender (Co-Chair), F. Bertola, J. Binney, M. Dickinson, D. Koo, Y. Mellier, M. Pettini, A. Renzini (Co-Chair), and R. Sancisi. The Local Organizing Committe included G. Hubert, E. Pancino and C. Stoffer, whom we warmly thank for their smooth and efficient organization. We also acknowledge the exquisite hospitality of the Istituto Veneto di Scienze, Lettere ed Arti, on whose premises the early registration of the participants and the welcome reception took place. Finally, we would like to thank F. Nisii and the whole staff of the Venice International University for their most helpful cooperation before and during the meeting.

Garching,   *Ralf Bender and Alvio Renzini*
May 2002

Contents

List of Participants

Ames, Susan
Radioastronomisches Institut
der Universität Bonn
sames@astro.uni-bonn.de

Axon, David
University of Hertfordshire
dja@star.herts.ac.uk

Balcells, Marc
Instituto de Astrofísica de Canarias
balcells@ll.iac.es

Barton Gillespie, Elizabeth
University of Arizona
bgillespie@as.arizona.edu

Baugh, Carlton
University of Durham,
Department of Physics
c.m.baugh@durham.ac.uk

Bender, Ralf
University Observatory, Munich
bender@usm.uni-muenchen.de

Bergeron, Jacqueline
Institut d'Astrophysique de Paris
bergeron@iap.fr

Bertola, Francesco
Università di Padova,
Dip. di Astronomia
bertola@pd.astro.it

Bettoni, Daniela
Osservatorio Astronomico di Padova
bettoni@pd.astro.it

Bissantz, Nicolai
University of Basel,
Astronomical Institute
Nicolai.Bissantz@unibas.ch

Brinchmann, Jarle
University of Oxford
jarle@astro.ox.ac.uk

Bunker, Andrew
University of Cambridge,
Institute of Astronomy
bunker@ast.cam.ac.uk

Burkert, Andreas
MPI für Astronomie, Heidelberg
burkert@mpia-hd.mpg.de

Chary, Ranga-Ram
University of California, Santa Cruz
rchary@ucolick.org

Chatzichristou, Eleni
STScI, Baltimore
elenic@stsci.edu

Christian, Carol
STScI, Baltimore
carolc@stsci.edu

Cimatti, Andrea
Osservatorio Astrofisico di Arcetri
cimatti@arcetri.astro.it

Ciotti, Luca
Osservatorio Astronomico di Bologna
ciotti@bo.astro.it

Clewley, Lee
Imperial College, London
l.clewley@ic.ac.uk

Corbelli, Edvige
Osservatorio Astrofisico di Arcetri
edvige@arcetri.astro.it

Corsini, Enrico Maria
Università di Padova,
Dip. di Astronomia
corsini@pd.astro.it

Courteau, Stéphane
University of British Columbia
courteau@astro.ubc.ca

Cristiani, Stefano
ST-ECF, Garching
scristia@eso.org

Cristobal, David
Instituto de Astrofísica de Canarias
dch@ll.iac.es

D'Onghia, Elena
Osservatorio Astronomico di Brera
elena@merate.mi.astro.it

Daddi, Emanuele
Università di Firenze
edaddi@arcetri.astro.it

Danese, Luigi
SISSA, Trieste
danese@sissa.it

Davies, Roger
University of Durham,
Department of Physics
roger.davies@durham.ac.uk

De Breuck, Carlos
Institut d'Astrophysique de Paris
debreuck@iap.fr

de Jong, Roelof
Steward Observatory,
University of Arizona
rdejong@as.arizona.edu

De Rijcke, Sven
University of Basel,
Astronomical Institute
sven.derijcke@rug.ac.be

Dejonghe, Herwig
Universiteit Gent, Sterrenkundig Obs.
Herwig.Dejonghe@rug.ac.be

Dekel, Avishai
The Hebrew Univerity, Jerusalem
dekel@astro.huji.ac.il

Dickinson, Mark
STScI, Baltimore
med@stsci.edu

Drory, Niv
University Observatory, Munich
drory@usm.uni-muenchen.de

Dunlop, James
University of Edinburgh,
Institute for Astronomy
jsd@roe.ac.uk

Eisenhardt, Peter
JPL/Caltech
prme@kromos.jpl.nasa.gov

Erickson, Lance
Embry-Riddle University, Florida
erickson@db.erau.edu

Falomo, Renato
Osservatorio Astronomico di Padova
falomo@pd.astro.it

Ferguson, Henry
STScI, Baltimore
ferguson@stsci.edu

Fontana, Adriano
Oss. Astronomico di Roma,
Monteporzio
fontana@mporzio.astro.it

Fort, Bernard
Institut d'Astrophysique de Paris
Fort@iap.fr

Fosbury, Bob
ST-ECF, Garching
rfosbury@eso.org

Franceschini, Alberto
Università di Padova,
Dip. di Astronomia
franceschini@pd.astro.it

Franco Balderas, Alfredo
UNAM, Mexico
alfred@astroscu.unam.mx

Franx, Marijn
Leiden Observatory
franx@strw.LeidenUniv.nl

Gavazzi, Giuseppe
Università di Milano
giuseppe.gavazzi@mib.infn.it

Genzel, Reinhard
MPI für extraterrestrische Physik,
Garching
genzel@mpe-garching.mpg.de

Gerhard, Ortwin
University of Basel,
Astronomical Institute
gerhard@astro.unibas.ch

Gottloeber, Stefan
Astrophysikalisches Institut Potsdam
sgottloeber@aip.de

Haiman, Zoltan
Princeton University
zoltan@astro.princeton.edu

Hill, Gary
McDonald Observatory,
University of Texas
hill@astro.as.utexas.edu

Iliev, Ilian
Osservatorio Astrofisico di Arcetri
iliev@arcetri.astro.it

Illingworth, Garth
University of California, Santa Cruz
gdi@ucolick.org

Jarvis, Matt
Leiden Observatory
jarvis@strw.leidenuniv.nl

Jetzer, Philippe
University of Zurich
jetzer@physik.unizh.ch

Keeton, Chuck
Steward Observatory,
University of Arizona
ckeeton@as.arizona.edu

Kelm, Birgit
Osservatorio Astronomico di Bologna
kelm@bo.astro.it

Kleyna, Jan
University of Cambridge,
Institute of Astronomy
kleyna@ast.cam.ac.uk

Kochanek, Christopher
Harvard Smithsonian Center
for Astrophysics
ckochanek@cfa.harvard.edu

Koo, David
Lick Observatory
koo@ucolick.org

Labbe, Ivo
Leiden Observatory
ivo@strw.leidenuniv.nl

Londrillo, Pasquale
Osservatorio Astronomico di Bologna
londrillo@bo.astro.it

MacArthur, Lauren
University of British Columbia
lauren@astro.ubc.ca

Magliocchetti, Manuela
SISSA, Trieste
maglio@sissa.it

Magorrian, John
University of Durham,
Department of Physics
John.Magorrian@durham.ac.uk

Masegosa, Pepa
Instituto de Astrofisica de Andalucia
pepa@iaa.es

Mayer, Lucio
University of Washington, Seattle
mayer@astro.washington.edu

McLure, Ross
University of Oxford
rjm@astro.ox.ac.uk

Mellier, Yannick
Institut d'Astrophysique de Paris
mellier@iap.fr

Metcalf, Ben
University of Cambridge,
Institute of Astronomy
bmetcalf@ast.cam.ac.uk

Milvang-Jensen, Bo
University of Nottingham
ppxbm@nottingham.ac.uk

Moeller, Ole
Kapteyn Astronomical Institute,
Groningen
ole@astro.rug.nl

Moorwood, Alan
ESO, Garching
amoor@eso.org

Nipoti, Carlo
Osservatorio Astronomico di Bologna
carlo@aspera.bo.astro.it

Pahre, Michael
Harvard Smithsonian Center
for Astrophysics
mpahre@cfa.harvard.edu

Pancino, Elena
ESO, Garching
epancino@eso.org

Papovich, Casey
STScI, Baltimore
papovich@stsci.edu

Pizzella, Alessandro
Università di Padova,
Dip. di Astronomia
pizzella@pd.astro.it

Pozzetti, Lucia
Osservatorio Astronomico di Bologna
lucia@bo.astro.it

Primack, Joel
University of California, Santa Cruz
joel@ucolick.org

Prochaska, Jason
Carnegie Observatories, Pasadena
xavier@ociw.edu

Rao, Sandhya
University of Pittsburgh,
Dept. of Physics & Astronomy
rao@everest.phyast.pitt.edu

Renzini, Alvio
ESO, Garching
arenzini@eso.org

Rich, Michael
University of California, Los Angeles
rmr@ciel.astro.UCLA.EDU

Rigopoulou, Dimitra
MPI für extraterrestrische Physik,
Garching
dar@mpe.mpg.de

Rinaldi, Giovanna
Università di Padova,
Dip. di Astronomia
rinaldi@pd.astro.it

Rocca-Volmerange, Brigitte
Institut d'Astrophysique de Paris
rocca@iap.fr

Romanowsky, Aaron
School of Physics & Astronomy,
University of Nottingham
aaron.romanowsky@nottingham.ac.uk

Rosati, Piero
ESO, Garching
prosati@eso.org

Sancisi, Renzo
Osservatorio Astronomico di Bologna
sancisi@bo.astro.it

Scarlata, Claudia
Università di Padova
scarlata@stsci.edu

Schreier, Ethan
STScI, Baltimore
ejs@stsci.edu

Seitz, Stella
University Observatory, Munich
stella@usm.uni-muenchen.de

Shaver, Peter
ESO, Garching
pshaver@eso.org

Simard, Luc
Steward Observatory,
University of Arizona
lsimard@as.arizona.edu

Smail, Ian
University of Durham,
Department of Physics
ian.smail@durham.ac.uk

Steinmetz, Matthias
Steward Observatory,
University of Arizona
msteinmetz@as.arizona.edu

Stern, Daniel
JPL/Caltech
stern@zwolfkinder.jpl.nasa.gov

Swaters, Rob
Carnegie Institution of Washington
swaters@dtm.ciw.edu

Tacconi, Linda
MPI für extraterrestrische Physik,
Garching
linda@mpe.mpg.de

Thatte, Niranjan
MPI für extraterrestrische Physik,
Garching
thatte@mpe.mpg.de

Tornatore, Luca
Università di Trieste,
Dip. di Astronomia
tornatore@ts.astro.it

Treu, Tommaso
Caltech, Pasadena
tt@astro.caltech.edu

Treves, Aldo
Università dell'Insubria, Como
treves@uni.mi.astro.it

Tully, Brent
Institute for Astronomy,
Univ. of Hawaii
tully@willick.ifa.hawaii.edu

Turnshek, David
Univ. of Pittsburgh,
Dept. of Physics & Astronomy
turnshek@quasar.phyast.pitt.edu

van den Bosch, Frank
MPI für Astrophysik, Garching
vdbosch@mpa-garching.mpg.de

Verheijen, Marc
University of Wisconsin - Madison
verheyen@astro.wisc.edu

Vernet, Joël
Osservatorio Astrofisico di Arcetri
Vernet@arcetri.astro.it

Weiner, Benjamin
Lick Observatory
bjw@ucolick.org

Wilkinson, Mark
University of Cambridge,
Institute of Astronomy
markw@ast.cam.ac.uk

Willott, Chris
University of Oxford
cjw@astro.ox.ac.uk

Yamada, Toru
National Astronomical Observatory
of Japan, Tokyo
yamada@optik.mtk.nao.ac.jp

Yan, Lin
SIRTF Science Center, Caltech
lyan@ipac.caltech.edu

Zamorani, Giovanni
Osservatorio Astronomico di Bologna
zamorani@astbo3.bo.astro.it

The Mass of the Milky Way

Konrad Kuijken

Kapteyn Institute, PO Box 800, 9700 AV Groningen, The Netherlands

Abstract. The Mass distribution of the Galaxy is reasonably well understood. Out to a radius of at least 200kpc, the enclosed mass increases roughly linearly with radius at about $10^{10} M_\odot$ per kpc. The inner mass distribution is inconsistent with what we know to be present in stars and gas. Recent evidence indicates that the potential gets quite round at large radii.

1 Introduction

In spite of the difficulties inherent in trying to compile a global picture of our Galaxy from the inside, the Milky Way is still one of the best-studied galaxies, about whose mass distribution we know more than in most other cases. This review is an attempt to highlight the recent developments in our understanding of the mass of the Milky Way.

It is now widely accepted that dark matter plays a dominant role in the universe. On the scales of galaxies the influence of dark matter is still important, but less so than on larger scales: the density parameter Ω drops from 1 at large scales to about 0.3 in galaxy groups and clusters, while baryons can represent no more than 0.03 and the visible material in galaxies accounts for only 0.01. The inner parts of galaxies span the interesting range where dark matter and baryons exert comparably strong gravitational fields.

2 Rotation Curve

The best measures of galaxy masses are still the rotation curves of cold hydrogen disks, which often extend to larger radii than the bulk of the stellar mass. The rotation velocity traces directly the strength of the radial gravitational acceleration in the disk plane, which is quite closely related to the amount of mass enclosed (Fig. 1). In other galaxies, rotation curves and other mass measures, such as gravitational lensing (e.g., Fisher et al. 2000) agree. Moreover, gravitational lenses provide some of the most direct evidence that the shapes of mass distributions and of light distributions can differ substantially (e.g., Koopmans et al. 1998).

In the case of our Galaxy, measuring the rotation curve at large radii is complicated by the fact that our reference frame is moving with the rotation itself. It is impossible by means of radial velocity measurements alone to detect a solid-body rotation (all relative distances stay the same in that case), so only

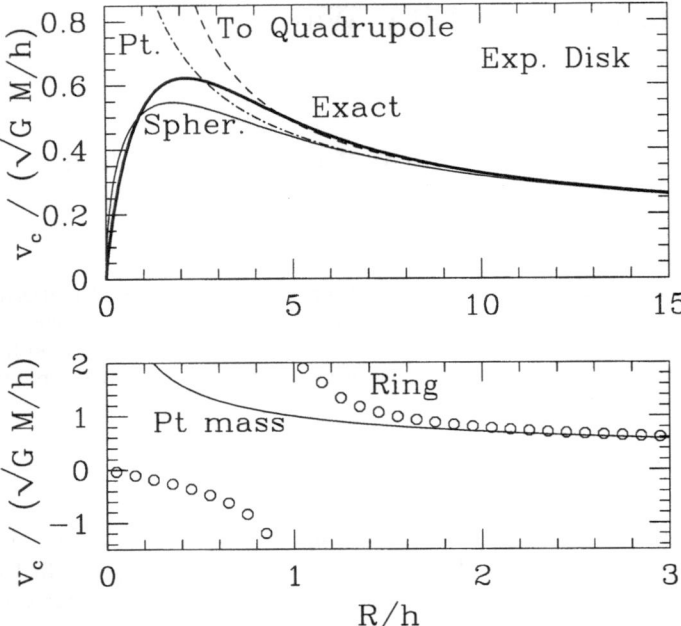

Fig. 1. *Top*: The rotation curve of an exponential disk, and various approximations to it. *Thin solid line*: same radial enclosed mass distribution $M(< R)$, but distributed spherically. *Dashed line*: far-field multipole expansion up to quadrupole. *Dash-dot line*: far-field monopole expansion. At large radii all agree well, illustrating that the rotation curve is a reasonably accurate mass tracer, not very dependent on the shape of the mass distribution. *Bottom*: As an extreme case, comparison between the rotation curve due to a thin ring compared to that of a point mass

deviations from solid-body are detectable. Such differential rotation curves have now been measured quite well from the galactic gas disk (see Merrifield 1992 for a review), and from deep radial velocity surveys of stars (e.g. Metzger 1998).

The solar rotation speed around the center of the Galaxy is best measured by means of proper motions of nearby stars with respect to an absolute reference frame, or by the reflex motion of Sgr A* with respect to distant qso's. Both techniques agree very well at an angular speed ('Oort $A - B$') of the solar neighbourhood of about 220km/s / 8kpc (Mignard 2000; Backer & Sramek 1999). To turn this number into a rotation speed we need the distance to the center of the Galaxy, which is still uncertain at the 10% level.

While these complications mean that the rotation curve of the Galaxy is less well-measured than that of external systems, it is nevertheless clear that the rotation curve is more or less flat out to a distance of 20kpc. There is no evidence for a keplerian fall-off at large radii that would indicate an edge to the mass distribution.

3   Distant Satellites

The gravity field in the outer reaches of the Milky Way may be probed by tracing the dynamics of distant satellites. Recently Wilkinson and Evans (1999) compiled data for 30 such objects, including 6 with well-measured proper motions. The latter allows them to constrain the orbital structure of the satellite population, one of the main uncertainties in turning a collection of measured velocities into a mass estimator. Though one would like to have more data, particularly to test whether the orbital structure in the inner few 10s of kpc (where proper motions are available) is the same at 100kpc radius, where the strongest mass constraint comes from, this study represents the best measurements of the gravity field beyond the HI disk that we have. Wilkinson & Evans conclude that the mass of the Galaxy is about $2 \times 10^{12} M_\odot$ inside a radius of 200kpc. Combined with the rotation curve results, a good rule of thumb is that the enclosed mass in the Galaxy is close to $10^{10} M_\odot$ per kpc of radius, at least as far out as 200kpc from the center.

4   Local Group Dynamics

Unlike rotation curves which provide an indication of the shape and extent of the mass distribution inside a galaxy, integral constraints on total masses can be derived from the dynamics of galaxies relative to each other. The local group provides a nice example, where the dynamics of the two dominant galaxies (M31 and our Galaxy) as well as a host of smaller objects (satellites of the two giants, as well as smaller local group members) can be modelled. All galaxies are assumed to have started out in a common Hubble expansion, and to have deviated from that under the influence of each other's gravity – orbital solutions are then constructed which reproduce the current configuration and relative motions (including proper motions where available) of the Local Group.

All studies coincide that the total mass of the local group is of the order of $3 \times 10^{12} M_\odot$. It is difficult to make these analyses very precise because of a number of uncertainties. The age of the universe enters as it determines how much time is available for deviations from Hubble expansion to grow; galaxies will evolve in mass and extent as time passes, so they cannot be treated as isolated systems of constant mass; the mass ratio between our Galaxy and M31 needs to be input into the model; and there may have been early-time mergers; some local group galaxies may yet have to be identified. A nice review of the local group dynamics is given by Zaritsky (1990).

The local group dynamics results are thus nicely consistent with what we know of the inner mass distribution of the Milky Way itself.

5   How Much Mass Resides in the Disk / How Flat Is the Halo?

Galaxies are not WYSIWYG. The Milky Way is no different: if the flat rotation curve we measure were to be realized as a disk-like mass distribution, it would have the mass profile of a Mestel disk, with surface mass density $\Sigma(R) = V^2/2\pi GR$. At the solar radius the surface mass density would have to be $185 M_\odot/pc^2$, significantly higher than all recent measurements (see Kuijken 1999 for a review). Much of the mass in the Galaxy that is responsible for the rotation curve lies outside the disk plane.

Locally, the gravitational potential is quite flattened (isopotentials have an E4 shape at 1.1kpc from the disk plane, Kuijken & Gilmore 1991), indicating that the disk has an important influence. But at larger radii, where the halo should dominate, the flaring of the HI layer indicates only a moderate halo flattening (axis ratio of iso-density surfaces 0.8, Olling & Merrifield 2000), though this result is quite sensitive to the adopted Galactic constants.

A different constraint on the shape of the halo comes from tracing substructure in the old spheroid population. More and more evidence for such substructure is now being found, probably corresponding to remnants of tidally disrupted satellites. These streams can be very long-lived, especially when viewed in phase space, and they provide a wonderful constraint on the shape of the gravitational potential. The nicest example to date is that of the Sgr dwarf, whose tidal tails have now been identified right around the sky, forming quite a narrow great circle traced by carbon stars. Ibata et al. (2001) have modelled the distribution and velocities of these stars, and conclude that only in a quite spherical halo (axis ratio 0.9 or higher) would the tidal streams have remained as planar as is observed. The galactic halo appears to get rounder at larger radii.

6   Local Halo Density and Kinematics

Locally, how much dark matter is there? This question is of interest to those trying to detect the dark halo directly, be it through particle physics experiments, or astronomical surveys. In spite of the uncertainties in modelling the Galaxy, the local halo density turns out to be quite robust to different assumptions of halo-to-disk ratio, or disk scale length, and lies between 0.004 and 0.01 $M_\odot/pc^3$. The dominant driver of the uncertainty is the local circular speed, which is this calculation has been taken to lie between 180 and 220 km/s.

In the absence of any detection of the dark matter, there is a great deal of modelling freedom as far as the halo kinematics are concerned. The halo could have zero angular momentum, or it could rotate pro- or retrograde. The orbits could be preferentially radial, or tangential, or isotropic. Many equilibrium models are possible (e.g., Kuijken & Dubinski 1995), and exploring this model freedom is particularly important for the interpretation of microlensing results to the Magellanic Clouds.

7   Conclusions

The galaxy remains a tantalizing object to study, in spite of the unique complications involved. There is no reason to expect this to change soon, far from it: with new astrometric missions such as GAIA it will be possible to obtain the complete stellar velocity field of the galaxy, to eliminate once and for all the uncertainties in the Galactic constants, and to take the study of phase-space mapping of satellite debris streams to a new level.

Nevertheless it is useful to keep a few cautionary remarks in mind. All analyses described so far try to model a Galaxy in equilibrium. Particularly at large radii this may be dangerous: galaxies generically have quite irregular outskirts, with disk warps that are often asymmetric (Garcia-Ruiz et al. 2002), and disk kinematics or morphologies that are lop-sided. Also ellipticals can have complex outer regions (see e.g. the spectacular deep image of M87 by Weil et al. 1997). When exploring the outermost regions of galaxies in search for mass estimators, non-equilibrium effects may well prove to be the limiting factor.

References

1. Backer, D. C. & R. A. Sramek: Ap.J. **524**, 805 (1999)
2. Fischer, P. et al.: A.J. **120**, 1198 (2000)
3. Garcia-Ruiz, I., Sancisi, R. & Kuijken, K.: in preparation
4. Ibata, R., G. F. Lewis, M. Irwin, E. Totten, & T. Quinn: Ap.J. **551**, 294 (2001)
5. Koopmans, L. V. E., A. G. de Bruyn, & N. Jackson: MNRAS **295**, 534 (1998)
6. Kuijken, K.: ApSS **267**, 217 (1999)
7. Kuijken, K. & J. Dubinski: MNRAS **277**, 1341 (1995)
8. Kuijken, K. & G. Gilmore: Ap.J.Lett **367**, L9 (1991)
9. Merrifield, M. R.: A.J. **103**, 1552 (1992)
10. Metzger, M. R., J. A. R. Caldwell, & P. L. Schechter: A.J. **115**, 635 (1998)
11. Mignard, F.: A&A **354**, 522 (2000)
12. Olling, R. P. & M. R. Merrifield: MNRAS **311**, 361 (2000)
13. Weil, M. L., J. Bland-Hawthorn, & D. F. Malin: Ap.J. **490**, 664 (1997)
14. Wilkinson, M. I. & N. W. Evans: MNRAS **310**, 645 (1999)
15. Zaritsky, D.: ASP Conf. Ser. 10: Evolution of the Universe of Galaxies, 51 (1990)

Distant Field BHB Stars
and the Mass of the Milky Way

Lee Clewley and Steve Warren

Blackett Laboratory, Imperial College of Science Technology and Medicine,
Prince Consort Rd, London SW7 2BW

Abstract. We present a new calculation of the mass of the Milky Way derived from radial velocities and distances measured from a sample of faint $16 < B < 20$ field blue horizontal branch (BHB) stars. This study aims to reduce the uncertainty in the measured mass of the Milky Way by increasing by a factor of five the number of halo objects at Galactocentric distances r > 30 kpc.

1 The Survey

We have embarked on a study of BHB stars in the halo with the aim of significantly increasing the number of distant halo tracers. The motivation for this work is to improve on the large uncertainty in the best estimate of the mass of the Milky Way. The analysis of Wilkinson & Evans [1] calculate the mass within 50 kpc to be $5.4^{+0.2}_{-3.6} \times 10^{11} M_\odot$. A principal source of uncertainty arises from the number of halo objects in the data set - there are currently only 27 known satellite galaxies and globular clusters at Galactocentric distances > 20 kpc. We aim to increase this number by isolating halo BHB stars.

Our targets are selected by B-V colour in a UBV two-colour plot with colours $0.0 < (B - V)_0 < 0.2$. These A-type stars include not only luminous field BHB stars, of absolute magnitude $M_V \sim 0.7$, but also contaminating stars that are some two magnitudes fainter. BHB stars are standard candles, so their distances can be measured. This is why they are valuable dynamical tracers. The contaminants, the so-called blue stragglers, are less distant, so it is crucial to isolate the two populations.

2 Isolating BHB Stars in the Halo

Kinman, Suntzeff, and Kraft [2] studied this problem in detail. They analysed three surface gravity indicators, each plotted against colour. Two use spectrophotometry to measure the steepness and size of the Balmer jump, the third compares the Balmer widths of the lines. Unfortunately spectrophotometry is not practical for remote BHB stars at r > 30 kpc. This is because the method requires $S/N \sim 40\text{Å}^{-1}$, which at $V = 19$ would require some two hours integration on a 4m telescope in dark time. Instead, we use three techniques: (a) we have refined an older technique that compares Balmer line width as a function of colour; (b) we have developed a routine that looks at the shape of the Balmer

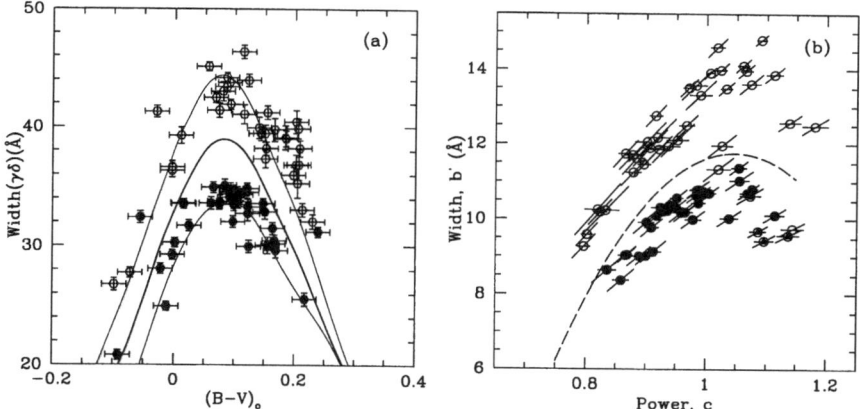

Fig. 1. The separation of halo BHB stars (filled) and blue stragglers (open) for the KSK data using our methods. Plot (a) is our Balmer width measurements. The curves are derived from Kurucz [3] model atmospheres for log g = 3.0, 3.5 and 4.0 and [Fe/H] = -1. Plot (b) shows the shape parameters of the Balmer line fit. Parameter b is the scale width and c is the curvature. The error bars are the axis of the 1-σ error ellipse

line and (c) by examining the strength of the Calcium II K lines we remove stars of high metallicity. Figure 1 shows the effectiveness of method (a) and (b) for the Kinman, Suntzeff, and Kraft [3] data. There is one interloper in Figure 1(a) and two in Figure 1(b). These stars have high metallicity measured by method (c) and are eliminated from the sample.

We run Monte Carlo simulations on these data to simulate the data typical of our faint star survey. We find that with an accuracy on the B-V colour < 0.03, and using spectra of S/N $\geq$ 20 (integrated over 2 Å), we can reliably classify any halo A star. At this S/N the spectra are of high enough quality for measurement of the radial velocity, with an accuracy of 15 km/s. This then provides a practical way of identifying remote tracers to determine the mass of the Milky Way halo.

3   Estimating the Mass

We refer to Wilkinson et al. (this volume) and [1] for details of the theoretical approach. Using this code we estimate the mass within 50 kpc using a preliminary sample of 37 BHB stars and the 27 satellite galaxies and globular clusters to be $5.36^{+0.1}_{-3.5} \times 10^{11} M_\odot$ and the total mass to be $2^{+3.8}_{-1.8} \times 10^{12} M_\odot$.

References

1. Wilkinson M.I., Evans N. W., MNRAS, 310, 645, (1999)
2. Kinman T.D., Suntzeff N.B., Kraft R.P., AJ, 108, 1722, (1994)
3. Kurucz R., CD-ROM No.13, Smithsonian Astrophysical Observatory (1993)

A Pattern Speed in the Galaxy's OH/IR Stars

Victor P. Debattista[1], Ortwin Gerhard[1], and Maartje N. Sevenster[2]

[1] Astronomisches Institut, Universität Basel, Venusstrasse 7, CH-4102 Binningen, Switzerland
[2] RSAA/MAAAO, RSAA/MSSSO, Cotter Road, Weston ACT 2611, Australia

1 Introduction

The Milky Way Galaxy (MWG) contains both a bar and spirals. The pattern speed/rotation frequency, Ω_p, of these components have been measured with a variety of models. For the bar, models have found values $40 \lesssim \Omega_p \lesssim 60$ km s^{-1} kpc^{-1} (Binney et al. 1991; Fux 1999; Englmaier & Gerhard 1999; Weiner & Sellwood 1999; Dehnen 1999; Bissantz et al. 2002). The spiral arm Ω_p is even more uncertain, with values in the range $13.5 \lesssim \Omega_p \lesssim 27$ km s^{-1} kpc^{-1} reported (Morgan 1990; Amaral & Lépine 1997; Mishurov & Zenina 1999).

A model-independent method, based on the continuity equation, for measuring pattern speeds in external galaxies has been developed by Tremaine & Weinberg (1984). This method can be extended to the MWG (Kuijken & Tremaine 1991; Debattista et al. 2002). For discrete tracers in the MWG, the Tremaine-Weinberg (TW) method is contained in the following expression:

$$
\Delta V \equiv \Omega_p R_0 - V_{\mathrm{LSR}} \equiv \frac{\mathcal{K}}{\mathcal{P}} - u_{\mathrm{LSR}} \frac{\mathcal{S}}{\mathcal{P}}
$$
$$
= \frac{\sum_i f(r_i) v_{r,i}}{\sum_i f(r_i) \sin l_i \cos b_i} - u_{LSR} \frac{\sum_i f(r_i) \cos l_i \cos b_i}{\sum_i f(r_i) \sin l_i \cos b_i} \tag{1}
$$

where R_0 is the Sun-MWG center distance, V_{LSR} is the tangential velocity of the local standard of rest (LSR), u_{LSR} is the radial velocity of the LSR, $f(r_i)$ is the observational detection probability (which need not be known), v_r is the heliocentric radial velocity of a discrete tracer, and (l, b) are its Galactic coordinates. Eqn. 1 assumes only one pattern speed and a low amplitude for any rapidly growing structure. Moreover, the tracer population needs to be sampled uniformly; one such survey is the ATCA/VLA OH 1612 MHz survey (Sevenster et al. 1997a,b & 2001), covering $|l| \le 45°$ and $|b| \le 3°$.

2 Pattern Speed of the OH/IR Population

We extracted from the ATCA/VLA OH 1612 MHz survey a sample of ~ 250 OH/IR stars which are relatively old (older than 0.8 Gyr) and bright (flux density greater than 0.16 Jy). These selection criteria give OH/IR stars between 4 and 10 kpc away from the Sun. We applied the TW analysis of Eqn. 1 to this sample, obtaining the results shown in Fig. 1 (Debattista et al. 2002). Note that the

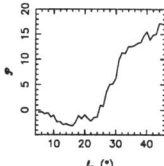 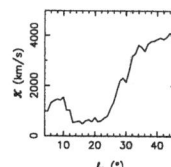 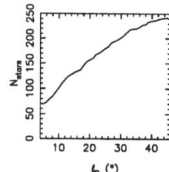 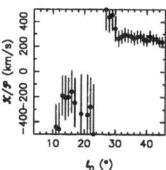

Fig. 1. The TW analysis for the OH/IR stars for changing l_0 (the maximum $|l|$ in Eqn. 1). From left to right are $\mathcal{P}$, $\mathcal{K}$, the number of stars and the resulting $\mathcal{K}/\mathcal{P}$

value of $\mathcal{K}/\mathcal{P}$ has converged, within the errors, for $|l| > 30°$. Simple tests with models show that a measurement of ΔV with a sample of this size should give an average accuracy of $\sim 17\%$ and always better than 40%, when the asymmetry signal is as large as the one we find. From re-sampling experiments, we find $\Delta V = 252 \pm 41$ km s^{-1}. If we assume $V_{\mathrm{LSR}}/R_0 = 220/8$ km s^{-1} kpc^{-1} (from SgrA* motion, Backer et al. 1999; Reid et al. 1999 and Cepheid proper motions, Feast & Whitelock 1997) and $u_{\mathrm{LSR}} = 0$ (from SgrA* HI absorption spectrum, Radhakrishnan et al. 1980), we obtain $\Omega_p = 59 \pm 5$ km s^{-1} kpc^{-1}. We estimate systematic error to be ~ 10 km s^{-1} kpc^{-1}.

The signal we found is concentrated close the plane ($|b| \leq 1°$) and at large longitude, suggesting a spiral is responsible for it (possibly the Scutum arm). The high Ω_p suggests this spiral arm is driven by the bar. Alternatively, the non-axisymmetric structure involved is an inner ring (Sevenster & Kalnajs 2001).

References

1. Amaral L. H., Lépine J. R. D., 1997, MNRAS, 286, 885
2. Backer D. C., Sramek R. A. 1999, ApJ, 524, 805
3. Binney J., Gerhard O. E., Stark A. A., et al. 1991, MNRAS, 252, 210
4. Bissantz N., Englmaier P., Gerhard O. 2002, MNRAS, *submitted*
5. Debattista, V. P., Gerhard, O. & Sevenster, M. N. 2001, MNRAS, *in press*
6. Dehnen W. 1999, ApJ, 524, L35
7. Englmaier P., Gerhard O. E. 1999, MNRAS, 304, 512
8. Feast M., Whitelock P. 1997, MNRAS, 291, 683
9. Fux R. 1999, A&A, 345, 787
10. Kuijken K., Tremaine S. 1991, *Dynamics of Disc Galaxies*, ed. B. Sundelius (Göteborg: Sweden) pg. 71
11. Mishurov Y. N., Zenina I. A. 1999, A&A, 341, 81
12. Morgan S. 1990, PASP, 102, 102
13. Radhakrishnan V., Sarma V. N. G. 1980, MNRAS, 85, 249
14. Reid M. J., Readhead A. C. S., Vermeulen R. C., et al. 1999, ApJ, 524, 816
15. Sevenster M. N., Kalnajs A. 2001, AJ, 122, 885
16. Sevenster M. N., Chapman J. M., Habing H. J., et al. 1997a, A&AS, 122, 79
17. Sevenster M. N., Chapman J. M., Habing H. J., et al. 1997b, A&AS, 124, 509
18. Sevenster M. N., van Langevelde H., Chapman J. M., et al. 2001, A&A, 366, 481
19. Tremaine S., Weinberg M. D. 1984, ApJ, 282, L5
20. Weiner B. J., Sellwood J. A. 1999, ApJ, 524, 112

Dark Matter in the Local Group: The Dark Haloes of the Milky Way, M31 and the Draco dSph

Mark I. Wilkinson[1], Jan T. Kleyna[1], N. Wyn Evans[2], and Gerard Gilmore[1]

[1] Institute of Astronomy, Madingley Road, Cambridge CB3 0HA
[2] Theoretical Physics, 1 Keble Road, Oxford, OX1 3NP, UK

Abstract. Mass estimates of the dark matter haloes of the Milky Way and M31 based on the latest observational data sets of tracer objects outside 20kpc are presented. The most likely masses are found to be $2.0^{+3.8}_{-1.8} \times 10^{12} M_\odot$ and $1.0^{+1.8}_{-0.6} \times 10^{12} M_\odot$, respectively. We conclude that there is no dynamical evidence that M31 is more massive than the Milky Way. The results of a detailed study of the internal dynamics of the Draco dwarf spheroidal (dSph) galaxy based on a new data set of stellar velocities are presented, providing the first clear evidence of an extended dark matter halo in Draco. We obtain a best-fit mass within 3 core radii of $8^{+3}_{-2} \times 10^7 M_\odot$ implying a V-band M/L ratio of $440 \pm 240 M_\odot / L_\odot$. Both tidal disruption and modified gravity (MOND) models are shown to be unsatisfactory explanations of the data.

1 The Milky Way and M31

Despite its importance as a cosmological parameter, the total mass of the Milky Way has remained poorly constrained by observations due to the small number of dynamical tracers at large radii. Further, the distant dSph Leo I has thwarted attempts to obtain a conclusive estimate of the mass of the Milky Way halo, as excluding it can affect the mass estimate by as much as a factor of 5. Mass estimates of the Milky Way (e.g. [12], [10]) have steadily increased over the past decade as more data were obtained. The most recent estimate [23] found a mass of $1.9 \times 10^{12} M_\odot$ and a scale length of ~ 170kpc. This estimate was not strongly sensitive to the presence or absence of Leo I. Recently, the proper motions for the distant globular clusters Pal 13 and NGC 7006 have been obtained ([22], [2]), bringing to eight the number of objects with full space motions. The other major improvement to the data set has been the addition of a sample of 37 halo Blue Horizontal Branch stars (Clewley & Warren, this volume) making a total of 64 distant tracers.

We model the halo of the Milky Way using a truncated, flat rotation curve (TF) halo potential (e.g. [23]). The potential-density pair is given by

$$\rho(r) = \frac{M}{4\pi} \frac{a^2}{r^2(r^2 + a^2)^{3/2}} \qquad \psi(r) = v_0^2 \log\left(\frac{\sqrt{r^2 + a^2} + a}{r}\right) \qquad (1)$$

where a and M are the scale-length and total mass of the halo, respectively, and the velocity normalisation v_0 is chosen so that the amplitude of the halo

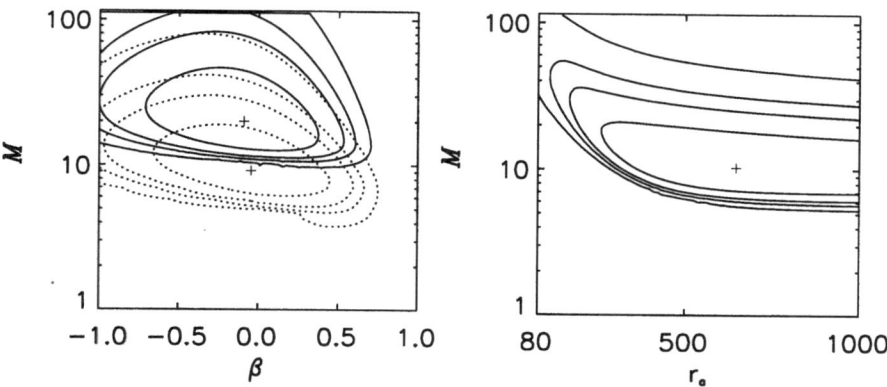

Fig. 1. Left: Likelihood contours for the total mass M of the Milky Way halo (in units of $10^{11}M_\odot$) and the velocity anisotropy β. Results including (solid curves) and excluding (dotted curves) Leo I are shown. Contours are at heights of 0.32, 0.1, 0.045 and 0.01 of peak height and the most likely values are indicated by plus signs. Right: Likelihood contours for the total mass M of the M31 halo (in units of $10^{11}M_\odot$) and the velocity anisotropy radius r_a. Contour heights as in left panel

rotation curve at the solar radius is $220\,\mathrm{km\,s^{-1}}$. The number density of satellites is fitted by a power law of index 4, and 2-integral distribution functions of the form $F(E, L) = L^{-2\beta}f(E)$ are constructed, where β is the Binney [1] anisotropy parameter and the function $f(E)$ is given in [23]. Models with $\beta = 0$ have isotropic velocity distributions, while models with positive (negative) β are radially (tangentially) anisotropic.

We use Bayes' theorem to obtain likelihood contours in the plane of the parameters M (or, equivalently, a) and β. The solid (dotted) contours in the left panel of Figure 1 show the likelihood contours obtained including (excluding) Leo I from the data set. From the Figure, the velocity distribution is approximately isotropic, although there is a considerable range of β values. The most likely total masses are $2 \times 10^{12}M_\odot$ including Leo I and $9.1 \times 10^{11}M_\odot$ excluding Leo I, corresponding to a values of 178kpc and 81kpc, respectively. The discrepancy between the two estimates is greater than that reported by [23]. This is due to changes in the existing proper motion estimates and emphasises the crucial importance of obtaining accurate proper motions for the Local Group satellites.

A detailed error analysis [23] shows that the main contributions to the uncertainty in the mass are the large errors in the proper motions and the small size of the data set. Assuming Leo I to be bound to the Milky Way, the best estimate for the total mass of the halo is $2.0^{+3.8}_{-1.8} \times 10^{12}M_\odot$. The mass within 50 kpc is $5.4^{+0.18}_{-3.5} \times 10^{11}M_\odot$.

Until recently, mass estimates of M31 were based either on the motions of the distant dwarf galaxies surrounding M31 or on its closer planetary nebulae and globular cluster populations. A recent mass estimate based on the complete data set of objects outside a projected radius of $\sim$ 20kpc, suggested that while the mass of the inner $\sim$ 30kpc of the M31 halo is larger than the equivalent mass in the Milky Way, the total mass of the halo may be smaller [3]. This conclusion was reinforced by the inclusion in the data set of radial velocities for 5 newly discovered dwarf satellites of M31 [4].

We model the halo of M31 using a TF potential and represent the number density of satellite galaxies by a TF density profile with appropriate choice of a. We build Osipkov-Merritt type DFs $F(E, L) \equiv F(Q)$ where $Q = E - L^2/2r_a^2$. At radii inside r_a the velocity distribution is isotropic, while at large radii it becomes radially anisotropic. The GC and PNe data sets are modelled using power law densities and exponential distribution functions which include rotation as there is evidence that both systems are slowly rotating ([6], [16]). For the GCs and PNe we have only projected positions, and so we must project the DFs along the line of sight [3].

The right panel of Figure 1 shows the likelihood contours in the plane of total mass and anisotropy radius. As in the case of the Milky Way, the velocity distribution is roughly isotropic ($r_a \sim$ 650 kpc). The most likely mass is $1.0^{+1.8}_{-0.6} \times 10^{12} M_\odot$, corresponding to a scale length a of 90 kpc. Given the large uncertainties in the estimates of both the Milky Way and M31 masses it is not possible to make precise statements about their relative masses. However, the dynamical evidence supports a picture in which both haloes have similar masses, rather than the standard Local Group picture in which M31 is $30 - 50\%$ more massive than the Milky Way (e.g. [13]).

The error bars on the total masses of both the Milky Way and M31 will be reduced significantly following the astrometric missions SIM and GAIA, which will obtain proper motions for all the Local Group satellites. This will reduce the uncertainty in the total masses to $\sim$ 20% [23]. On a shorter time scale, the mass estimates within 50kpc will be improved in the case of the Milky Way by increasing the size of the data set of BHB stars (Clewley & Warren, this volume) and for M31 by increasing the size of the GC and PNe data sets [21].

2   The Draco dSph

Dwarf spheroidal galaxies are the smallest dark matter dominated stellar systems in the universe (e.g. [13]) and, as such, they are vital laboratories for studying the properties of the dark matter. A new data set of 159 velocities for giants (V< 19.5) at large radii in the Draco dSph [8] has made it possible for the first time to investigate the radial variation of the velocity dispersion profile of a highly dark matter dominated dSph. The upper left panel of Figure 2 shows that the dispersion profile of Draco is flat or slowly rising in the outer parts.

2.1   Dark Matter Halo Models

We model the light distribution of Draco using a Plummer sphere with core radius $r_0 = 232\text{pc} = 9.7'$ [24]. The lower left panel of Figure 2 shows the mass

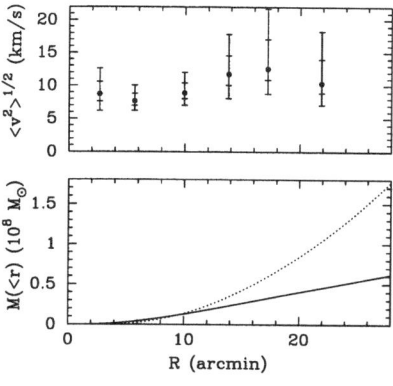
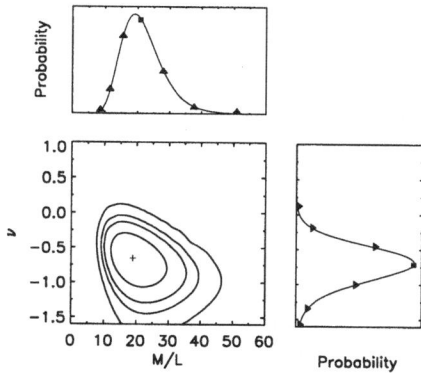

Fig. 2. Left, Top: Line of sight velocity dispersion as a function of projected radius R in the Draco dSph. Tick marks show the 1 and 2σ errorbars assuming a binary fraction of 40%. Left, bottom: Three-dimensional mass profiles for Draco obtained using the Jeans equations. Right: Likelihood contours for Draco's M/L and the anisotropy parameter ν assuming the validity of MOND. The contours are at enclosed two-dimensional χ^2 probabilities of 0.68, 0.90, 0.95 and 0.997. Projections of the contours onto the two parameter axes are also shown. The median of each distribution is represented by a square and the triangles show the 1σ, 2σ and 3σ limits

estimates obtained from the dispersion profile using the Jeans' equations and assuming velocity isotropy. Depending on whether the profile is assumed to be flat (solid curve) or linearly rising (dotted curve) the estimated mass within $3r_0$ (the region probed by our data) ranges from $6.3 - 18.0 \times 10^7 M_\odot$. These estimates imply M/L ratios of $350 - 1000$, where the V-band luminosity of Draco is $1.8 \times 10^5 L_{\odot,\mathrm{V}}$ [7].

Wilkinson et al. present a 2-parameter family of halo models for dSph galaxies which we use to model our Draco data set [24]. The halo potential $\psi(r)$ is given by

$$\psi(r) = \begin{cases} \psi_0(1 + r^2/r_0^2)^{-\alpha/2} & \text{if } \alpha \neq 0, \\ -(v_0^2/2)\log\left[1 + r^2/r_0^2\right] & \text{if } \alpha = 0. \end{cases} \tag{2}$$

The parameter α determines the mass distribution of Draco's halo, and ranges from models in which mass-follows-light ($\alpha = 1$) to models in which the visible dwarf lies in the approximately harmonic core of a much larger halo ($\alpha = -2$). When $\alpha = 0$ the rotation curve of the halo is asymptotically flat. The distribution functions for Plummer density profiles in the potentials (2) are derived in [24] and contain a second parameter ν which measures the velocity anisotropy. Isotropic models have $\nu = 0$ and radially (tangentially) anisotropic models have $\nu > 0$ ($\nu < 0$). A Bayesian analysis of our data set (see Kleyna et al., this volume and [8]), assuming uniform priors in both model parameters yields a value of $8^{+3}_{-2} \times 10^7$ for the mass contained within $3r_0$, implying a V-band M/L ratio of 440 ± 240 (see Figure 1 of Kleyna et al., this volume). Our data are sufficient to rule out both mass-follows-light ($\alpha = 1$) and harmonic core ($\alpha = -2$) models at the 2.5σ level.

2.2   Alternatives to Dark Matter I: Tides

A number of authors have questioned the assumption of virial equilibrium which underlies determinations of the M/L ratio for dSph galaxies based on stellar velocities (e.g. [11]), and have suggested that what we observe as dSph galaxies are actually the tidally disrupted remnants of bound stellar systems. However, in the case of Draco, three pieces of evidence argue strongly against tides having played an important role in its evolution. First, Draco is almost round on the sky ($\varepsilon = 0.29$), and has a very regular appearance. This is unlikely to be due to chance alignment of an extended tidal remnant. Estimates of the proper motion of Draco suggest that it is currently at or near pericentre [9]. N-body simulations show that, near pericentre, tidal remnants are elongated along the orbit (e.g [19]), and therefore line of sight elongation would imply that the orbit is similarly aligned. The observed proper motions are sufficient to rule out such alignment within $45°$ with 98% confidence [9]. Second, a new surface brightness profile for Draco based on SDSS data [17] shows no evidence for tidal tails around Draco, and no break in the profile which might indicate tidal truncation. Kleyna et al. [9] discuss the tidal radius of Draco in detail based on the available orbit information and conclude that it is significantly larger than the light distribution. Thirdly, N-body simulations have found that the characteristic signature of tidal disruption is an apparent rotation of the remnant about its minor axis (in 3 dimensions). As Draco is close to pericentre, in projection this apparent rotation would be expected about an axis perpendicular to the projection of its orbit. However, Kleyna et al. [9] find only a statistically insignificant rotation signal which, if real, is about an axis which is in 3σ disagreement with the axis expected for apparent rotation due to tides. Finally, the lack of alignment between the major axis of Draco and the projection of its orbit suggests that any flattening of the stellar distribution is not due to tides.

We therefore conclude that Draco is unlikely to be a tidal remnant.

2.3   Alternatives to Dark Matter II: MOND

MOdified Newtonian Dynamics (MOND) was proposed as an alternative explanation for the flatness of spiral galaxy rotation curves [14]. A number of authors have investigated whether MOND could also explain the internal dynamics of dSph galaxies ([5], [15]). However, these studies were based only on data from the central regions of dSphs and did not yield unambiguous results. We performed a full Bayesian analysis of our velocity data set within the MOND paradigm assuming Draco to be an isolated, spherical stellar system. The assumption of isolation is justified by noting that for $\sim$ 90% of our stars the gravitational attraction of the Milky Way disk is smaller than the internal accelerations due to other Draco stars. Representing the slightly flattened light distribution of Draco by a spherical model has greater impact when applying MOND since all the mass is assumed to be in the stars. However, this effect is not significant as our spherical model underestimates the mass interior to radius r by $\leq 20\%$, resulting in M/L ratios which are at most 20% overestimated.

The right panel of Figure 2 presents likelihood contours in the plane of the stellar M/L ratio and the anisotropy ν of the velocity distribution. The data suggest a velocity distribution that is slightly tangentially anisotropic. More significantly the most likely value of the M/L ratio is 19 with a 3σ lower limit of 8.6. This is significantly higher than would be expected for Draco based on studies of Galactic globular clusters [20] which suggest that the upper limit on the allowed M/L values is about 4.3. We conclude that the flat/rising velocity dispersion profile in the outer parts of the Draco dSph presents a serious challenge to the standard MOND formulation.

References

1. Binney, J., 1981, MNRAS, 196, 455
2. Dinescu D. I., Majewski S. R., Girard T. M., Cudworth K. M., 2001, AJ, 122, 1916
3. Evans, N. W., Wilkinson M. I., 2000 MNRAS, 316, 929
4. Evans, N. W., Wilkinson M. I., Guhathakurta P., Grebel E. K., Vogt S. S., 2000, ApJL, 540, 9
5. Gerhard O.E., 1994, in Proc. of the ESO/OHP Workshop, "Dwarf Galaxies", ed. G. Meylan & P. Prugniel (Garching:ESP), 335
6. Huchra J. P., Stauffer J., van Speybroeck L., 1982, ApJL, 259, 57
7. Irwin M.J., Hatzidimitriou D., 1995, MNRAS, 277, 1354
8. Kleyna J.T., Wilkinson M.I., Evans N.W.,Gilmore G., Frayn C.M., 2002a, MNRAS, in press, astro-ph/0109450
9. Kleyna J.T., Wilkinson M.I., Evans N.W.,Gilmore G., 2002b, ApJL, 563, 115
10. Kochanek C., 1996, ApJ, 457, 228
11. Kroupa P., 1997, New Astronomy, 2, 139
12. Little B., Tremaine S. D., 1987, ApJ, 320, 493
13. Mateo M., 1998, ARA&A, 36, 435
14. Milgrom M., 1983, ApJ, 270, 365
15. Milgrom M., 1995, ApJ, 455, 439
16. Nolthenius R., Ford H. C., 1987, ApJ, 317, 62
17. Odenkirchen M., et al. 2001, AJ, 122, 2538
18. Oh K.S., Lin D.N.C., Aarseth S.J., 1995, ApJ, 442, 142
19. Parmentier G., Gilmore G., 2001, A&A, 378, 97
20. Perrett K. M. et al., 2000 AAS, 196, 2803
21. Siegel M. H., Majewski S. R., Cudworth, K. M., Takamiya M., 2001, AJ, 121, 935
22. Wilkinson M. I., Evans N. W., 1999, MNRAS, 310, 645
23. Wilkinson M.I., Kleyna J.T., Evans N.W., Gilmore G., 2002, MNRAS, astro-ph/0109451

First Evidence for an Extended Dark Halo in the Draco Dwarf Spheroidal

Jan Kleyna[1], Mark Wilkinson[1], N. Wyn Evans[2], and Gerard Gilmore[1]

[1] Institute of Astronomy, Madingley Road, Cambridge, CB3 OHA, UK
[2] Theoretical Physics, 1 Keble Road, Oxford, OX1 3NP, UK

Abstract. We present a new velocity data set in the Draco dwarf spheroidal, containing the velocities of 159 giant stars extending to large radii. We fit this data set with a family of models parameterised by the orbital anisotropy and the dark matter halo scale. We are able to rule out a mass follows light model at the 2.5σ significance level; we also find that Draco possesses a dark halo with a density profile producing a near-flat rotation curve.

1 Introduction

The stellar velocity dispersions of several Local Group dwarf spheroidal galaxies (dSphs) are much larger than would be expected for self-gravitating systems of stars. In the past, several authors have obtained mass to light ratios in excess of $100 M_\odot/L_\odot$ for a number of dSphs, based on data in the core and usually assuming a mass follows light King model for the stellar distribution.

In this work, we present a new stellar velocity data set inside the Draco dSph, consisting of 159 Draco member stars extending to a radius of $\sim 25'$ [1,2]. We obtained stellar spectra using the WYFFOS multifibre spectrograph, which permits ~ 80 objects to be observed at once. We selected target giant branch stars from a good quality CCD photometric survey, thereby suppressing most of the Galactic foreground. We measured velocities by cross-correlating a synthetic template spectrum with the stellar Ca triplet.

2 Models

An important problem in modeling stellar systems is the degeneracy between mass and anisotropy. A system's velocity dispersion may rise with radius either because of a massive halo, or because the stellar orbits become more tangential with distance from the center. To address this degeneracy, we have developed a set of two parameter (α, ν) dynamical models with a Plummer light profile closely resembling Draco's observed profile [3]. The parameter α is an exponent describing the shape of the dark matter halo: $\alpha = 1$ represents a system in which the mass follows the light, and $\alpha = -2$ describes a system embedded deep in the uniform density core of an extended halo. The parameter ν is an logarithmic measure of the anisotropy, with $\nu > 0$ representing radially anisotropic systems, and $\nu < 0$ corresponding to tangentially anisotropic ones.

For each value of α, ν and projected radius, the models predict a line profile. We convolve this line profile with a Gaussian representing our velocity errors and with a plausible binary velocity distribution, and treat the convolved profile as a probability distribution from which our observed velocities are drawn. We then maximise the probability of observing the observed velocity data set over the space of α, ν.

Figure 1 shows the likelihood contours resulting from fitting our Draco data to this family of models. We rule out both mass follows light ($\alpha = 1$) and an extended core model ($\alpha = -2$) at the $2.5\,\sigma$ level, and find that the best fit yields a uniform rotation curve (isothermal) halo. Our best fit mass within 3 core radii is $8^{+3}_{-2} \times 10^7\,M_\odot$ and the V-band mass to light ratio is $440 \pm 240\,M_\odot/L_\odot$.

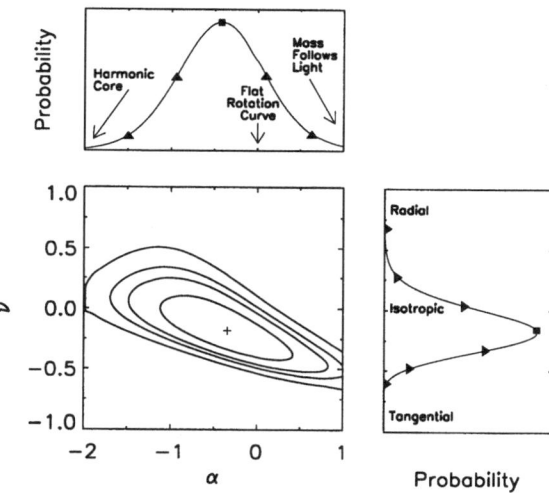

Fig. 1. Likelihood contours for fitting stellar velocities in Draco with a two parameter (halo exponent α and anisotropy ν) model. The top and right panels are the probability distributions in α and ν, respectively

Additionally, we determine that the MOdified Newtonian Dynamics (MOND) alternative to dark matter requires Draco to have an implausibly large mass to light ratio (see Wilkinson *et. al.*, this volume)

References

1. J. Kleyna, M. Wilkinson, N.W. Evans, G. Gilmore: ApJ, **563**, L115 (2001)
2. J. Kleyna, M. Wilkinson, N.W. Evans, G. Gilmore, C. Frayn: MNRAS, accepted [astro-ph/0109451] (2002)
3. M. Wilkinson, J. Kleyna, N.W. Evans, G. Gilmore: MNRAS, accepted [astro-ph/0109450] (2002)

The Inner Halo of the Draco dSph

John Magorrian

Dept. of Physics, University of Durham, South Road, Durham DH1 3LE, England

In projection, Draco is close to round and shows no obvious indications of interactions [3]. Virial mass estimates [1] suggest that it is dark-matter dominated. Here I try to constrain the distribution of this dark matter. I model the galaxy as a spherical, collisionless system in equilibrium, located 72 kpc from the sun.

1 Data

My models fit the galaxy's projected light distribution [3] and radial velocities of individual stars. I use the kinematics published in [1] and [2], yielding 326 velocity measurements of 186 Draco member stars. A handful of these stars are clearly members of binary systems. Strictly speaking, such binaries violate my collisionless assumption, but I can still use the velocity of the binary's centre-of-mass as a respectable tracer of the galaxy's DF.

To allow for potential binary contamination in the kinematics of the other stars, I assume that *every* star is a member of circular binary whose centre of mass moves at projected velocity v and whose observed member is moving relative to this with peak projected velocity $u > 0$. Single (i.e., non-binary) stars in this model will have $u \simeq 0$. After many applications of Bayes' theorem, a good description of the kinematics of the full ensemble of 186 stars is that u is distributed as a Gaussian with zero mean and dispersion $1.6\,\mathrm{km\,s^{-1}}$. Then, having a set of velocity measurements for any one star, I use this prior in u to obtain the probability distribution of the supposed circular binary's projected centre-of-mass velocity v.

2 Model

Having only 186 stellar velocities, I assume a restricted form for the DF – $f(E, L^2) = L^{-2\beta} f_0(E)$ – with constant anisotropy β. Given the surface brightness distribution [3], an assumed mass distribution $\rho(r)$ and some choice of anisotropy parameter β, I solve for $f_0(E)$ numerically. It is then straightforward to calculate the probability density that a star at radius R will have projected velocity v, and, from this, the likelihood $\Pr(D|\rho\beta)$ of the kinematical data D in the model (ρ, β). From Bayes' theorem, the probability of the model is

$$\Pr(\rho\beta|D) \propto \Pr(D|\rho\beta) \Pr(\rho) \Pr(\beta).$$

I take $\Pr(\beta)\mathrm{d}\beta = \mathrm{d}\beta/[(1 - \beta)^{1/2}(2 - \beta)]$.

3   Results

Figure 1 shows various projections of $\Pr(\rho\beta|D)$ for models assuming densities

$$\rho(r) = \frac{M_{300}}{4\pi(3-a)} \left(\frac{300\,\mathrm{pc}}{r}\right)^a,$$

with prior $\Pr(\rho)$ flat in $\log M_{300}$ and a. There is weak evidence for tangential anisotropy ($\beta < 0$): this could be an artifact of my assumption of spherical symmetry. The best-fitting models have $a \simeq 0.5$, but CDM-like haloes with $a \geq 1$ cannot be ruled out with the present kinematics.

These power-law models provide much better descriptions of the data than models in which mass follows light, with bookmakers' odds of 10^4 to 1.

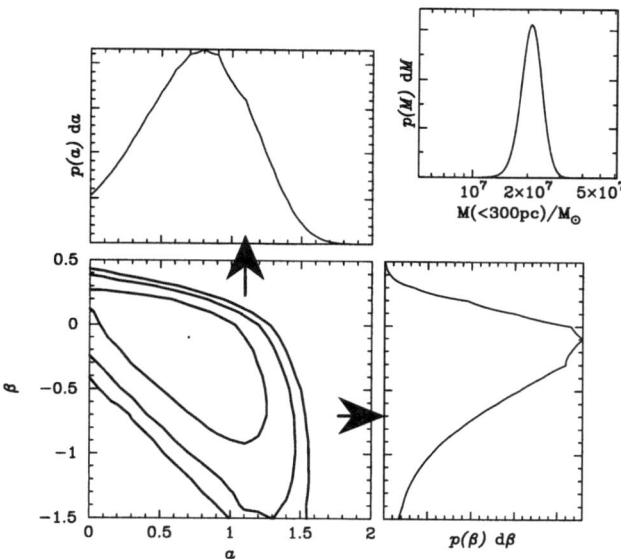

Fig. 1. Posterior probability distribution of (M_{300}, a, β). The panel on the bottom left shows the 68%, 90% and 95% confidence levels of $\Pr(a, \beta|D)$. The marginalized distributions $\Pr(a|D)$ and $\Pr(\beta|D)$ are plotted in the neighbouring panels. The top right panel shows $\Pr(M_{300}|D)$

References

1. T.E. Armandroff, E.W. Olszewski, C. Pryor: AJ **110**, 2131 (1995)
2. J.T. Kleyna, M.I. Wilkinson, N.W. Evans, G. Gilmore: MNRAS, in press (astro-ph/0109450). See also these proceedings.
3. M. Odenkirchen et al: AJ **122**, 2538 (2001)

Mass/Light Variations with Environment

R. Brent Tully

University of Hawaii, Honolulu HI 96822 USA

Abstract. In the local part of the universe, a majority of the total mass is in only a small number of groups with short dynamical times. The Virgo Cluster dominates far out of proportion to its contribution in light. Ninety percent of the total mass is in groups with $12.5 < \log M/M_\odot < 15$.

1 Introduction

It would hardly be surprising if the relationship between mass and light varies with environment, even if structures form from an invariant baryon to dark matter fractionation. Dark matter could be more dispersed than the baryons that are manifested in stars or detectable gas. This possible differentiation would create a 'bias' between what is seen and what exists [1]. Recent modelling [2,3] suggests there could be a complex relationship between mass and light, that dark matter may be underrepresented by light at both extremes of low density and high density. Mass-to-light may grow with scale around galaxies to an asymptotic limit [4]. This paper presents observational evidence from the motions of galaxies that there are mass-to-light differences with environment.

2 Galaxy Groups

Dark halos extend beyond the observed baryonic components of galaxies so there cannot be a sufficient understanding of the distribution of dark matter from studies of the internal kinematics of galaxies. The motions of galaxies relative to each other provide a probe of the distribution of matter on the scales of the separations between objects.

This discussion will begin with a new look at old data. Masses can be estimated for groups of galaxies using the virial theorem. To be strictly applicable, the group should be relaxed. However, within a possible systematic of a factor of two, the virial theorem gives an estimate of group masses as long as the basic assumption is met that the group is bound.

Candidate groups can be found among the structures defined by a dendogram analysis [5]. Within a sample of N galaxies, the pair is selected with the extreme of an appropriate property, say, the largest product luminosity / separation^2. This pair is then considered as a single unit and the procedure is repeated. After $N-1$ steps, all galaxies in the sample are merged. At each merger step, a luminosity density can be characterized by the sum of the light of the components divided by the cube of the separation at this step. Now the timescale for separation of an overdensity from the cosmic expansion and its collapse depends on the inverse square root of the mass overdensity. If there is a constant relationship between mass and light, then luminosity is a standin for mass. Then, in a plot of the merger dendogram as a function of luminosity density there would be a discrete cut that would distinguish entities that could have collapsed within the age of the universe.

A constant relationship between mass and light remains to be demonstrated and it would not be appropriate to specify such a relationship since this is to be a product of the study. However an independent observable monitors the probability that a candidate group is bound: the crossing time - a characteristic dimension for the group divided by the velocity dispersion. Assume a cut on the luminosity density dendogram. If that cut is at a low density, there would be a lot of entities above the cut with crossing times longer than the age of the universe. If the cut is at a very high density, the entities that survive the cut would all have crossing times much shorter than the age of the universe. Like Goldilocks, we try to find a cut that is just right, characterized by crossing times that scatter up to, but not above, the age of the universe. The details of the choice of cut and consequent properties of a sample of groups within the Local Supercluster have been discussed [5]. The Nearby Galaxies Catalog [6] includes these group affiliations. This catalog also records entities as 'associations' if they satisfy a luminosity density threshold an order of magnitude lower than the cut that specifies groups. This more lax prescription saves a more extended list of group candidates, a matter of interest later in the discussion.

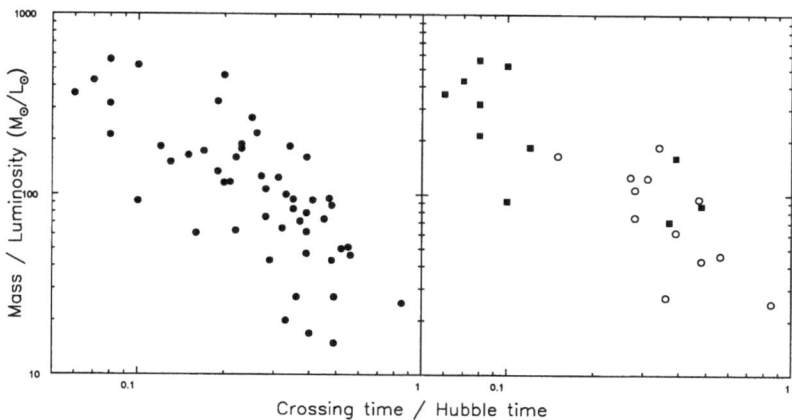

Fig. 1. Correlation between dynamical time and M/L_B. *Left panel:* groups with at least 5 identified members. *Right panel:* solid squares identify groups with a majority of types E-S0-Sa; open circles identify groups with a majority of types Sab-Irr

A fresh look at this old material reveals that there are strong correlations between mass-to-blue-light ratio (M/L_B) and either crossing time or morphology. These correlations are seen in Figure 1 where M/L_B is plotted against crossing time for a sample of 49 groups with at least 5 members that lie within $25h_{75}^{-1}$ Mpc. In the panel at the right, only groups with at least 6 known members with $M_B < -17$ are considered (here, $H_o = 75$ km s^{-1} Mpc^{-1} and $h_{75} = H_o/75$).

Fig. 1 reveals that groups predominantly composed of early type galaxies tend strongly to have short crossing times and high M/L_B. Groups of predominantly late type galaxies inevitably have longer crossing times and lower M/L_B. The range in M/L_B values extends over more than a decade. The trend in Fig. 1 has too big an amplitude be explained by correlated errors in velocity dispersion in the calculation of M/L_B and crossing time. The correlation between group morphology and crossing time rests on independent information. The same information is seen in Figure 2. The symbols again denote group morphology. There is a trend with group mass, with a larger fraction of early type groups toward higher masses. The consequence is a roll off toward a shallow slope between mass and light at high mass. It is seen, though, that at intermediate masses around 10^{13} $M_\odot$ there is a substantial range in M/L_B values. The variance is strongly correlated with both crossing time and group morphology.

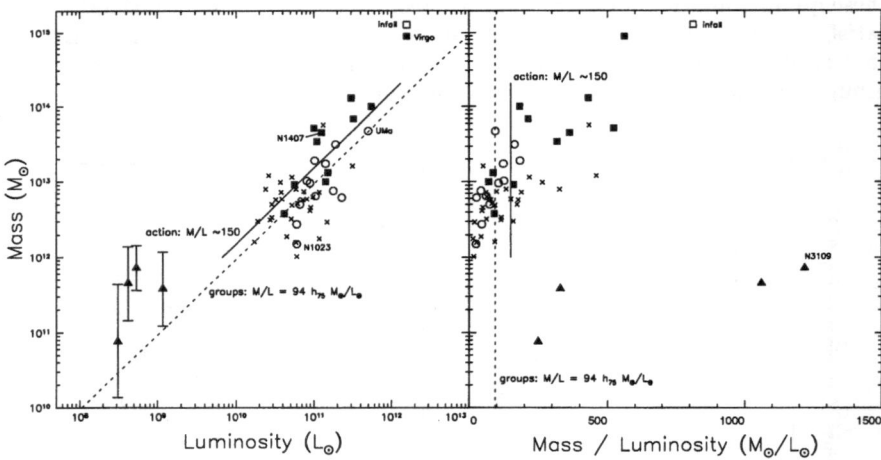

Fig. 2. In the left panel, group mass is plotted against group luminosity. In the right panel, the same mass is plotted against mass-to-light ratio. Filled squares and open circles represent the same groups seen in the right panel of Fig 1 and the crosses represent the smaller groups that fill out the left panel of Fig 1. The 4 triangles with error bars locate the groups of dwarfs discussed in section 4. The dashed lines indicate the mean M/L value found for groups [5], the solid lines indicate the mean field result from numerical action models [9], and the open box is based on the mass requirement for the Virgo Cluster that follows from the numerical action models

3   Numerical Action Modelling

The distribution of dark matter on scales larger than groups is recorded in the slosh of galaxies about Hubble flow on scales of megaparsecs to tens of megaparsecs. The general properties of the gravitational field can be recovered through the construction of plausible galaxy orbits with mixed boundary condition constraints [7,8]. It is to be appreciated that, while to date the specifics of individual orbits are poorly constrained by this modelling, mean mass densities are tightly constrained. In the general field, $M/L_B \sim 150\ M_\odot/L_\odot$ is found from this modelling [9] a result indicated by the solid lines in Fig. 2. This M/L_B value is 50% larger than the mean value derived from the group analysis which may be reasonable given the reference to scales larger than the domain of bound groups.

The action modelling of the Local Supercluster region raised an important issue regarding M/L_B variations with environment. There is specific information about the infall of galaxies toward the Virgo Cluster. Galaxies are entering the cluster with 1-D velocities in excess of 1500 km s^{-1} and the current zero-velocity shell with respect to the cluster is at about 28° radius from the cluster center. It is impossible to provide an adequate description of this infall with a mass assigned to the Virgo Cluster in accordance with the low M/L_B value found overall. The infall pattern requires a mass of $1.4 \times 10^{15} h_{75}^{-1}\ M_\odot$ and $M/L_B \sim 900 h_{75}^{-1}\ M_\odot/L_\odot$ [9]. These values are large but the mass relates to the cluster on a scale in excess of 2 Mpc radius. This datum is recorded by the open square in Fig. 2. The key constraints are (a) a high mass is required in the cluster to explain the high infall velocities in the immediate vicinity of the cluster, but (b) if there were a significant amount of additional mass between us and the cluster then we would have a higher retardation from Hubble flow.

4   Groups of Dwarfs

It was mentioned that the group analysis [5] also identified entities called associations that pass a luminosity density cut an order of magnitude fainter than the group cut. There are a small fraction of entities identified this way that deserve special attention. There are a few associations comprised of only dwarf galaxies which would easily have been called groups if only separations were considered (ie, luminosity not a factor). Velocity dispersions for these associations are extremely low, of order 20 km s^{-1}, so if these entities are bound then masses are modest, of order $10^{11}\ M_\odot$. However there is little light so M/L_B values range from ~ 250 to 1200 $M_\odot/L_\odot$.

These proposed groups have been discussed [10] in the context of a plausible galaxy formation scenario. The details of the formation mechanism will not be pursued here but the generic possibility is raised that there could be many low mass dark halos with little or no accompanying stars or gas. It would follow from this hypothesis that there could be environments with *some* star formation, enough to provide probes of the potential, but still so little that M/L_B is large.

The four candidate dwarf groups that are represented at the low mass end of Fig 2 are identified and details on their properties are given in reference [10].

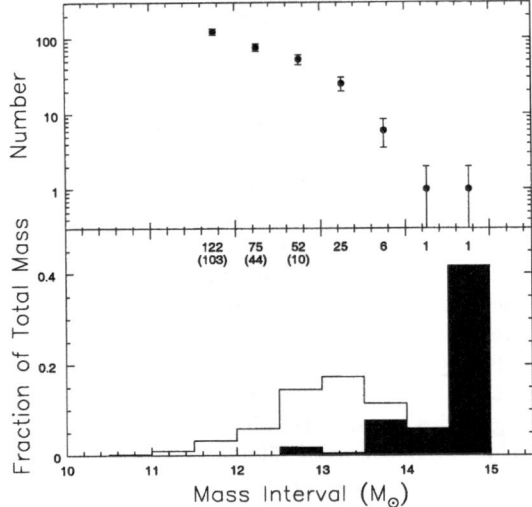

Fig. 3. *Lower panel:* fraction in half decade mass intervals of the total mass associated with local groups and galaxies. Filled histogram: 14 groups with crossing times less than $0.2H_0^{-1}$. Open histogram: all other groups and individual galaxies. Numbers of groups plus individual galaxies per bin are given across the top of the panel (numbers of individual galaxies in brackets). The sample is reasonably complete above $\log M = 11.5$. *Upper panel:* number of groups plus individual galaxies per mass interval

The critical issue is whether the proposed groups are bound because, if so, then M/L_B values are surely high. It is to be noted that these 4 candidate groups are all within 5 Mpc and within this distance beyond the Local Group (restricting to high latitudes) there are only 4 'normal' groups with luminous galaxies. There is the prospect that bound groups of dwarfs are numerically common, though they may contribute only a small fraction of the total mass of the universe. One could suspect that dark matter structures devoid of stars would also exist, which would be in line with the original 'biasing' idea. However the action analysis suggests that such structures would not contribute too much to the overall mass inventory.

5 Conclusions

Figure 3 shows the inventory of the clustered mass in the local region. Mass contributions are summed over all groups and individual galaxies within a distance of $25h_{75}^{-1}$ Mpc and with $|b| > 30°$ (distances based on a numerical action kinematic model). In the case of groups, masses come from application of the virial theorem. In the instances of pairs or triples where the observed dispersion in velocities is dominated by measurement uncertainties, or in the case of single galaxies, masses are inferred assuming $M/L_B = 100 M_\odot/L_\odot$, a round off of the mean result for groups represented by the dotted lines in Fig 2. The volume contains a total luminosity of $1.0 \times 10^{13} L_\odot$ in cataloged galaxies and a total mass of $2.1 \times 10^{15} M_\odot$ (only 10% of this mass total is inferred from the M/L assumption). Within this nearby volume, there should be reasonable completion down to $0.1 L_B^*$. The overall $M/L \sim 200 M_\odot/L_\odot$ is consistent with $\Omega_m \sim 0.2$ [9]. The mean density in this local region is $4.4 \times 10^{-30} h_{75}^2$ gm/cm^3, indicating this region has twice the cosmic mean density if $\Omega_m = 0.2$. Hence our census accounts for all the mass anticipated by the numerical action dynamical modelling. There are two striking points to note with Fig 3. One point is that 90% of the mass is in bound entities with $\log M > 12.5$. The other point is that the Virgo Cluster, with 15% of the light in this volume, has over 40% of the mass! Of course, the statistics above $\log M = 14$ are inadequate with this local sample (Fornax Cluster is the second massive entity).

The information provided by groups about the distribution of dark matter on scales of hundreds of kiloparsecs and the information provided by galaxy flows about dark matter on scales up to tens of megaparsecs give rise to a consistent picture. M/L_B values in dynamically evolved regions can be an order of magnitude higher than in the great majority of places that are dynamically young. The evidence from both galaxy motions and an inventory of groups suggests $M/L_B \sim 200 \ M_\odot/L_\odot$ overall, consistent with $\Omega_{matter} \sim 0.2$. Similar results are found from wide field weak lensing studies [11]. Within the high latitude, inner 25 Mpc region of the group sample, roughly 60% of the mass is in 14 low crossing time groups dominated by early galaxy types that contribute a quarter of the light. We seem to live in a curious universe where a majority of the clumped matter is in the modest percentage of locations with crossing times < 2 Gyr.

Groups containing familiar luminous galaxies all have masses above $\sim 10^{11}$ $M_\odot$. Candidate groups of only dwarf galaxies are identified with masses near this $10^{11} \ M_\odot$ limit. These associations are probably bound, whence they would contain mostly dark matter. It is reasonable to speculate that there are low mass halos without any stars or gas. However most of the mass of the universe seems to be in collapsed regions with $10^{12.5} - 10^{15} \ M_\odot$.

References

1. Kaiser, N. ApJ, 284, L9 (1984)
2. Blanton, M., Cen, R., Ostriker, J.P., & Strauss, M.A. ApJ, 522, 590 (1999)
3. Somerville, R.S. et al. MNRAS, 320, 289 (2001)
4. Bahcall, N.A., Lubin, L.M., & Dorman, V. ApJ, 447, L81 (1995)
5. Tully, R.B. ApJ, 321, 280 (1987)
6. Tully, R.B. Nearby Galaxies Catalog, Cambridge University Press (1988)
7. Peebles, P.J.E. ApJ, 344, L53 (1989)
8. Shaya, E.J., Peebles, P.J.E., & Tully, R.B. ApJ, 454, 15 (1995)
9. Tully, R.B., & Shaya, R.B. Evolution of Large-Scale Structure, astro-ph/9810298
10. Tully, R.B., Somerville, R., Trentham, N., & Verheijen, M. astro-ph/0107538
11. Wilson, G., Kaiser, N., & Luppino, G.A. (2001), astro-ph/0102396

On the Black Hole – Bulge Mass Relation in Active and Inactive Galaxies

R.J. McLure[1] and J.S. Dunlop[2]

[1] Oxford University, UK
[2] Institute for Astronomy, Edinburgh University, UK

Abstract. New virial black-hole mass estimates are presented for a sample of 72 AGN covering three decades in optical luminosity. Using a model in which the AGN broad - line region (BLR) has a flattened geometry, we investigate the $M_{bh} - L_{bulge}$ relation for a combined 90-object sample, consisting of the AGN plus a sample of 18 nearby inactive elliptical galaxies with dynamical black-hole mass measurements. It is found that, for all reasonable mass-to-light ratios, the $M_{bh} - L_{bulge}$ relation is equivalent to a linear scaling between bulge and black-hole mass. The best-fitting normalization of the $M_{bh} - M_{bulge}$ relation is found to be $M_{bh} = 0.0012 M_{bulge}$, in agreement with recent black-hole mass studies based on stellar velocity dispersions. Furthermore, the scatter around the $M_{bh} - L_{bulge}$ relation for the full sample is found to be significantly smaller than has been previously reported ($\Delta \log M_{bh} = 0.39$ dex). Finally, using the nearby inactive elliptical galaxy sample alone, it is shown that the scatter in the $M_{bh} - L_{bulge}$ relation is only 0.33 dex, comparable to that of the $M_{bh} - \sigma$ relation. These results indicate that reliable black-hole mass estimates can be obtained for high redshift galaxies.

1 Introduction

The correlation between black-hole mass and bulge luminosity is now well established for both active and inactive galaxies (Magorrian et al. 1998; Laor 1998). However, despite recent attention in the literature, the usefulness of the $M_{bh} - L_{bulge}$ relation as a black-hole mass estimator is at present severely limited due to its large scatter ($\simeq 0.5$ dex). Although the correlation between black-hole mass and stellar velocity dispersion for nearby inactive galaxies displays a much smaller scatter ($\simeq 0.3$ dex, Merritt & Ferrarese 2001a), it is clear that a $M_{bh} - L_{bulge}$ correlation with reduced scatter would be highly desirable, given the extreme difficulty in obtaining stellar velocity dispersions for high redshift galaxies.

This conference proceeding presents the main results of a new study (McLure & Dunlop 2002) in which we investigate the black hole - bulge mass relation using a 90-object sample comprised of 72 AGN (53 QSOs and 19 Seyfert 1s) and 18 nearby quiescent ellipticals with dynamically determined black-hole mass estimates. Those interested in the details of our analysis, particularly the flattened geometry model adopted for the calculation of the virial black-hole mass estimates, are referred to McLure & Dunlop (2002).

2 The Black-Hole Mass – Bulge Luminosity Relation

In Fig 1 absolute R–band bulge magnitude is plotted against black-hole mass for the 72 objects in the AGN sample. Also shown is absolute R–band bulge magnitude plotted against dynamically-estimated black-hole mass for our nearby inactive elliptical galaxy sample. Two aspects of Fig 1 are worthy of immediate comment. Firstly, as was shown by McLure & Dunlop (2001) and by Laor (1998 & 2001), it can be seen that bulge luminosity and black-hole mass are extremely well correlated, with $r_s = -0.77$ (7.3σ). Secondly, it is clear that the AGN and nearby inactive galaxy samples follow the same $M_{bh} - L_{bulge}$ relation over > 3 decades in black-hole mass, and > 2.5 decades in bulge luminosity. This second fact strongly supports the conclusions of Dunlop et al. (2001) and Wisotzki et al. (2001), that the host-galaxies of powerful quasars are normal massive ellipticals drawn from the bright end of the elliptical galaxy luminosity function. Thirdly, there can be seen to be no systematic offset between the Seyfert 1 and quasar samples, reinforcing the finding of McLure & Dunlop (2001) that, contrary to the results of Wandel (1999), the bulges of Seyfert galaxies and QSOs form a continuous sequence which ranges from $M_R(\text{bulge}) \simeq -18$ to $M_R(\text{bulge}) \simeq -24.5$. If we adopt an integrated value of $M_R^\star = -22.2$ (Lin et al. 1998), then this implies that the $M_{bh} - L_{bulge}$ relation holds from $L_{bulge} \simeq 0.01 L^\star$, all the way up to objects which constitute some of the most massive ellipticals ever formed; $L_{bulge} \simeq 10 L^\star$.

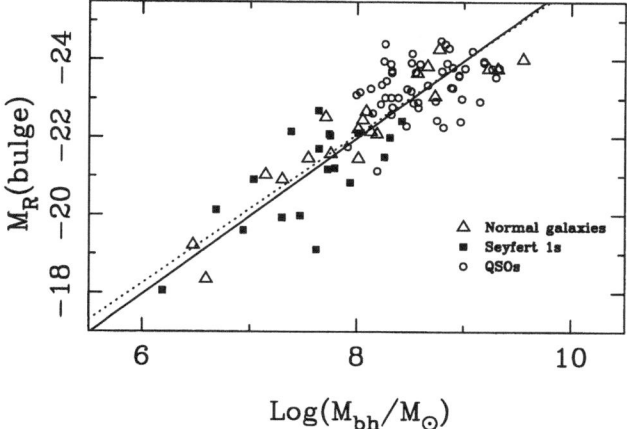

Fig. 1. Absolute R–band bulge magnitude versus black-hole mass for the full 90-object sample. The black-hole masses for the 72 AGN are derived from their Hβ line-widths under a disc-like BLR model (see McLure & Dunlop 2002). The black-hole masses of the inactive galaxies (triangles) are dynamical estimates as compiled by Kormendy & Gebhardt (2001). Also shown is the formal best-fit (solid line) and the best-fitting linear relation (dotted line)

The best-fit to the full 90-object sample has the following form:

$$\log(M_{bh}/\,\mathrm{M}_\odot) = -0.50(\pm0.02)M_R - 2.96(\pm0.48) \tag{1}$$

and is shown as the solid line in Fig 1. The scatter around this best-fitting relation is only $\Delta M_{bh} = 0.39$ dex, an uncertainty factor of < 2.5. The reduced scatter found here in comparison to previous studies is due to two factors. Firstly, all of the bulge luminosities used in this study are derived from full two-dimensional modelling of high resolution data, the majority of which is from HST. The second factor is the inclination corrections to the black-hole mass estimates provided by our flattened-geometry BLR model. Both of these aspects are discussed in detail in McLure & Dunlop (2002).

Given that the 18 objects in the nearby inactive galaxy sample have actual dynamical black-hole mass estimates, it is obviously of interest to quantitatively test how consistent the $M_{bh} - L_{bulge}$ relation for these objects is with the fit to the full, AGN dominated, sample. The best-fit to the inactive galaxy sub-sample alone, has the following form:

$$\log(M_{bh}/\,\mathrm{M}_\odot) = -0.50(\pm0.05)\mathrm{M_R} - 2.91(\pm1.23) \tag{2}$$

which can be seen to be perfectly consistent with the best-fit to the full sample in terms of both slope and normalization. Indeed, the best-fitting relations for the full sample, quasar sample, Seyfert galaxy sample and the nearby inactive galaxy sample are all internally consistent, and display comparable levels of scatter. This is a remarkable result given that it implies that the combined bulge/black hole formation process was essentially the same throughout the full sample, which as well as featuring both active and inactive galaxies, includes galaxies of both late and early-type morphology.

2.1   The Linearity of the Black Hole – Bulge Mass Relation

In our previous study (McLure & Dunlop 2001) of a sample of 45 AGN we found that $M_{bh} \propto M_{bulge}^{1.16\pm0.16}$, and therefore concluded that there was no evidence that the $M_{bh} - M_{bulge}$ relation was non-linear. In contrast, evidence for a non-linear relation was recently found by Laor (2001). In his $V-$band study of the black hole to bulge mass relation in a 40-object sample (15 PG quasars, 16 inactive galaxies and 9 Seyfert galaxies) Laor found a best-fitting relation of the form $M_{bh} = M_{bulge}^{1.54\pm0.15}$, which is clearly inconsistent with linearity. However, in order to determine the $M_{bh} - M_{bulge}$ relation it is obviously necessary to convert the measured bulge luminosities into masses, via an adopted mass-to-light ratio. The form of this mass-to-light ratio affects the derived slope of the $M_{bh} - M_{bulge}$ relation in the following way. If the mass-to-light ratio is parameterized as $M/L \propto L^\alpha$, then the resulting slope (γ) of the $M_{bh} - M_{bulge}$ relation is given by $\gamma = \frac{-2.5\beta}{1+\alpha}$, where β is the slope of the $M_{bh} - L_{bulge}$ relation (Eqn 1).

Here we choose to adopt the derived $R-$band mass-to-light ratio for the Coma cluster from Jørgensen, Franx & Kjærgaard (1996), which has $\alpha = 0.31$.

With this mass-to-light ratio our best-fitting $M_{bh} - L_{bulge}$ relation transforms to a $M_{bh} - M_{bulge}$ relation of the following form:

$$M_{bh} \propto M_{bulge}^{0.95 \pm 0.05} \tag{3}$$

It can immediately be seen that from our results there is no indication that the scaling between black hole and bulge mass is non-linear.

In order to calculate the bulge mass of the objects in his sample, Laor (2001) adopted a $V-$band mass-to-light ratio of $M_{bulge} \propto L_{bulge}^{1.18}$ (Magorrian et al. 1998), which is significantly different from our chosen mass-to-light ratio. However, irrespective of this, our new best-fit to the slope of the $M_{bh} - L_{bulge}$ relation ($\beta = -0.50 \pm 0.02$) of our new sample, which has a larger dynamic range in L_{bulge} than both the samples studied in McLure & Dunlop (2001) and Laor (2001), means that any disagreement about mass-to-light ratios cannot now alter the conclusion that the $M_{bh} - M_{bulge}$ relation is consistent with being linear. To demonstrate this we conclude by noting that even using the $M_{bulge} \propto L_{bulge}^{1.18}$ mass-to-light ratio adopted by Laor (2001), our best-fitting $M_{bh} - L_{bulge}$ relation is equivalent to $M_{bh} \propto M_{bulge}^{1.06 \pm 0.06}$, again, completely consistent with a linear scaling.

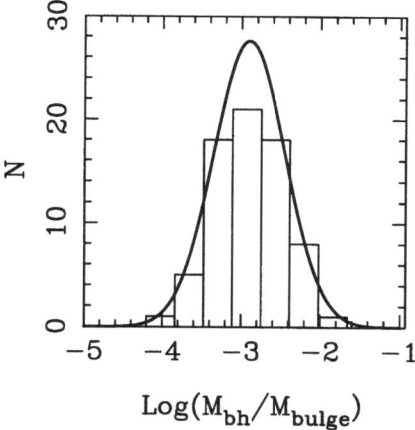

Fig. 2. Histogram of the ratio of black-hole mass to bulge mass for the 72-object AGN sample. Over-plotted for comparison is a gaussian with $\langle \log(M_{bh}/M_{bulge}) \rangle = -2.90$ and standard deviation 0.45 (see text for discussion)

2.2   The Normalization of the Black Hole – Bulge Mass Relation

Having established that the $M_{bh} - M_{bulge}$ relation is consistent with being linear, we now assume perfect linearity in order to establish the normalization of the

$M_{bh} - M_{bulge}$ relation. With the mass-to-light ratio adopted here, a linear scaling corresponds to enforcing a slope of -0.524 in the M_{bh} vs. M_R relation. Under this restriction the best-fitting relation has a normalization of $M_{bh} = 0.0012 M_{bulge}$, and can clearly be seen to be an excellent representation of the data (Fig 1). It is noteworthy that the normalization of $M_{bh} = 0.0012 M_{bulge}$ is identical to that determined by Merritt & Ferrarese (2001b) from their velocity dispersion study of the 32 inactive galaxies in the Magorrian et al. sample.

The closeness of the agreement between the M_{bh}/M_{bulge} ratios determined here with those determined by Merritt & Ferrarese is highlighted by Fig 2, which shows a histogram of the M_{bh}/M_{bulge} distribution for our 72-object AGN sample. The AGN M_{bh}/M_{bulge} distribution has $\langle \log(M_{bh}/M_{bulge}) \rangle = -2.87 \pm 0.06$ with a standard deviation of $\sigma = 0.47$. This is in remarkably good agreement with the Merritt & Ferrarese results, which were $\langle \log(M_{bh}/M_{bulge}) \rangle = -2.90$ and $\sigma = 0.45$. Finally, we note that the normalization of $M_{bh} = 0.0012 M_{bulge}$ agrees very well with the predictions of recent models of coupled bulge/black hole formation at high redshift (Archibald et al. 2001).

3   Bulge Luminosity versus Stellar Velocity Dispersion

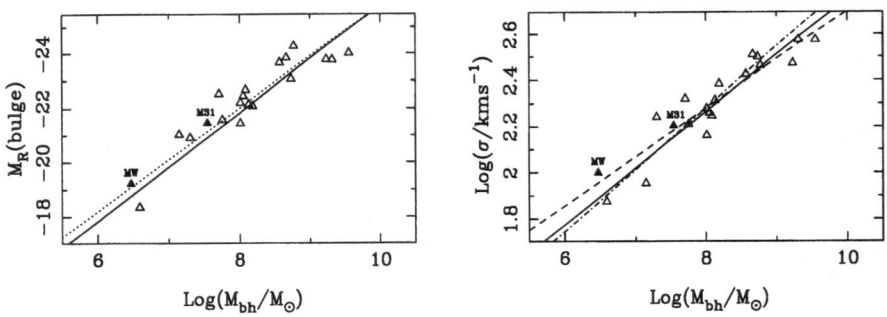

Fig. 3. Left-hand panel shows absolute R−band bulge magnitude versus dynamical black-hole mass estimate for our inactive galaxy sample. The solid line is the best-fitting relation ($M_{bh} \propto M_{bulge}^{0.95 \pm 0.09}$) and the dotted line is the best-fitting linear relation ($M_{bh} = 0.0012 M_{bulge}$). The right-hand panel is the same with bulge luminosity replaced by stellar velocity dispersion. The solid line is the best-fit ($M_{bh} \propto \sigma^{4.09}$), the dashed line is the Merritt & Ferrarese (2001a) relation ($M_{bh} \propto \sigma^{4.72}$), and the dot-dashed line is the Gebhardt et al. (2000) relation ($M_{bh} \propto \sigma^{3.75}$). The location of the Milky Way and M31 are indicated for the interest of the reader, although neither were included in the analysis

The quality of the fit to the inactive galaxy sample is illustrated by the left-hand panel of Fig 3, which shows the $M_{bh} - L_{bulge}$ relation for the inactive galaxy sample alone. Of particular interest is the scatter around this best-fit relation, given that it has been widely reported in the literature (eg. Merritt & Ferrarese 2001a, Kormendy & Gebhardt 2001) that the scatter around the $M_{bh} - L_{bulge}$ relation is significantly greater than that around the $M_{bh} - \sigma$ relation. However, in contrast, we find that the scatter around the $M_{bh} - L_{bulge}$ relation for our sample of nearby inactive galaxies, which excludes non E-type morphologies, is only 0.33 dex, in excellent agreement with the scatter around the $M_{bh} - \sigma$ relation (Merritt & Ferrarese 2001a).

To test this result further, in the right-hand panel of Fig 3, we investigate the $M_{bh} - \sigma$ relation for our nearby inactive galaxy sample. The scatter around the best-fit relation ($M_{bh} \propto \sigma^{4.09}$) is 0.30 dex, leading us to the conclusion that the intrinsic scatter around the $M_{bh} - L_{bulge}$ relation for *elliptical* galaxies is comparable to that in the $M_{bh} - \sigma$ relation.

4 Conclusions

The main conclusions of this study can be summarized as follows:

- The best-fitting $M_{bh} - L_{bulge}$ relation to the combined sample of 72 AGN and 18 nearby inactive elliptical galaxies is found to be consistent with a linear scaling between black hole and bulge mass ($M_{bh} \propto M_{bulge}^{0.95\pm0.05}$), and to have much lower scatter than previously reported ($\Delta \log M_{bh} = 0.39$ dex).
- The best-fitting normalization of the $M_{bh} - M_{bulge}$ relation is found to be $M_{bh} = 0.0012 M_{bulge}$, in excellent agreement with recent stellar velocity dispersion studies.
- In contrast to previous reports it is found that the scatter around the $M_{bh} - L_{bulge}$ and $M_{bh} - \sigma$ relations for nearby inactive elliptical galaxies are comparable, at only ~ 0.3 dex.

References

1. E.N. Archibald, et al: ApJ submitted, astro-ph/0108122
2. J.S. Dunlop, et al: MNRAS submitted, astro-ph/0108397
3. I. Jørgensen, M. Franx, P. Kjærgaard: MNRAS **280**, 167 (1996)
4. K. Gebhardt, et al: ApJ **539**, 13 (2000)
5. J. Kormendy, K. Gebhardt: submitted, astro-ph/0105230
6. M. Lacy, et al: ApJ **551**, L17 (2001)
7. A. Laor: ApJ **505**, L83 (1998)
8. A. Laor: ApJ **553**, 677 (2001)
9. H. Lin, et al: ApJ **464**, 60 (1996)
10. J. Magorrian, et al: AJ **115**, 2285 (1998)
11. R.J. McLure, J.S. Dunlop: MNRAS **327**, 199 (2001)
12. R.J. McLure, J.S. Dunlop: MNRAS in press, astro-ph/0108417
13. D. Merritt, L. Ferrarese: ApJ **547**, 140 (2001a)
14. D. Merritt, L. Ferrarese: MNRAS **320**, L30 (2001b)
15. A. Wandel: ApJ **519**, L39 (1999)
16. L. Wisotzki, B. Kuhlbrodt, K. Jahnke: astro-ph/0103112

The 2MASS Luminosity, Velocity, and Mass Functions of Galaxies

Michael A. Pahre, Chris S. Kochanek, Emilio E. Falco, and John P. Huchra

Harvard–Smithsonian Center for Astrophysics,
60 Garden Street, Cambridge, MA 02138, USA

Abstract. The velocity function (VF) of galaxies is analogous to the luminosity function (LF), but a more direct tool for comparing observations and theory. The 2MASS LF is calculated here for a local sample of $\sim$ 4000 galaxies, and the VF is derived from a $\sim$ 1000 galaxy subsample drawn from it. The observed VF follows a Schechter functional form. All current analytical, semi-analytical, and numerical galaxy formation models predict power-law (or nearly so) shapes for the VF, and hence cannot reproduce the observations. It remains an open challenge for *any* galaxy formation model to explain the observed VF.

1 Introduction

The last several decades have seen a substantial development in the study of galaxy statistical properties. Today, a number of different studies have generated luminosity functions (LF) of a few thousand or tens of thousands of galaxies at optical [13,1,4] and near-infrared [9,3] wavelengths. Galaxy luminosities are very difficult to calculate in theoretical models, however, due to the complicated physical processes required (baryonic collapse, star formation, extinction, bar instabilities, etc.). While some global properties of the observational LFs – like the shallow faint-end slope – might challenge many theoretical models, such a comparison between observations and theory might better be characterized as inconclusive due to the large uncertainties in the models.

Galaxy halo masses are much more easily calculated from cosmological simulations and analytical (or semi-analytical) models, but are more challenging than luminosities to measure observationally – hence the intellectual motivation for this workshop. On the observational side, the circular rotational velocity for disks and the central velocity dispersion provide indicators of galaxy halo mass, albeit ones that are far from perfect since both sample only the inner portions of the dark halo. On the theoretical side, the dark halo mass can be converted into the equivalent circular velocity at its virial radius, the maximum rotation velocity of the halo, or the maximum velocity after baryonic infall has occurred. None of these are exactly comparable to observational measurements, since they either sample large radii inaccessible to observations (virial radius) or are subject to systematic errors resulting from inadequate resolution in the models (maximum velocities). Nonetheless, a velocity or mass function of galaxies – analogous to the LF – promises to be a powerful tool which can provide a more direct comparison between the observational properties of galaxies and theoretical models.

Since it was first proposed as a tool [2], work on the velocity function (VF) has lagged substantially behind the LF [5,7,8,6]. Most of the previous work on the subject was motivated to explain the incidence of gravitational lensing, but its applications are arguably more general, since the VF offers a method to discriminate between luminosity evolving and merging processes of galaxy evolution. Those VF studies can really only be taken as preliminary investigations into the topic, since they have a number of key limitations: (1) Completely different samples are used to define the LF and the kinematic relations (Tully-Fisher [TF] for spiral galaxies, Faber-Jackson [FJ] for elliptical galaxies). (2) Many early-type galaxies show evidence for rotational support, such that the morphological galaxy sequence (from elliptical to spiral galaxy) is not a one-to-one mapping onto the dynamical galaxy sequence (from pressure to rotational support). (3) Spectroscopic classifications based on fiber spectrographs are typically systematically biased with respect to apparent magnitude [10]. And (4) observational and theoretical definitions of a quantity can be significantly different – such as how the theoretical models typically use the halo virial velocity, but observations trace the kinematics at much smaller radii.

2   Data

An observational measurement of the VF which overcomes these limitations is a challenge that can be met with current instrumentation and databases. In this paper, we present the VF for a galaxy sample based on the 2MASS galaxy catalog.

We start with a complete sample of ~ 4000 galaxies with $K_s \leq 11.25$ mag from the 2MASS catalog. Many galaxies have known redshifts from prior surveys, so we measured a modest number of new redshifts to achieve 100% completeness for the redshift survey. We then visually classified the morphologies of the entire sample using DSS and DSS2 images.

A substantial subsample of galaxies were further supplemented with central velocity dispersions and/or maximum rotational velocities, taken from the literature and then normalized onto common observational systems. More than 500 early-type galaxies have measurements of central velocity dispersions, and a similar number of late-type have maximum circular rotational velocities, thus this subsample represents $\sim 25\%$ of the parent LF sample. Approximately one-fourth (~ 130) of these early-type galaxies have maximum rotational velocities (major axis) in addition to the central velocity dispersions. We refer to the present work as the MALIGN (Masses And Luminosities of Infrared Galaxies Nearby) project, Phase I.

Future work (MALIGN, Phase II) will pursue the task in a more complete manner, by obtaining spatially-resolved internal kinematics of a complete sample of 2MASS selected galaxies, thereby providing a sample with a full understanding of various selection functions.

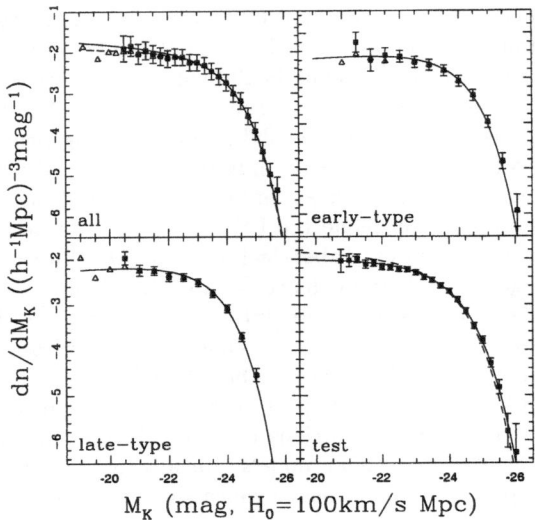

Fig. 1. (upper left) The K-band LF of galaxies from the 2MASS catalog, and broken down into **(upper right)** early- and **(lower left)** late-type LFs. From [9]

3 Measuring the VF Using Stellar and Gas Kinematics

The method of deriving the VF starts with the LF in its standard Schechter [15] form:

$$\frac{dn}{dL} = n_* \left(\frac{L}{L_*}\right)^\alpha \exp\left(-\frac{L}{L_*}\right) \tag{1}$$

which provides $n(L)$, while we wish to find $n(v)$. The LF for the 2MASS sample is shown in Fig. 1 [9], which has a relatively flat faint-end slope and is quantitatively consistent with the more distant 2dF K-band LF of [3].

The method used here is to adopt the two kinematic relations of the Tully-Fisher (spiral galaxies) and Faber-Jackson (elliptical galaxies):

$$\begin{aligned} L &= L_s v_c^{b_s} && \text{Tully} - \text{Fisher(spirals)} \\ &= L_e \left(\sqrt{2}\sigma_0\right)^{b_e} && \text{Faber} - \text{Jackson(ellipticals)} \end{aligned} \tag{2}$$

where $b_s \sim b_e \sim 4.0 \pm 0.2$ in the K-band, v_c is the rotational circular velocity, and $\sigma_0 = v_c/\sqrt{2}$ is the central velocity dispersion. These kinematic relations for the 2MASS subsamples are shown in Fig. 2. The full Fundamental Plane (FP) of elliptical galaxies is not used for two reasons: one, the sample is chosen in apparent magnitude, so the luminosity version of the scaling relation is required to account for any selection effects; and two, the FP has a luminosity density term (the mean surface brightness) which is not available for the photometric sample from the 2MASS catalog. Future work will address both issues. An additional

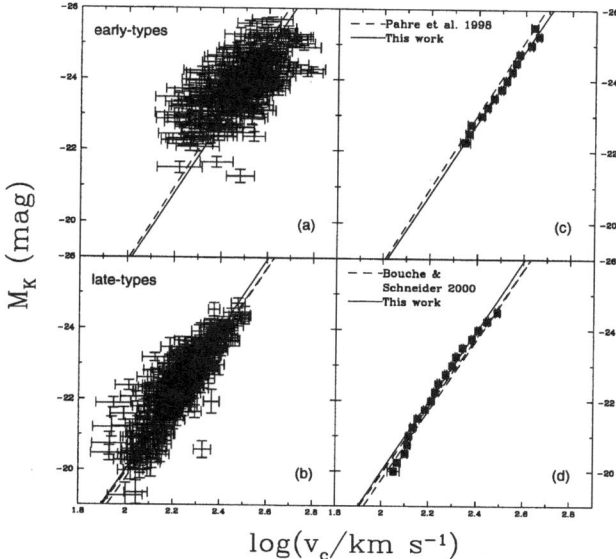

Fig. 2. The kinematical relations: (a) Faber-Jackson for early-types, and (b) Tully-Fisher for late-types. (c) and (d) are binned versions of (a) and (b), respectively

uncertainty in (2) occurs, since some of the early-type galaxies show partial rotational support. This is shown in Fig. 3. For the present work, we derive separate VFs with and without this early-type galaxy rotational support, and find a difference between the two only at the high velocity (mass) end.

Substituting (2) into (1) after normalizing the velocity by a characteristic velocity v_*:

$$\frac{L}{L_*} = \left(\frac{v}{v_*}\right)^b \tag{3}$$

yields an estimate of the velocity function:

$$\frac{dn}{d\log v} = \frac{dn}{dL}\left|\frac{dL}{d\log v}\right| = n'_* \ln(10) \left(\frac{v}{v_*}\right)^{b(1+\alpha)} \exp\left(-\left(\frac{v}{v_*}\right)^b\right), \tag{4}$$

which has a form similar to that of the Schechter function. The VF based on (4), calculated separately for early- and late-type galaxies (and then summed), is shown in Fig. 4. As was found for the K-band LF, this differential VF has a relatively flat low velocity (mass) end, meaning that most of the integrated galaxy mass comes from the massive galaxies near v_*. Including rotational support for the early-type galaxies shifts the high velocity end of the VF to the right by a small amount, but does not modify the low velocity end because it is flat.

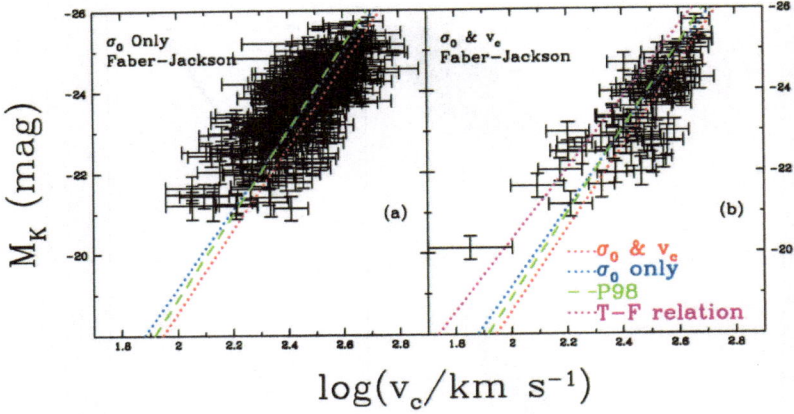

$$\log(v_c/\mathrm{km\ s^{-1}})$$

Fig. 3. The Faber-Jackson relation for early-type galaxies (**a**) without using any rotational support information [500 galaxies], and (**b**) using the rotational support information [130 galaxies]. The latter shows a small, systematic shift to the right

4   Comparison of the VF with Previous Observations

A comparison between our 2MASS derived VF and previous determinations [6] is shown in Fig. 4(**a**), along with a "standard gravitational lensing" determination [8]. Since our sample has complete LF information and a statistically treated subsample with kinematic information, our derivation of the VF preserves the covariance and allows us to estimate robust error bars. The scatter among the previous determinations is greater than the internal error of our new measurement. Furthermore, we have improved upon the first three of the four limitations of previous work identified in Section 1, hence our external systematic errors should be reduced significantly.

5   Comparison of the Velocity Function with Models

A comparison between our 2MASS derived VF and various model predictions (mostly from [6]) is shown in Fig. 4(**b**). None of the model predictions match the data. The fundamental problem is that all of the models – even those with the semi-analytical model treatment – predict roughly power-law VFs, while the observed VF follows the Schechter form.

Simple dark matter models without infall (see [6]) or adiabatic compression [11] under-predict the number of galaxies near to v_*. Clearly some amount of compression of the baryonic content is required to obtain the correct space densities of galaxies near v_*.

All models over-predict the number of high velocity galaxies, which is only partially surprising since those models do not distinguish between the massive dark matter halos that form galaxies and those that form groups of galaxies.

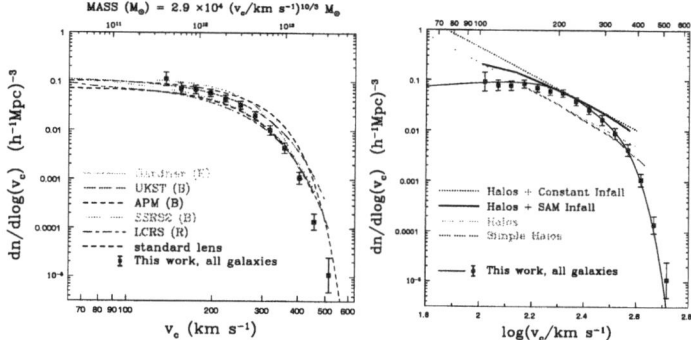

Fig. 4. (*left*) Comparison of the 2MASS VF with other observational determinations. The scatter in the previous work is largely due to large systematic uncertainties among various LFs, but some is no doubt also due to the heterogeneity of their galaxy samples. The conversion from the VF to the mass function is represented on the upper axis (see §6). (**right**) Comparison with theoretical model predictions. The observed VF follows a Schechter functional form. All current models, however, follow power-law (or nearly so) forms, and hence dramatically over-predict the observed VF at both the low- and high-velocity ends. From [14], plotted with various VFs and models from [6]

All models also over-predict the number of low velocity galaxies, which probably points to serious inadequacies in the input physics of the models. Low mass galaxies might be unstable to the energetic feedback effects of even one supernova, but these energetics are not fully treated in any of the models displayed.

6   The Mass Function of Galaxies

Following [12], we can investigate what would be required in order match the theoretical mass function to the observational velocity function. For approximately one decade in mass near to M_*, the conversion from observed velocity v_c into theoretical halo mass follows the power-law relation $v_c \propto M^{10/3}$. Applying this relation to the observational VF thus converts it into a MF (albeit only in relation to this particular model); this MF is plotted as the upper axis in Fig. 4 (**left**). This is one promising approach for studying the methodology of converting an observational VF into an "observed" halo MF. An alternate method for future investigations will be to estimate halo mass using the quantity $r_d v_c^2$ suggested by van den Bosch (these proceedings).

The VF provides a comparison between observations and theory that is more direct than the LF. All models explored to date fail to reproduce the observed VF presented here, which points to serious inadequacies in the models. It remains an open challenge for *any* galaxy formation model to explain the observed VF.

The authors wish to thank the SOC/LOC for organizing an exciting meeting. We thank A. Gonzalez for providing many of the models and observations plotted in Fig. 4. MAP acknowledges support at various stages of this project from NASA grants HF-01099.01-97A and NAG5-10777.

References

1. M. R. Blanton, et al.: AJ, 121, 2358 (2001)
2. S. Cole, N. Kaiser: MNRAS, 237, 1127 (1989)
3. S. Cole, et al.: MNRAS, 326, 255 (2001)
4. N. Cross, et al.: MNRAS, 324, 825 (2001)
5. M. Fukugita, Turner, E. L.: MNRAS, 253, 99 (1991)
6. A. H. Gonzalez, et al.: ApJ, 528, 145 (2000)
7. C. S. Kochanek:ApJ, 419, 12 (1993)
8. C. S. Kochanek: ApJ, 466, 638 (1996)
9. C. S. Kochanek, et al.: ApJ, 560, 566 (2001)
10. C. S. Kochanek, M. A. Pahre, E. E. Falco: ApJ, submitted (2002) (astro-ph/0011458)
11. C. S. Kochanek, M. White: ApJ, submitted (2002) (astro-ph/0102334)
12. C. S. Kochanek: in the proceedings of The Dark Universe (meeting at STScI, April 2-5, 2001) M. Livio, ed. (2002) (astro-ph/0108160)
13. H. Lin, et al.: ApJ, 464, 60 (1996)
14. M. A. Pahre, C. S. Kochanek, E. E. Falco: in preparation (2002)
15. P. Schechter: ApJ, 203, 297 (1976)

Estimating the Mass of Local Disk Galaxies from the NIR Luminosity

Giuseppe Gavazzi

Università degli Studi di Milano - Bicocca, P.zza delle Scienze 3, 20126 Milano, Italy

1 Introduction

The relation between the dynamical mass ($\propto V_{max}^2 \times r_{opt}$) (within the optical radius) and the luminosity of disk galaxies was studied by [6] using U, B, V and H band aperture photometry of several hundred galaxies at z=0. They found that the slope of the $L(\lambda) \propto M^{\alpha(\lambda)}$ relation increases with wavelength, $\alpha(\lambda) = 0.69(U), 0.76(B), 0.82(V), 1.00(H)$. In other words only the NIR luminosity is in direct proportionality with the dynamical mass of disk systems, or their M/L_H is nearly constant. Beside offering a cheap mass estimator, useful for example for estimating the mass of galaxies at intermediate redshifts, the $M \propto L_H$ relation has important implications on our understanding of the structure and perhaps of the evolution of galaxies.

2 The Updated $M \propto L_H$ Relation

An updated version of the $M \propto L_H$ relation for spiral galaxies is given in the left panel of Fig.1. It results from a NIR surface photometry of nearly 1500 galaxies selected in the Virgo cluster and in the Coma supercluster analyzed in [7],[8], providing better total H magnitudes (extrapolated to infinity) and a more objective mass estimate ($\propto V_{max}^2 \times r_e$). The resulting relation is marginally steeper $\alpha = 1.07(H)$ than in [6], but still consistent with a slope of one.
The NIR imaging survey was extended to cover early-type as well as spiral galaxies. A $M \propto L_H$ relation was obtained also for E+S0 galaxies (see right panel of Fig.1). I should spend some words illustrating how we constructed this relation. M was derived following these steps:

- The observed H band radial light profiles were deprojected, assuming spherical symmetry.
- Assuming M/L=k within galaxies, the radial mass profiles were derived from the light profiles.
- Assuming isotropy of the velocity dispersion tensor and hydrostatical equilibrium, the radial profile of the velocity dispersion was derived.
- Integrating this profile along the line of sight within a standard aperture, the central velocity dispersion σ_c (model) was obtained.
- By adjusting M/L such that σ_c (observed) = σ_c (model), M as plotted in the right panel of Fig.1 was finally obtained.

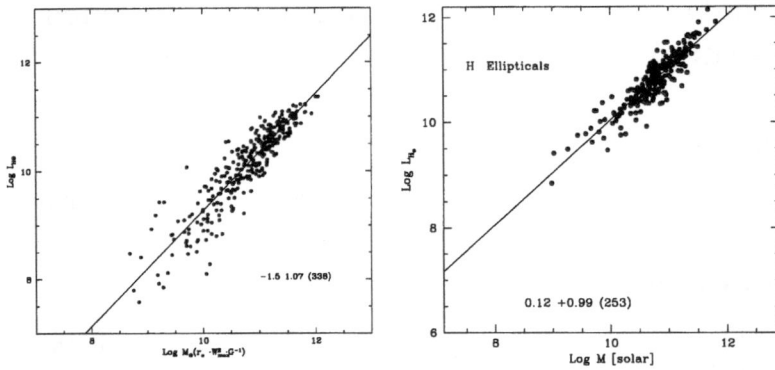

Fig. 1. The $M \propto L_H$ relation for spiral (left) and elliptical (right) galaxies, as obtained from surface photometry

The mass so derived for elliptical galaxies is found in direct proportionality with L_H.

The contradiction between the above statement and the existence of the tilt of the Fundamental Plane (see [5]) or k3 space (see [2]), even in their NIR versions, (see [12]) resides on the assumption of homology (σ_c (observed) $\propto \sigma$ (virial)), which we believe is broken in elliptical systems ([14]) because the ratio of the two velocities is found to vary systematically along the luminosity sequence. Whereas M/L varies (producing the tilt of the FP) if M is computed using σ_c (observed), M/L remains constant if M is computed using σ (virial) as obtained from the model. In conclusion, the H luminosity provides a good first-order estimate of the dynamical mass of galaxies, irrespective of their morphological types, allowing us to use hereafter L_H as synonymous of mass.

3   The Mass Dependence of the NIR Structural Parameters

The NIR light profiles of 1500 galaxies were fitted with de Vaucouleurs (Bulge), exponential (Disk) and Bulge+Disk profiles (see [7],[8]). The frequency of the profile decompositions as a function of the system H luminosity (Mass) is given in Fig.2. The fraction of pure de Vaucouleurs profiles increases with the luminosity, opposite to that of exponential disks. It is remarkable that this pattern is independent of the Hubble type, indicating that a global parameter, such as the total system mass, governs the shape of galaxies, as traced by their old stars. Thus the observed variety of morphological types is not only regulated by the angular momentum, as more commonly assumed.

From the imaging material the NIR light concentration index C_{31} was obtained (see [9]). This is a model-free morphological parameter which assumes values > 7

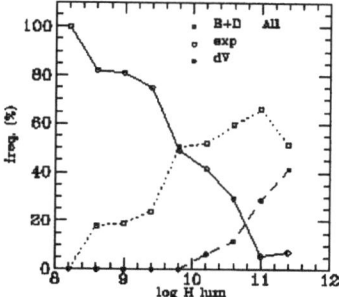

Fig. 2. The frequency of the NIR light-profile decompositions as a function of H luminosity, including E and S galaxies

if central cusps (bulges) are present or low values ($C_{31} \sim 3$) in exponential disks. C_{31} is found to depend non-linearly on the system mass (see bottom-right panel of Fig.3). Similarly, the global galaxy colors are known to depend on luminosity (see top-right panel of Fig.3). Altogether it appears that three photometric parameters: the "shape" (C_{31}), the mass (L_H) and the color of galaxies are necessary and sufficient to fully characterize the variety of galaxies, without invoking the Hubble classification, as discussed in [9].

4   The Mass Dependence of the Star Formation History

With the aim of constraining the star formation history (SFH) of galaxies in the cluster nearest to us, we recently undertook a spectro-photometric survey of the Virgo cluster ([11]). So far we obtained 125 spectra spanning a broad range in luminosity and Hubble type. The spectroscopic data were combined with photometry taken over the broadest possible wavelength baseline, including UV (2000 Å), optical (UBV) and infrared (JHK) photometry. The spectral energy distributions (SED) so obtained cover the domain 2000–22000 Å. The SEDs were fitted with Bruzual & Charlot [4] population synthesis models assuming a Salpeter IMF, an age T=13 Gyr and letting the metallicity and the time scale τ of the assumed SFH as free parameters. We adopted a star formation history with delayed-exponential form "a la Sandage" which mimics the SFH first proposed by [13] (see left panel in Fig.5). Fig.4 shows the fit of the templates SEDs with B&C models which were obtained averaging photometry and spectra of several galaxies in 10 bins of Hubble type. It appears that the whole Hubble sequence can be modeled assuming increasingly delayed star formation histories of longer duration (τ).

B&C models were also fitted to the SEDs of the individual (spiral) galaxies and the resulting relationship between τ and the H luminosity is given in right panel of Fig.5. For L_H spanning 4 decades, τ varies between 3 and 20 Gyr. The most luminous spiral galaxies have experienced a "short" burst of star formation in their early history, whereas the less luminous systems have long and shallow star formation histories.

5   Conclusions

The point I wanted to stress in my contribution is that the more we look at galaxies, the more we find that their global mass plays a fundamental role in determining their structural properties. With increasing mass we find that:

- Their shape becomes progressively more "cuspy" (old star are more concentrated toward the center producing significant bulges). This is true for both Elliptical and spiral galaxies.
- Their color becomes progressively redder. In ellipticals this is most likely the result of a metallicity sequence ([1]), whereas in spirals it reflects a genuine "ageing" of the stellar population with increasing mass ([9], [3]).
- Their star formation history becomes progressively consistent with shorter star formation events shifted toward earlier epochs.

Altogether it seems that massive galaxies appear "older" than their corresponding low mass counterparts. This evidence is to be taken into serious considerations and might contribute disentangling between models of galaxy formation and evolution.

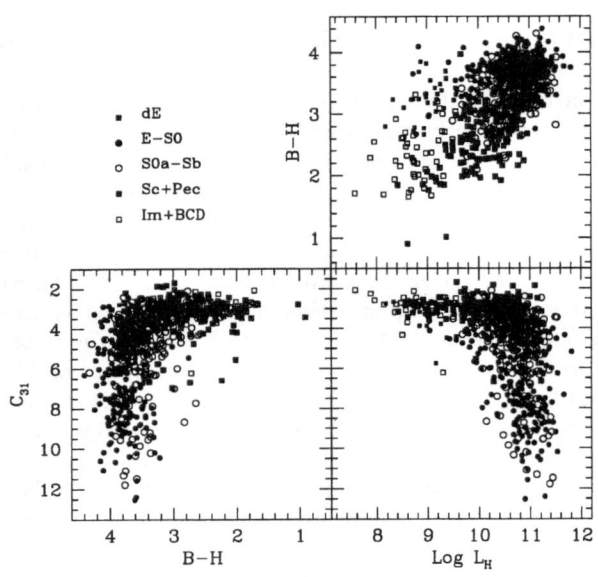

Fig. 3. The Color-C_{31}-H luminosity relation

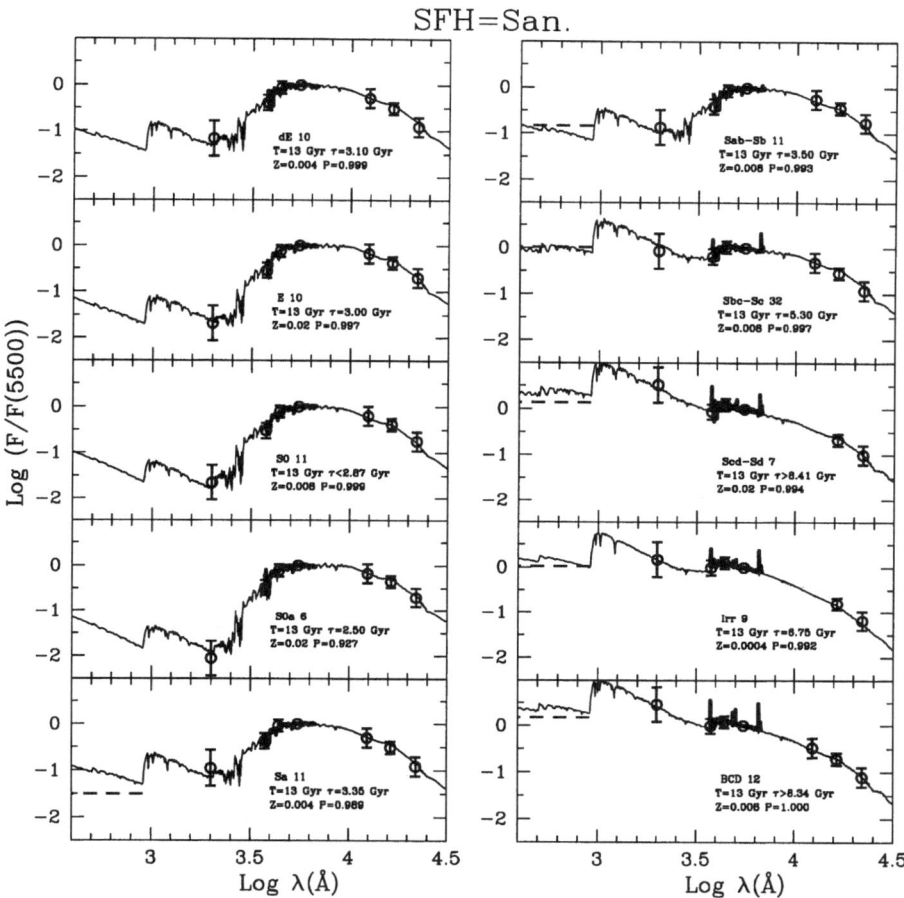

Fig. 4. The template SEDs, obtained averaging the photometry (dots) and the spectra (thick line) of 125 galaxies in the Virgo cluster, grouped in 10 bins of Hubble type. The SEDs are fitted with B&C models (thin line), assuming an age T=13 Gyr, and letting Z and τ as free parameters

Acknowledgements

Christian Bonfanti, Alessandro Boselli, Paolo Franzetti, Gerry Sanvito, Marco Scodeggio and Stefano Zibetti are warmly acknowledged for their precious contributions to this work. This work is based on observations taken at the ESO, OHP, TIRGO, and TNG observatories.

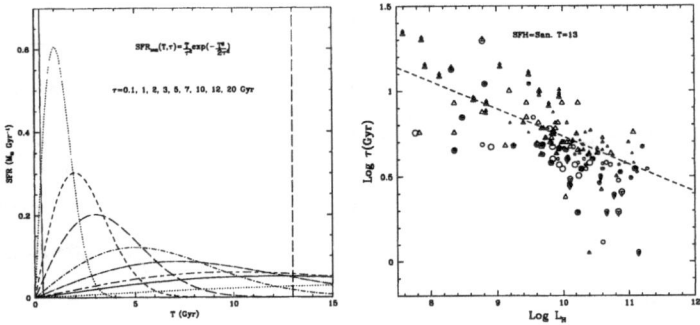

Fig. 5. Left panel: the time evolution of a star formation event a la "Sandage" is plotted for 8 values of τ. This SFH is characterized by a steep rise followed by an exponential decay. It becomes progressively shallower and the position of the peak is delayed in time with increasing τ. Right panel: the dependence of the fitted τ on L_H for spiral galaxies

References

1. N. Arimoto, Y. Yoshii, A&A, 173, 23 (1987)
2. D. Berstein, R. Bender, S. Faber, R. Nolthenius, AJ, 114, 1365 (1997)
3. A. Boselli, G. Gavazzi, J. Donas, M. Scodeggio, AJ, 121, 753 (2001)
4. G., Bruzual & S. Charlot, ApJ, 405, 538 (1993)
5. S. Djorgowski & M. Davis, ApJ, 313, 59 (1987)
6. G. Gavazzi, D. Pierini, A. Boselli, A&A, 312, 397 (1996)
7. G. Gavazzi, P. Franzetti, M. Scodeggio, A. Boselli, D. Pierini, A&A, 361, 863 (2000)
8. G. Gavazzi, S. Zibetti, A. Boselli, P. Franzetti, M. Scodeggio, S. Martocchi, A&A, 372, 29 (2001)
9. M. Scodeggio, G. Gavazzi, A. Boselli, P. Franzetti, S. Zibetti, D. Pierini, A&A (2002) (in press).
10. G. Gavazzi, C. Bonfanti, G. Sanvito, A. Boselli, M. Scodeggio, (2002) (submitted to AJ)
11. G. Gavazzi, A. Boselli, C. Bonfanti, G. Sanvito, M. Scodeggio, G. Stasinska, J. Lequeux, & R. Kennicutt, (2002) (submitted to A&A).
12. M. Pahre, S. Djorgowski, R. de Carvalho, AJ, 116, 1591 (1998)
13. A. Sandage, A&A, 161, 89 (1986)
14. S. Zibetti, G. Gavazzi, P. Franzetti (2002) (in preparation)

Galaxy Masses: Disks and Their Halos

Stacy McGaugh

Department of Astronomy, University of Maryland, College Park, MD 20742

Abstract. I review what we currently do and do not know about the masses of disk galaxies and their dark matter halos. The prognosis for disks is good: the asymptotic rotation velocity provides a good indicator of total disk mass. The prognosis for halos is bad: cuspy halos provide a poor description of the data, and the total mass of individual dark matter halos remains ill-constrained.

1 Disk Masses

The great regularity of the Tully-Fisher relation [32] has long been thought to originate from a strong mass-velocity relation and a near constancy of mass-to-light ratio. The latter requires a fair but not unreasonable amount of regularity in stellar populations. Put simply,[1]

$$L \sim \mathcal{M} \sim V^a \,.$$ (1)

There have long been indications (Sancisi 1995, private communication) that this simple scaling may fail at low luminosities. This has become more clear as data have improved [22],[29]. This breakdown of the Tully-Fisher relation might arise because of the chaotic star formation histories of low mass galaxies, or as a result of a breakdown in the underlying mass-velocity relation. Another possibility is that optical luminosity ceases to trace mass because stars cease to be the dominant mass component in these disks [15].

It has now become clear that this last possibility is in fact the case. Low mass galaxies are often dominated by gas rather than stars. If instead of luminosity or stellar mass, we plot disk (star + gas) mass against the flat rotation velocity, a nice mass-velocity relation is recovered over many orders of magnitude (Fig. 1). This 'Baryonic Tully-Fisher Relation' (BTF) is [21]

$$\mathcal{M}_d = \mathcal{A}V_f^b \,,$$ (2)

for which the data in Fig. 1 give

$$\mathcal{A} = 50\ \mathcal{M}_\odot\ \mathrm{km}^{-4}\,\mathrm{s}^4$$
$$b = 4.0 \pm 0.1 \,.$$ (3)

---

[1] "Stacy, you're a genius! ... [!] ... when it comes to pepper grinders" (van den Bosch 2001, private communication).

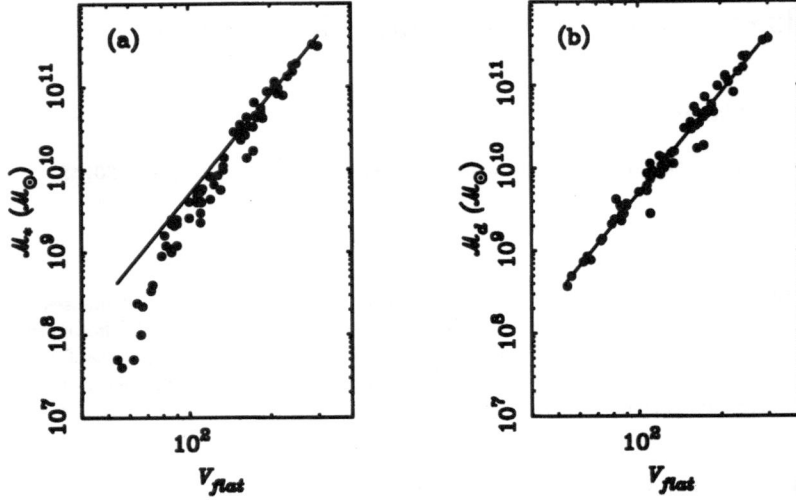

Fig. 1. The Tully-Fisher relation expressed in terms of (**a**) stellar mass and (**b**) baryonic disk mass (the sum of stars and gas). The luminous Tully-Fisher relation holds well for galaxies dominated by stars, but breaks down for low mass galaxies where the gas mass can often exceed the stellar mass (**a**). The sum of stars and gas provides a better correlation (**b**): the asymptotic flat rotation velocity is a good indicator of disk mass [21]. The data shown here are taken from a large compilation of high quality data [27]. Consequently, the scatter is greatly reduced from that in [21]. The galaxies shown here are drawn from the full range of disk Hubble types: mostly Sc, Sd, Sm, Im, but also a few Sa and Sb galaxies are present. The intrinsic scatter is small, with room only for scatter in the stellar mass-to-light ratio due to variations in star formation histories (probably not in the IMF), and scatter due to the modest ellipticities of disks [1]

The normalization of the BTF is rather uncertain: formally acceptable values fall in the range $34 < \mathcal{A} < 85$. The precise value of the slope has been modestly controversial: $b = 4.0$ was given by [21] while $b = 3.5$ was found by [2]. This difference can be traced to different assumptions about the (rather goofy [14]) distance to the UMa cluster for which some of the better rotation curve [38] and photometric data [33] exist. As the distance increases, the gas mass increases faster than the stellar mass (as D^2 and as D, respectively). This boosts the total mass of gas dominated galaxies by a larger factor than star dominated galaxies. Since these reside at opposite ends of the relation, the slope tips to shallower values with increasing D. Nevertheless, the population models of [2] are consistent with a slope of $b = 4.0$ (Fig. 2). While the calibration of the BTF can always be improved, it already provides an excellent indicator of disk mass. Moreover, continuity between gas-rich and star-rich galaxies constrains stellar population mass-to-light ratios. The favored values are reasonable in terms of population synthesis models (Fig. 2), but unpleasantly heavy for cuspy dark matter halos.

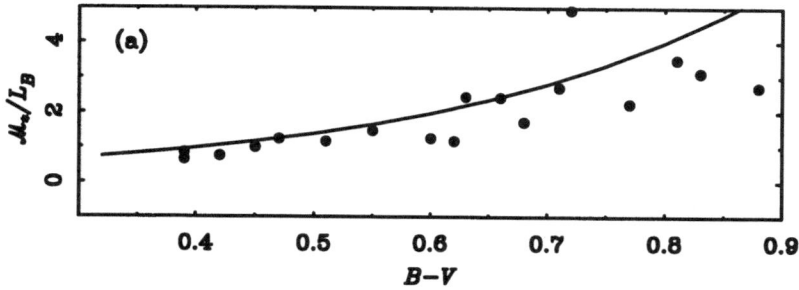

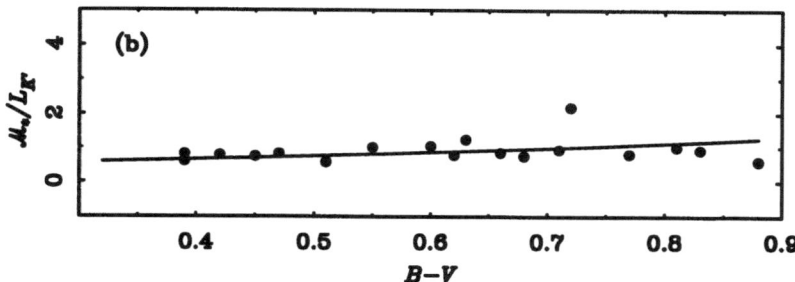

Fig. 2. The stellar mass-to-light ratios in (a) the B-band and (b) the K'-band predicted by a slope 4 BTF for the UMa galaxies [37], [33], [38]. These are plotted as a function of $B - V$ color, together with the Bruzual & Charlot, Salpeter IMF model from [2] (the first model in their table 4). The population synthesis models are in good agreement with the BTF, indicating that we have a good handle on $\mathcal{M}_*/L$ and disk masses

2   Dark Matter Halos

Rotation curves, by themselves, can only give a lower limit on the total halo mass: that enclosed by the last measured point. However, if the functional form of the halo were known, it might be possible to provide some constraint by fitting the observations to the known form. The NFW halo paradigm [23],[27] which has arisen from cosmological N-body simulations in principle gives a way to do this.

Unfortunately, if not surprisingly, observed rotation curves never extend far enough to constrain the circular velocity at the virial radius, V_{200} [20]. There is a great deal of degeneracy between the concentration c and V_{200}. An example is given in Fig. 3, which shows how difficult it can be to distinguish between fits with NFW halos of rather different parameters.

Matters are made worse by the general failure of the NFW form to provide a good description of the data. The data just don't look like NFW halos. Statist-

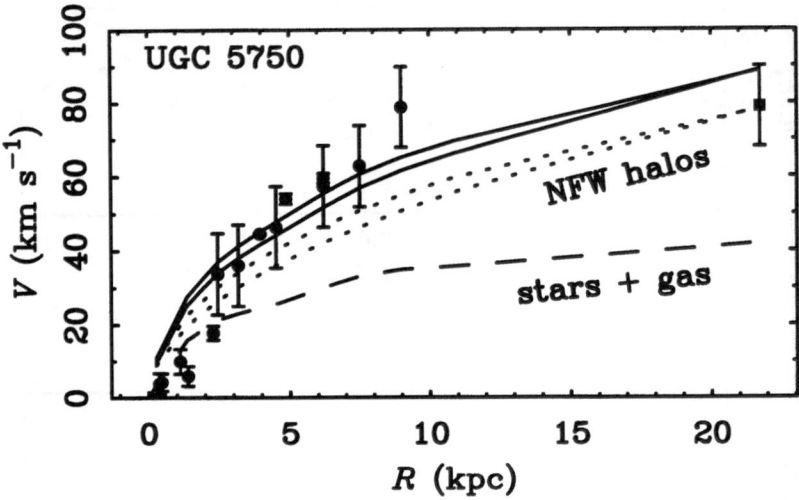

Fig. 3. An example of an NFW halo fit to the LSB galaxy UGC 5750 [11]. The best fit parameters in this case are $c = 1.9$ and $V_{200} = 117$ for $\mathcal{M}_*/L_R = 1.4\ \mathcal{M}_\odot/L_\odot$. Another tolerable fit with $c = 0.2$ and $V_{200} = 300$ is also shown (lower dotted curve) to illustrate the degeneracy between parameters. Though many models sort of fit, their concentrations are implausibly low for ΛCDM

ically, halos with constant density cores are almost always preferred over those with cusps [11]. This is most clear in the best resolved cases [10].

The most important systematic concern at this point is not observational. Resolution has improved by an order of magnitude [31], [20], [9] over the original 21 cm data for LSB galaxies [36], [12]. The NFW shape has not become apparent as the data have improved. Instead, the systematics pointed out by [18],[19] as problematic for CDM (independent of the cusp issue) have only become more clear. Concern over slit mispositioning [30] are misplaced: independent observers reproduce one anothers' results [20],[9]. While there are certainly cases in which the error bars are large enough to allow an NFW fit, isothermal fits are inevitably better. Simply changing the size of the error bars won't change this: a systematic change in the shapes of ~ 50 high resolution rotation curves is required. One can certainly imagine ways in which this might happen [30], but it is extremely unlikely that any of these ideas apply to real data, let alone to **all** of the data from various independent sources.

The most serious issue is in the mass models: stars have mass. Even in the limit of zero stellar mass, which is the most favorable to the NFW case, isothermal halos are statistically preferred [11],[9]. The situation only becomes more grim if stars are allowed to have mass. Though LSB galaxies are dark matter dominated down to small radii, plausible $\mathcal{M}_*/L$ models do require that *some* of the velocity be attributed to luminous mass. This pulls the inferred dark matter distribution further away from the expected cusp slope.

We are hardly unique in reaching these conclusions, which are shared by **all** published analyses of high resolution long slit Hα data [6],[8],[9],[10],[11],[13],[26]. High resolution Fabry-Perot [4],[39],[25] and CO [5] data are also inconsistent with cuspy halos, as are a variety of data for the Milky Way itself [3]. The only analyses which are favorable to NFW are those of low resolution data with large error bars [34],[35]. When the error bars are large, any model can be driven through them. Though it has not been emphasized, constant density cores provide as good or better fits even in these cases.

The isothermal halo form, while effective, is an extremely flexible fitting function which lacks the motivation of the NFW halo form. So one might persists that the NFW fits are still more appropriate in that they can be related to cosmology. Standard ΛCDM makes a clear prediction [27] for what the concentrations of dark matter halos should be: $c = 9$ for $\Omega_m h = 0.2$. Scatter about this value should be modest – the largest estimate [7] finds a lognormal distribution with $\sigma_c = 0.18$. The median observed concentration is $c = 6.4$ [11] which is different from the standard ΛCDM prediction by many σ. The problem with NFW halos is not just a matter of getting fits to individual galaxies, but also of understanding how the observed concentrations can be so low. These low concentrations would be tolerable in a very low density universe with $\Omega_m h \approx 0.12$ [17]. Even then there exists a significant tail of very low concentration ($c < 4$) galaxies which simply should not exist for any plausible cosmology.

The debate over halo profiles, while contentious, misses the real point. Many halo profiles are nominally viable because they have lots of degenerate free parameters. Mass modeling is a bit like fitting a high order polynomial to a few data points: the line goes through the data, but means nothing. One would prefer to have a minimal parameter description of the data. Such a prescription exists [16]. It has long been noted that there is a strong coupling between mass and light.[2] Oddly, this coupling persists for dark matter dominated LSB galaxies. One needs only a single parameter per galaxy, the stellar mass-to-light ratio, in order to fit the rotation curve in comparable or greater detail than can be matched by many-parameter halo models. The mass-to-light ratio in the K-band is close enough to constant that one can make a good zero parameter prediction with such data [28]. Until we come to terms with this observed phenomenology, debating the cusp slope of dark matter halos is rather akin to debating the number of angels that can dance on the head of a pin.

Acknowledgements: I am most grateful to Vera Rubin, Erwin de Blok, and Albert Bosma for their work on the issues discussed here. I would also like to thank Renzo Sancisi, Marc Verheijen, Rob Swaters, and Frank van den Bosch for many lively and stimulating conversations. No doubt, I have yet to hear the end of it! The work of SSM is supported in part by NSF grant AST9901663.

---

[2] Renzo's Rule: when you see a feature in the light, you see a corresponding feature in the rotation curve.

References

1. D.R. Andersen, M.A. Bershady, L.S. Sparke, J.S. Gallagher, E.M. Wilcots: ApJ, **551**, L131 (2001)
2. E.F. Bell, R.S. de Jong: ApJ **550**, 212 (2001)
3. J.J. Binney, N.W. Evans: MNRAS, submitted (2002)
4. S. Blais-Ouellette, P. Amram, C. Carignan: AJ **121**, 1952 (2001)
5. A.D. Bolatto, J.D. Simon, A. Leroy, L. Blitz: ApJ in press (2002)
6. A. Borriello, P. Salucci: MNRAS **323**, 285 (2001)
7. J.S. Bullock, T.S. Kolatt, Y. Sigad, R.S. Somerville, A.V. Kravtsov, A.A. Klypin, J.R. Primack, A. Dekel: MNRAS **321**, 559 (2001)
8. S. Côté, C. Carignan, K.C. Freeman: AJ **120**, 3027 (2000)
9. W.J.G. de Blok, A. Bosma: A&A submitted (2002)
10. W.J.G. de Blok, S.S. McGaugh, A. Bosma, V.C. Rubin: ApJ **552**, L23 (2001)
11. W.J.G. de Blok, S.S. McGaugh, V.C. Rubin: AJ **122**, 2396 (2001)
12. W.J.G. de Blok, S.S. McGaugh, J.M. van der Hulst: MNRAS **283**, 18 (1996)
13. E. D'Onghia, these proceedings
14. W.L. Freedman et al.: ApJ **553**, 47 (2001)
15. K.C. Freeman: 'Historical Introduction'. In: *The Low Surface Brightness Universe, IAU Colloquium 171*, ed. J.I. Davies, C. Impey, S. Phillipps (Astronomical Society of the Pacific, San Francisco 1999) pp. 3-8
16. S.S. McGaugh: 'Dynamical Constraints on Disk Galaxy Formation'. In: *Galaxy Dynamics: from the Early Universe to the Present*, ed. F. Combes, G.A. Mamon, V. Charmandaris (Astronomical Society of the Pacific, San Francisco 2000) pp. 153-160
17. S.S. McGaugh, M.K. Barker, W.J.G. de Blok: ApJ, submitted (2002)
18. S.S. McGaugh, W.J.G. de Blok: ApJ, **499**, 41 (1998)
19. S.S. McGaugh, W.J.G. de Blok: ApJ, **499**, 66 (1998)
20. S.S. McGaugh, V.C. Rubin, W.J.G. de Blok: AJ **122**, 2381 (2001)
21. S.S. McGaugh, J.M. Schombert, G.D. Bothun, W.J.G. de Blok: ApJ **533**, L99 (2000)
22. L.D. Matthews, W. van Driel, J.S. Gallagher 1998: AJ **116**, 1169
23. B. Moore, T. Quinn, F. Governato, J. Stadel, G. Lake: MNRAS **310**, 1147 (1999)
24. J.F. Navarro, C.S. Frenk, S.D.M. White: ApJ **490**, 493 (1997)
25. P. Palunas, T.B. Williams: AJ **120**, 2884 (2000)
26. P. Salucci: MNRAS **320**, L1 (2001)
27. R.H. Sanders, S.S. McGaugh: ARA&A in press (2002)
28. R.H. Sanders, M.A.W. Verheijen: ApJ **503**, 97 (1998)
29. J.M. Stil: PhD Thesis, Leiden University (1999)
30. R.A. Swaters, these proceedings
31. R.A. Swaters, B.F. Madore, M Trewhella: ApJ **531**, L107
32. R.B. Tully, J.R. Fisher: A&A **54**, 661 (1977)
33. R.B. Tully, M.A.W. Verheijen, M.J. Pierce, J. Huang, R.J. Wainscoat: AJ **112**, 2471 (1996)
34. F.C. van den Bosch, B.E. Robertson, J.J. Dalcanton, W.J.G. de Blok: AJ, 119, 1579 (2000)
35. F.C. van den Bosch, R.A. Swaters: MNRAS **325**, 1017 (2001)
36. J.M. van der Hulst, E.D. Skillman, T.R. Smith, G.D. Bothun, S.S. McGaugh, W.J.G. de Blok: AJ **106**, 548 (1993)
37. M.A.W. Verheijen: PhD thesis, University of Groningen (1997)
38. M.A.W. Verheijen, R. Sancisi: A&A **370**, 765 (2001)
39. B.J. Weiner, J.A. Sellwood, T.B. Williams: ApJ **546**, 931 (2001)

Dark Matter in the Center of LSB and Dwarf Galaxies from Hα Observations

E. D'Onghia[1], D. Marchesini[2], G. Chincarini[3,4], C. Firmani[4], E. Molinari[4], P. Conconi[4], and A. Zacchei[5]

[1] Università degli Studi di Milano, Via Celoria 16, 20133, Milano, Italy
[2] SISSA, Via Beirut 4, 34014 Trieste, Italy
[3] Università degli Studi di Milano-Bicocca, Piazza dell'Ateneo Nuovo 1, 20126, Milano, Italy
[4] Osservatorio Astronomico di Brera-Merate, Via Bianchi 46, 23807 Merate (LC), Italy
[5] Telescopio Nazionale Galileo (TNG)

Abstract. This work is focused on the preliminary results of the observations of Hα rotation curves for some of the late type dwarf and LSB galaxies carried out at the TNG telescope. From our analysis we find good agreement between our Hα data and the HI observations taken from the literature, concluding that the HI rotation curves for these galaxies suffer very little from beam smearing. A preliminary analysis of our data rules out the CDM model in the inner regions of these galaxies.

1 Introduction

Cold Dark Matter (CDM) models provide a solid framework capable to explain most of the properties of the universe at large scales. Using N-body simulations it was showed that virialized haloes are well described by an universal density profile, the Navarro, Frenk, & White model (1997; hereafter NFW). This profile diverges at the center showing a cuspy core: $\rho \propto 1/r$. Recent high-resolution N-body simulations show that as the numerical resolution is increased, the resulting dark matter density profile goes as $\rho \propto 1/r^{1.5}$ (Moore et al. 1999). These profiles are in conflict with the observations: rotation curves of late-type dwarf and LSB galaxies seem to rule out singular halo profiles. Indeed, the rotation curves of these galaxies call for a finite central density (soft core), in conflict with the cuspy cores predicted by the model. Note that LSB and dwarf galaxies are systems strongly dark matter dominated, so their rotation curves are good tracers of the underlying dark halo gravitational potential and therefore good candidates to explore the innermost shape of the dark matter distribution. Recently, however, some authors have challenged the existence of soft cores in centers of galaxies measuring Hα rotation curves of late-type dwarf and LSB galaxies. Basically they find two results: HI rotation curves for most of the galaxies are affected by beam-smearing and a good agreement with concentration values predicted by the NFW model (van den Bosch and Swaters 2001).

1.1   Observations

We have carried out Hα rotation curves of dwarf and LSB galaxies at the TNG. The spatial resolution in the central regions is a factor 10 better than the one of the HI data published in literature for the same galaxies. Observations of late-type dwarf and LSB galaxies were carried out using the instrument Dolores (scale = 0.275 Åpxl^{-1}, dispersion = 0.8 Åpxl^{-1}, λ ∈ (6200, 7800) Å).

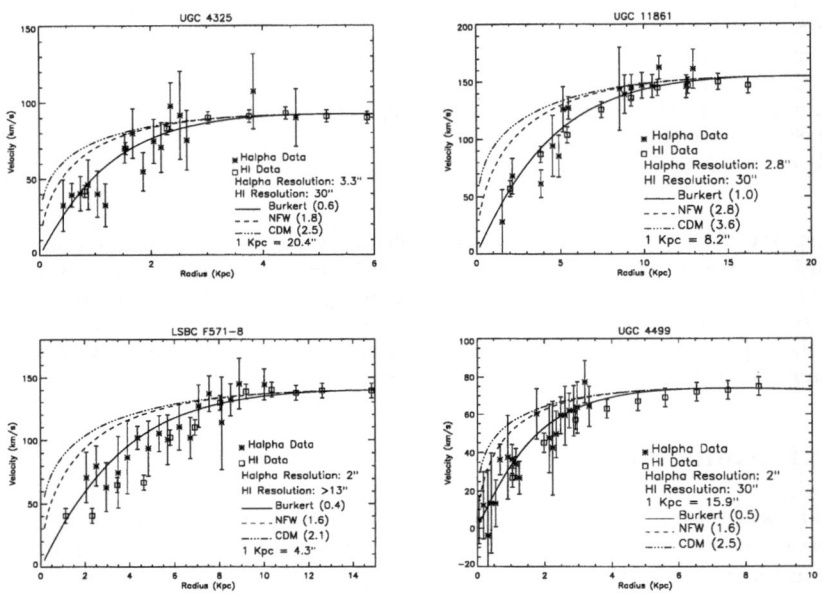

Our Hα data (stars in the figures) are plotted together with HI data (from literature, empty squares). As one can see, the HI data suffer of very little beam smearing. The solid lines are the best fits obtained by using the Burkert's profile; the dashed lines are NFW profile and the dotted-dashed lines are Moore's profile (CDM model). The numbers in parenthesis represent the normalized χ^2 for each fit. For each galaxy is stated the spatial resolution for both the Hα and the HI data. For all the four galaxies, the Burkert's profile is the best fit, while the Moore's profile does not match data. New observations with high resolution new technology grating (VPHG, Conconi et al. 2001) are being planned.

2   Conclusions

Our preliminary analysis of Hα rotation curves seems to rule out the CDM scenario in the inner regions of LSB and dwarf galaxies, showing evidence of soft cores for most of the galaxies.

References

1. de Blok, W.J.G., McGaugh, S.S., Rubin V.: ApJ, in press, (2001)
2. Moore,B., Quinn,T., Governato,F., Stadel,J., Lake,G.: MNRAS, 310, 1147 (1999)
3. Navarro, J.F., Frenk, C.S., White, S.D.M: ApJ, 490, 493 (1997)
4. van den Bosch, F.C., Swaters, R.A.: MNRAS, 325, 1017 (2001)
5. Conconi, P. et al.: SPIE Meeting, San Diego, U.S.A. (2001)

Galaxy Systems in Low Density Environments: The NGC 4756 System

Birgit Kelm[1], Paola Focardi[1], Ruth Grützbauch[2], Werner W. Zeilinger[2], and Roberto Rampazzo[3]

[1] Universitá di Bologna, Bologna, I
[2] Institut für Astronomie der Universität Wien, A
[3] Osservatorio Astronomico di Brera, Milano, I

Abstract. Isolated compact structures are investigated to identify possible links between their characteristics and the presence of an AGN.

1 Introduction

The study of isolated and dense groups of galaxies permits to assess the effects of strong interactions on morphology and physical properties of galaxies such as induced star formation and AGN-type phenomena [1], [2]. Despite the prediction that a fast merging evolution should affect dense systems observations of Hickson Compact Groups (HCGs) indicate that only 7% of the galaxies are presently merging [3]. Moreover though dwarf and low-luminosity AGNs appear to be frequent, bright AGNs are hosted in only few of the member galaxies [4], [5], [6]. A model assuming galaxies in CGs to constitute virialized systems embedded within a common massive halo may help to reduce the discrepancies between observed and predicted galaxy-galaxy interaction rate [7].

2 The NGC 4756 System

We report preliminary results for the NGC 4756 group which is part of a larger survey on physical properties of galaxies in low density environments. It is an extremely compact, isolated system projected onto a rich background cluster (ACO 1631). It includes 5 bright ($M_B \leq$-18) physical members within $200h^{-1}$ kpc and no further ($M_B \leq$-18) neighbours within a $600h^{-1}$kpc radius region. So far group membership is controversial. Garcia [8] classifies NGC 4756 as a 3 member group while Giuricin et al. [9] attribute 12 to 14 members to the group. New deep imaging and spectroscopy data obtained with EFOSC II at the ESO 3.6m telescope reveal a complex structure for the group. The dominant member is NGC 4756, a bright elliptical (E3) with 2 fainter companions in the close neighbourhood. A subgroup consisting of 4 bright members is found SW of NGC 4756 to which another 2 faint members are associated.

Deep imaging reveals no signatures of ongoing or past interactions for NGC 4756. The subgroup, consisting of an elliptical (IC 829) and 3 spirals (MCG-2-33-35, MCG-2-33-36 and MCG-2-33-38) reveals however peculiarities typical of interacting systems. MCG-2-33-36 displays an extremely blue arm extending toward IC 829 whose spectrum shows typical signatures of enhanced star formation. MCG-2-33-38 is a tidally distorted disk galaxy whose spectrum shows AGN-type signatures. Strong [OIII], [NII] and [SII] emission combined with weak, broad $H\alpha$ emission suggests an intermediate Seyfert type [10].

Dynamical mass estimates indicate that the group has a mass between 2.88×10^{12} and 3.45×10^{13} $M_\odot$, according to prescriptions in Bahcall & Tremaine [11] and Heisler et al. [12]. Preliminary analysis of the the X-ray emission from the ASCA data archive locates the two maxima at the position of the Seyfert galaxy within the sub-group and in correspondence to NGC 4756. Because of the large PSF, clear identification of a diffuse component encompassing the whole system is uncertain. We attribute most of the diffuse emission in the area around NGC 4756 to ACO 1631.

3   Discussion

The NGC 4756 system resembles HCG 40. Both CGs include a dominant elliptical closed to a subgroup displaying on-going merging as well as Seyfert activity. This kind of configuration is quite atypical for bright elliptical CG members, generally appearing surrounded by an "isotropic" galaxy distribution. We suggest that the bright elliptical might enhance the rate of galaxy-galaxy interaction in the subgroup and eventually trigger an AGN phenomenon. The mechanism acting could be analogous to the one suggested by Bekki [13] to provide a physical explanation for the relationship between cluster starburst galaxies and cluster substructures.

References

1. B. Moore et al.: Nature, **379**, 613 (1996)
2. J.E. Barnes: Nature **338**, 123 (1989)
3. S.E. Zepf: ApJ **407**, 448 (1993)
4. R. Coziol, A.L.B. Ribeiro, R.R. de Carvalho, H.W. Capelato: ApJ **493**, 563 (1998)
5. B. Kelm, P. Focardi, G.G.C: Palumbo: A&A. **335**, 912 (1998)
6. M. Shimada et al.: AJ **119**, 266 (2000)
7. A.I. Zabludoff & J.S. Mulchaey: ApJ **496**, 39 (1998)
8. A.M. Garcia et al.: A&AS **98**, 7 (1993)
9. G. Giuricin et al.: ApJ **543**, 178 (2000)
10. D.E. Osterbrock: ApJ **249**, 462 (1981)
11. J.N. Bahcall & S. Tremaine: ApJ **244**, 805 (1981)
12. J. Heisler, S. Tremaine, J.N. Bahcall: ApJ **298**, 8 (1985)
13. K. Bekki: ApJ **510**, L15 (1999)

The Masses of Dwarf Elliptical Galaxies: First Results from an ESO Large Program

H. Dejonghe[1], S. De Rijcke[12], G. Hau[3], and W. W. Zeilinger[4]

[1] Universiteit Gent, Krijgslaan 281, S9, B9000 Gent, Belgium
[2] Universität Basel, Venusstrasse 7, 4102 Binningen, Switzerland
[3] ESO, Casilla 19001, Vitacura, Santiago, Chile
[4] Universität Wien, Türkenschanzstrasse 17, A1180 Wien, Austria

1   Introduction

Dwarf ellipticals are of particular interest for several reasons. Firstly, they dominate in numbers the nearby universe, and can therefore righteously claim to be a class on their own. In this contribution, we will restrict ourselves to those galaxies that can be found at the faint end of the luminosity function of elliptical galaxies ($M_B > -18$) and that have a regular-elliptical appearance (hence excluding Blue Compact Dwarfs or dwarf Irregulars). In particular, we will concentrate on photometric characterisations of the dEs as a class, and on the internal dynamics. Secondly, the morphological, dynamical and chemical properties of dEs should provide important clues to their formation, and in any case constrain theoretical models. Almost certainly, this knowledge will also be relevant for understanding the formation mechanisms of their larger brethren. In this connection, the effects of the environment, and the relation of dEs to the bulges of spirals or even globulars come to mind. From a broader perspective, cosmological theories, in particular the CDM paradigm, expect generally that dark matter should be present ([14],[4]). It is obviously of great interest to check this prediction. Thirdly, a great deal of attention has gone to theoretical models of the evolution of dEs. The role of dark matter has not been explored to a great extent, but the discussion up to now rather focussed on the role of SN driven galactic winds. The basic idea is that the associated mass loss would render the galaxy rounder and eventually also produce a more anisotropic orbital structure. If, on the other hand, the dE would turn out to be too heavy to lose a substantial fraction of its mass through galactic winds, one would expect that the metal-rich gas gets expelled towards the outskirts, hence (1) triggering star formation at larger radii, with a corresponding effect on the photometric profile, and (2) a subsequent redistribution of the metal-rich gas, with a corresponding flat or slowly rising metallicity and an outward reddening as a result ([7],[17],[16]). All this can be tested by studying the photometry, internal kinematics and line strengths of dEs, which was the purpose of the ESO Large Program.

As for the existing data, it must be pointed out that it is extremely difficult to obtain reliable kinematical data with telescopes that are smaller than the 8m class. Hence, only for the photometry does there exist a solid result, namely that all ellipticals, from cDs to dEs, follow rather well the Sérsic photometric profile

$I(r) = I_0 \exp[-(r/r_0)^{1/n}]$ with reasonably convincing correlations $n(M)$, $I_0(M)$ and $r_0(M)$, M being total absolute magnitude. On the other hand, for only a few – relatively bright – dEs are projected velocities and/or velocity dispersion profiles available ([2], [3], [6], [11],[12], [15]). There are no data for faint dEs. Also, the existing data do not seem to be very consistent. It would seem from them that brighter dE's have $(V/\sigma)^* < .4$, i.e. that they are supported by pressure rather than by rotation. As to the central velocity dispersions σ_0, a few tens more are published ([16],[5]). Mass-to-light ratios resulting from them are unreliable however, because σ_0 may be sensitive to the dynamics of the nucleus (e.g. bright dEs may show a cold central cusp), and confidence intervals cannot be given. For reviews of the results of the pre-8m era, we refer to [13] and [10].

Table 1. Photometric and kinematic properties of the objects for which the observations have been fully reduced. The listed values for σ are not the central σ_0, but weighted averages over the central region.

name	type	M_R	M_B	$\langle \mu_R \rangle$	R_e (")	R_e (kpc)	$\overline{\sigma}$ (km/s)	v_{max} (km/s)
NGC 5044 group								
FS29	dE5	−18.71	−18.08	20.84	9.0	1.57	44.5±3.6	60.0±6.0
FS76	dE1	−17.92	−16.68	19.98	4.2	0.73	59.0±7.7	15.0±4.0
FS131	dE5N	−18.76	−17.48	20.56	8.1	1.41	67.7±4.3	20.0±4.0
NGC 5898 group								
dw2	dE6	−17.27	/	20.82	5.9	0.80	42.1±3.2	20.0±7.0
Fornax								
FCC136	dE2N	−18.47	−17.40	21.38	13.4	1.19	39.3±2.4	10.0±3.5
FCC150	dE2	−17.45	−16.50	20.59	5.9	0.52	32.6±3.3	0.0±7.0
FCC245	dE1N	−17.16	−16.20	22.18	10.7	0.95	37.0±4.7	0.0±7.0
FCC266	dE0N	−17.19	−16.30	21.11	6.6	0.59	25.9±7.2	0.0±3.5
FCC288	dE7/dS0	−17.85	−16.62	21.28	9.7	0.86	39.3±2.8	50.0±7.0

2   The Observations

We compiled a sample of dEs, with the intention to (1) cover all ellipticities and (2) to investigate the role of the environment. We chose objects in a dense cluster environment (center of Fornax cluster), in a cluster/group environment (NGC5044 group, outskirts of Fornax cluster) and in small group environment (NGC5898 group). At this point, we can report on 18 nights on VLT + FORS2. We obtained VRI imaging with typically sub-arcsecond seeing in the magnitude ranges $-20 < M_R < -16$, $-18 < M_B < -15$ and long-slit spectroscopy ($\lambda\lambda 790-930$nm around CAII triplet, $\sigma_{instr} \approx 30$ km/s) along apparent major and minor axes, for dEs with $-19 < M_R < -17$. The typical total integration times are 5-8 hrs per position angle, and the typical radial extent of the spectra is 1-2 R_e. We have complete data sets on 9 dEs so far. In table 1, a number of key parameters are listed. We note that FS29 and FS76 are clearly oblate rotators. FS131 has an extreme peanut shape, while FCC288 has disky isophotes.

3   Preliminary Results

In figure 1, we present the photometry of a superset of the galaxies listed in table 1. There is a clear positive trend of the Sérsic parameter n as a function of M_R, confirming the results of [13]. The positive correlation $(B - R)(M_R)$ is also very convincing, supporting the scenario in which more massive dwarfs keep their processed material longer.

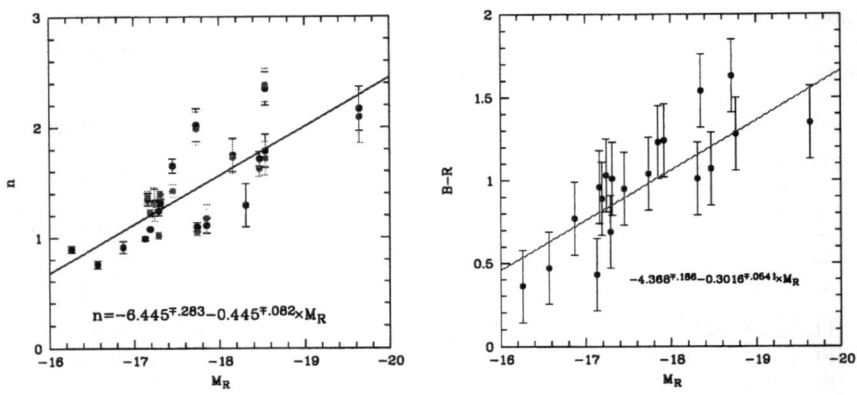

Fig. 1. Left panel: Sérsic parameter n versus absolute R magnitude. Black dots: n from R photometry, gray dots: n from I photometry. Right panel: $B - R$ color versus absolute R magnitude for the same sample

As for the kinematic properties, we already noted that fast rotating dwarfs do exist. Any formation scenario will have to take this into account. However, interpreting small number statistics, only a minority of the dEs (of the order of 15%) are flattened by rotation. There is also no obvious trend of v/σ with absolute magnitude, contrary to the case of the normal ellipticals. As can be seen in figure 2, there exists a reasonably tight Faber-Jackson type relation for dEs. However, there seems to be no obvious trend of v_{max} with absolute magnitude, or if there is one, the scatter is large, again contrary to the case of the normal ellipticals. As for the dynamical modeling, two galaxies have been modelled extensively so far. Details can be found in [8] and [9]. The main results of the modeling are estimates for the M/L and the mass inside $1.5R_e$ within a 90% confidence level, which we obtained on the basis of a large number of dynamical models. We find for FS29: $4.4 < (M/L)_B = 6.3 < 7.3$ or $2.6 < M/10^9 = 3.7 <$

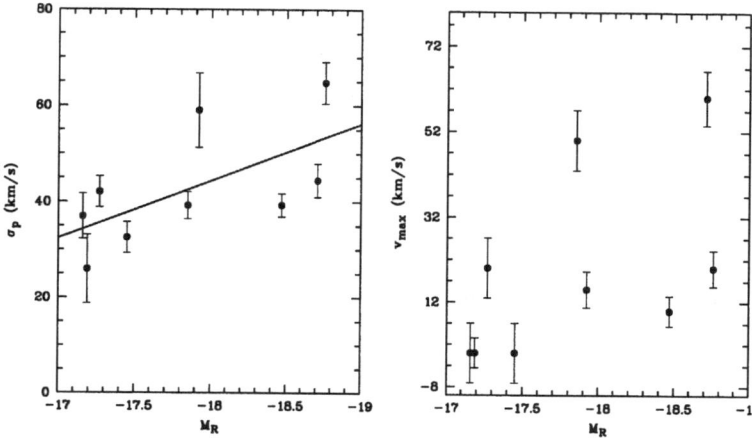

Fig. 2. Left panel: weighted central velocity dispersion versus absolute R magnitude. Right panel: maximum rotation velocity versus absolute R magnitude

4.3, and for FS76: $3.2 < (M/L)_B = 7.7 < 9.1$ or $1.2 < M/10^9 = 2.9 < 3.4$. Of all the models run so far, the best models are the ones with a dark halo. There is no indication that for dEs the issues raised by [1] are a major concern.

References

1. M. Baes, H. Dejonghe: this volume
2. R. Bender, J.-L. Nieto: Astron. & Astr. **239**, 97 (1990)
3. R. Bender, A. Paquet, J.-L. Nieto: Astron. & Astr. **246**, 349 (1991)
4. A. Burkert: Ap. J. **447**, L25 (1995)
5. R. Bender, D. Burstein, S.M. Faber: Ap. J. **399**, 462 (1992)
6. D. Carter, E.M. Sadler: Mon. Not. R. Astr. Soc. **245**, 12 (1990)
7. A. Dekel, J. Silk: Ap. J. **303**, 39 (1986)
8. S. De Rijcke, H. Dejonghe, G. Hau, W.W. Zeilinger: Ap. J., **559**, L21, (2001)
9. S. De Rijcke, H. Dejonghe: this volume
10. H.C. Ferguson, B. Bingeli: Ann. Rev. Astron. Astroph. **6**, 67 (1994)
11. E.V. Held, J. Mould, R. de Zeeuw: Astron. J. **100**, 415 (1990)
12. E.V. Held, T. de Zeeuw, J. Mould, A. Picard: Astron. J. **103**, 851 (1992)
13. H. Jerjen, B. Bingeli: Astron. J. **116**, 2873 (1998)
14. J. Navarro, C. Frenk, S.D.M. White: Ap. J. **462**, 563 (1996)
15. M. Mateo, E. Olszewski, D. Welch, P. Fisher, W. Kunkel: Astron. J. **102**, 914 (1991)
16. R.C. Peterson, N. Caldwell: Astron. J. **105**, 1411 (1993)
17. Y. Yoshii, N. Arimoto: Astron. & Astr. **188**, 13 (1987)

Estimating the Masses of Dwarf Ellipticals: First VLT Results

Sven De Rijcke[1,2], Herwig Dejonghe[1], Werner W. Zeilinger[3], and George T. K. Hau[4]

[1] Sterrenkundig Observatorium, Universiteit Gent, Belgium
[2] Astronomisches Institut, Universität Basel, Switzerland
[3] Institut für Astronomie, Universität Wien, Austria
[4] ESO, Chile

1 Introduction

The surface brightness profiles of both dwarf ellipticals (dEs) and normal ellipticals (Es) can be approximated by a Sérsic law, suggesting that Es and dEs are similar objects. But whereas Es are more likely to be rotationally flattened when they are fainter, dEs were found not to rotate. This is the famous *kinematic dichotomy* between Es and dEs [5]. The lack of rotation in dEs is interpreted as evidence for significant mass-loss and/or tidal heating. dEs may have lost their gas content, expelled by a global SN-driven wind [3]. Alternatively, small disk galaxies in a cluster or group environment undergo galaxy harassment, making them change their morphology [8,9]. Both scenarios lead to spheroidal, slowly rotating dE-like objects.

2 An ESO Large Program: The Dynamics of dEs

We obtained deep, high resolution major and minor axis spectra ($\lambda\lambda 7900 - 9300$Å) and VRI photometry with Kueyen-FORS2 (VLT–UT2) from a sample of 14 dEs in the Fornax Cluster and the NGC5044 and NGC5898 Groups. Kinematic information is available out to 1.5-2R_e. Dynamical models, fitted to the spectra or to the kinematics, serve to determine the orbital structure and to obtain reliable mass-estimates for these dEs. This drastic expansion of the existing data-set (for only 6 dEs was kinematic information available in the literature at the start of this program [1,2,7,10]) will help to check the reality of the kinematic dichotomy.

Here, we present the results of the dynamical modeling of FS29 and FS76, two dEs in the NGC5044 Group ($D = 36$ Mpc for $H_0 = 75$ km/s/Mpc) [6]. A dynamical model consists of a gravitational potential, which determines the stellar orbits, and a distribution function (DF), a function of the isolating integrals of motion, which distributes the stars over all possible orbits. Depending on the flattening of the galaxy, a spherical potential (with integrals E, L and L_z) or an axisymmetric one (with integrals E and L_z) is adopted. Deprojection yields the spatial luminosity density $\rho_{\text{lum}}(R, z)$. We parameterize the total mass density, including dark matter, as $\rho_{\text{tot}}(R, z) = A\rho_{\text{lum}}(R, z)(1 + Bm^\alpha)$ with $m = \sqrt{R^2 + (z/q)^2}$. We choose $\alpha = 1$ while A and B are constrained by the data. The potential follows from Poisson's equation. For some potentials, the DF would have to be negative somewhere in phase-space in order to adequately fit the data. The modeling technique [4] then finds the next best thing : the best fitting model that still has a positive DF. If the χ^2-value of a model is too high, it can be rejected at a certain confidence level. In this way, one can assess the range of possible mass-distributions. The model with the lowest χ^2 is obviously considered to be the best-fit model.

3   Results for FS29

This is a non-nucleated dE5 ($M_R = -18.71$, $1R_e = 9.0'' = 1.57$ kpc). Kinematic information is available out to $2R_e$ along the major axis and $0.5R_e$ along the minor axis. It is one of the fastest rotating dEs discovered so far ($(v/\sigma)^* = 1.07 \pm 0.20$, consistent with an isotropic oblate rotator). Axisymmetric models were fitted to the surface brightness and the major and minor axis kinematics. Self-consistent models ($B = 0$) can be rejected at the 90% confidence level. The best fitting model has $B = 4.0$ and the mass inside a $1.5\,R_e$ sphere is $M(1.5\,R_e) = 3.7^{+0.6}_{-1.1} \times 10^9 M_\odot$, equivalent to $\left(\frac{M}{L}\right)_I = 3.3^{+0.5}_{-1.0}$. The orbital structure is isotropic.

4   Results for FS76

FS76 is a non-nucleated dE1 ($M_R = -17.92$, $1R_e = 4.2'' = 0.73$ kpc). Kinematic information is available out to $1.5R_e$ along the major axis and $1.25R_e$ along the minor axis. FS76 is also a fast rotator ($(v/\sigma)^* = 0.61 \pm 0.25$). Spherical models were fitted to the surface brightness and the major and minor axis spectra. Self-consistent models cannot be ruled out but models where mass does not follow light ($B > 0$) provide a much better fit : the best model has $B = 0.4$ and $M(1.5\,R_e) = 2.9^{+0.5}_{-1.7} \times 10^9 M_\odot$, corresponding to $\left(\frac{M}{L}\right)_I = 5.4^{+1.0}_{-3.2}$. The model is slightly tangentially anisotropic ($\beta \approx -2$ at large radii).

References

1. Bender R. *et al.*, 1992, AJ, 103, 851
2. Bender R. & Nieto J.-L., 1990, AA, 239, 97
3. Dekel A. & Silk J., 1986, ApJ, 303, 39
4. De Rijcke S. & Dejonghe H., 1998, MNRAS, 298, 677
5. Ferguson C. & Binggeli B., 1994, AARv, 6, 67
6. Ferguson C. & Sandage A., 1990, AJ, 100, 1
7. Geha *et al.*, 2001, astro-ph/0107010
8. Mayer L. *et al.*, 2001, astro-ph/0103430
9. Moore B. *et al.*, 1998, 495, 139
10. Peterson R. & Caldwell N., 1993, AJ, 105, 1411

Dynamical Masses of Elliptical Galaxies

Ortwin Gerhard

Astronomisches Institut, Universität Basel, Venusstrasse 7, 4102 Binningen, Switzerland

Abstract. Recent progress in the dynamical analysis of elliptical galaxy kinematics is reviewed. Results reported briefly include (i) the surprisingly uniform anisotropy structure of luminous ellipticals, (ii) their nearly flat (to $\sim 2R_e$) circular velocity curves, (iii) the Tully-Fisher and $M/L - L$ relations and the connection to the Fundamental Plane, and (iv) the large halo mass densities implied by the dynamical models.

1 Introduction

Elliptical galaxies are the most massive galaxies and are generally found in dense environments. In the context of hierarchical models, they represent an advanced step in the galaxy formation process. Their mass distributions and dark halo properties, while not yet as well-understood observationally as those of spiral galaxies, are of considerable interest. X-ray and gravitational lensing studies imply mass-to-light ratios $M/L \sim 100$ on scales of $\sim 100\,\mathrm{kpc}$ for the most massive ellipticals containing hot gas atmospheres. Here the new XMM and Chandra satellite data will lead to mass measurements of unprecedented resolution and accuracy.

In the central $\sim 2R_e$, mass distributions $M(r)$ may be determined from absorption-line profile measurements and dynamical models. Planetary nebulae and globular clusters can be used as discrete velocity tracers to $\sim 4-5R_e$ (typical effective radii R_e are in the range 3-10 kpc). In less massive, gas-poor systems these may give the main constraints on $M(r)$. In gas-rich ellipticals where $M(r)$ can be accurately determined from the X-ray gas emission, the stellar-kinematic data will constrain the orbit structure best.

The following sections give a brief review of the dynamical analysis of the stellar-kinematic data, and the results obtained so far on mass distributions in ellipticals and on their dynamical family properties.

2 Dynamical Mass Estimation

As is well known, velocity dispersion profile measurements (and streaming velocities, if the galaxy rotates) do not suffice to determine the distribution of mass with radius, due to the degeneracy with orbital anisotropy. Only with very extended measurements can a constant M/L model be ruled out (e.g., Saglia et al. 1993), but even then the detailed $M(r)$ remains undetermined. Absorption line profile shapes (giving line-of-sight velocity distributions $L(v)$, LOSVD for short), however, contain additional information with which this degeneracy can largely be broken. Simple spherical models are useful to illustrate this (Gerhard 1993): at large radii, radial orbits are seen side-on, resulting in a peaked LOSVD (positive Gauss-Hermite parameter h_4), while tangential orbits lead to a flat-topped or double-humped LOSVD ($h_4 < 0$). Similar considerations can be made for edge-on or face-on disks (Bender 1990, Magorrian & Ballantyne 2001) and spheroidal systems (Dehnen & Gerhard 1993).

One may think of the LOSVDs constraining the anisotropy, after which the Jeans equations can be used to determine the mass distribution. However, the gravitational potential influences not only the widths, but also the shapes of the LOSVDs (see illustrations in Gerhard 1993). Furthermore, eccentric orbits visit a range of galactic radii and may therefore broaden a LOSVD near their pericentres as well as leading to outer peaked profiles. Thus, in practise, the dynamical modelling to determine the orbital anisotropy and gravitational potential must be done globally, and is typically done in the following steps:

(0) choose geometry (spherical, axisymmetric, triaxial);

(1) choose dark halo model parameters, and set total luminous plus dark matter potential Φ;

(2) write down a composite distribution function (DF) $f = \Sigma_k a_k f_k$, where the f_k can be orbits, or DF components such as $f_k(E, L^2)$, with free a_k;

(3) project the f_k to observed space, $p_{jk} = \int K_j f_k d\tau$, where K_j is the projection operator for observable P_j, and τ denotes the line-of-sight coordinate and the velocities;

(4) fit the data $P_j = \Sigma_k a_k p_{jk}$ for all observables P_j simultaneously, minimizing a χ^2 or negative likelihood, and including regularization to avoid spurious large fluctuations in the solution. This determines the a_k, i.e, the best DF f, given Φ, which must be $f > 0$ everywhere;

(5) vary Φ, go back to (1), and determine confidence limits on the parameters of Φ.

Such a scheme was employed, e.g., using orbits in spherical potentials by Rix et al. (1997) and Romanowsky & Kochanek (2001); using DF components in spherical potentials by Gerhard et al. (1998), Saglia et al. (2000), and Kronawitter et al. (2000); with orbits in axisymmetric geometry by Cretton et al. (2000) and Gebhardt et al. (2000); and with DF components in axisymmetry by Matthias & Gerhard (1999) and Statler et al. (1999). The modelling techniques used to constrain black hole masses from nuclear kinematics and dark halo parameters from extended kinematics are very similar.

Line-profile shape parameter measurements are now available for many nearby ellipticals (e.g., Bender, Saglia & Gerhard 1994), but those reaching to $\sim 2R_e$ are still scarce (see Kronawitter et al. 2000). Modelling of the mass profiles of ellipticals from such data has been done for some two dozen round galaxies in the spherical approximation, and for a few cases using axisymmetric three-integral models.

Typically the models that fit individual galaxies best imply small to modest amounts of dark matter within $2R_e$, but there are ellipticals which are very well represented by constant M/L models out to these radii. Most results obtained so far are from spherical models for round ellipticals; their mass distributions and radially anisotropic orbit structure are discussed in the next section. The effects of intrinsic deviations from sphericity and of embedded, near-face-on disks have been discussed by Kronawitter et al. (2000), Magorrian & Ballantyne (2001), and Gerhard et al. (2001). Axisymmetric models exist only for very few ellipticals. In NGC 1600 (E3.5, Matthias & Gerhard 1999), NGC 2300 (E2, Kaeppeli 1999), NGC 2320 (E3.5, Cretton, Rix & de Zeeuw 2000), and NGC 3379 (E1, Gebhardt et al. 2000) radially anisotropic structure at $\sim 0.5R_e$ has been inferred from three-integral models along the major axis, similar to that found in the round galaxies, but with less anisotropy on the minor axis.

Because the dark matter fraction inside $2R_e$ is still modest, and the orbit structure in the outer main bodies of ellipticals is not well-constrained by data that end at $2R_e$, it will be important to include discrete velocity data from planetary nebulae (PN) or globular clusters (GC) that reach to larger radii. Such data can be included directly in the above modelling scheme, using a likelihood maximization (Romanowsky & Kochanek 2001), or can be used a posteriori to differentiate between dynamical models at radii beyond the absorption line data on which these are based (Saglia et al. 2000). The number of ellipticals with such data is still small, but is expected to be growing rapidly thanks to the special purpose PN spectrograph (Douglas et al. 2002) and ongoing programs with globular cluster samples around ellipticals. Typically one may expect a few hundred discrete velocities per galaxy from these programs. Using slitless spectroscopy at the VLT Méndez et al. (2001) succeeded in measuring 535 PN radial velocities around the E5 galaxy NGC 4697. Detailed modelling of these data is still in progress, but simple models for this relatively low luminosity elliptical galaxy are consistent with constant M/L out to $\sim 3R_e$, in agreement with Dejonghe et al. (1996).

Two particularly interesting cases are the central galaxies of the Fornax and Virgo clusters, NGC 1399 and M87. From a comparison of dynamical models for the absorption line kinematics (Saglia et al. 2000) with the mass distribution obtained from ASCA data (Ikebe et al. 1996), it appears that PN (Arnaboldi et al. 1994) and GC velocities (Richtler et al. 2001) are just in the right radial range to allow a study of the transition between the potential of the central NGC 1399 galaxy and the potential of the Fornax cluster. Similarly, from a study of the stellar kinematics and the GC velocities around M87, and a comparison to the X-ray mass profile, Romanoswky & Kochanek (2001) find a rising circular velocity curve, and suggest that the potential of the Virgo cluster may already dominate at $r \sim 20\,\mathrm{kpc}$ from the center of M87.

3   Dynamical Family Properties

The discussion in this section is based on the work of Kronawitter et al. (2000) and Gerhard et al. (2001), who analyzed the line-profile shapes of a sample of 21 mostly luminous, slowly rotating, and nearly round elliptical galaxies in a uniform way, using spherical dynamical models. A similar study using three-integral axisymmetric models will be very worthwhile, but is still some time away. The sample of Kronawitter *et al.* includes a subsample with mostly new extended kinematic data, reaching to $\sim 2R_e$, and a subsample based on the less extended older data of Bender *et al.* (1994). Based on these data and on photometry, non-parametric spherical models were constructed from which circular velocity curves, anisotropy profiles, and radial profiles of M/L were derived, including confidence ranges. The main results from this study are as follows.

(1) The circular velocity curves (CVCs) of elliptical galaxies are flat to within $\simeq 10\%$ for $R \gtrsim 0.2R_e$ to at least $R \gtrsim 2R_e$, independent of luminosity (Fig. 1). The CVC is a convenient measure of the potential even though luminous elliptical galaxies do not rotate rapidly. Quantitatively, the median ratio of the circular velocity at the radius of the outermost kinematic data point, $v_c(R_{\mathrm{max}})$, to the maximum circular velocity of the respective best model, v_c^{max}, is 0.94, with 95% confidence ranges of $\sim \pm 0.1$. This argues against strong luminosity segregation in the dark halo potential.

(2) Despite the uniformly flat CVCs, there is a spread in the ratio of the CVCs from luminous and dark matter, i.e., in the radial variations of cumulative mass-to-light ratio. The sample includes galaxies with no indication for dark matter within $2R_e$, and others where the best dynamical models result in local M/L_Bs of 20-30 at $2R_e$. As in spiral galaxies, the combined rotation curve of the luminous and dark matter is flatter than those for the individual components ("conspiracy").

(3) Most of these ellipticals are moderately radially anisotropic, with average $\beta \equiv 1 - \sigma_\theta^2/\sigma_r^2 \simeq 0 - 0.35$, again independent of luminosity. The dynamical structure of ellipticals is therefore surprisingly uniform. The maximum circular velocity is accurately predicted by a suitably defined central velocity dispersion. Averaging σ within $0.1R_e$, $\sigma_{0.1} = 0.66v_c^{\mathrm{max}}$.

(4) Elliptical galaxies follow a Tully-Fisher (TF) relation with marginally shallower slope than spiral galaxies (see also Magorrian & Ballantyne 2001). At given circular velocity, they are about 1 mag fainter in B and about 0.6 mag in R, and appear to have slightly lower baryonic mass than spirals, even for the maximum M/L_B allowed by their kinematics (minimum dark halo models). The residuals from the TF (and Fundamental Plane, FP) relations do not correlate with dynamical anisotropy β.

(5) The luminosity dependence of M/L_B indicated by the tilt of the FP corresponds to a real dependence of dynamical M/L_B on L_B. The tilt of the FP is therefore not due to deviations from homology or a variation of dynamical anisotropy with L_B, although the slope of M/L_B versus L_B could still be influenced by photometric non-homology. The tilt can also not be due to an increasing dark matter fraction with L_B, unless (i) the most luminous ellipticals have a factor > 3 less baryonic mass than spiral galaxies of the same circular velocity, and (ii) the range of IMF is larger than currently discussed, and (iii) the IMF or some other population parameter varies systematically along the luminosity sequence

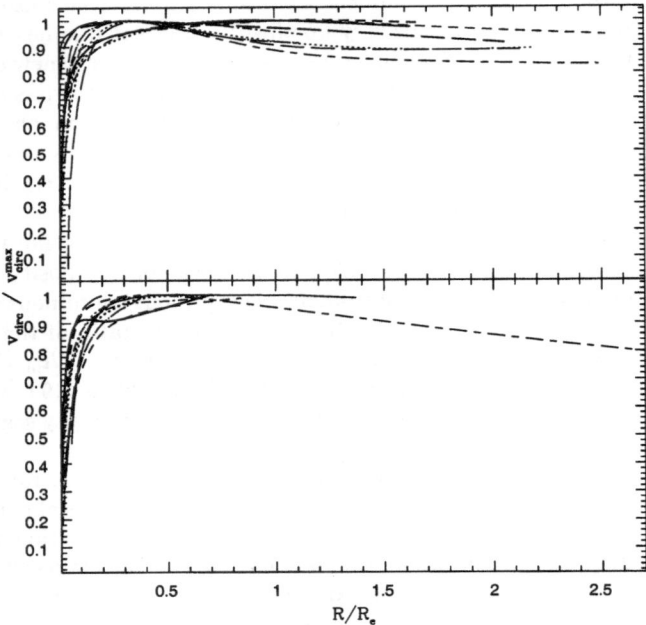

Fig. 1. "Best model" circular velocity curves of all galaxies from the Kronawitter *et al.* (2000) sample, plotted as a function of radius scaled by the effective radius R_e, and normalized by the maximum circular velocity. The upper panel shows the galaxies from the extended kinematics subsample, the lower panel the galaxies from the subsample with older data from Bender *et al.* (1994). From Gerhard *et al.* (2001)

such as to undo the increase of M/L_B expected from simple stellar population models for more metal-rich luminous galaxies. This seems highly unlikely.

(6) The tilt of the FP is therefore best explained as a stellar population effect. Population models show that the values and the change with L_B of the maximal dynamical M/L_Bs are consistent with the stellar population M/L_Bs based on published metallicities and ages, within the uncertainties of IMF and distance scale. Because of (5) above and because the observed correlation between age and luminosity is far too weak to explain the $M/L_B - L_B$ relation (Forbes & Ponman 1999), the main driver of the B-band tilt is therefore probably metallicity. This would not explain the observed K-band tilt, which must then be explained by a secondary population effect.

(7) The population models show that the dynamical models would have overestimated the luminous masses of these elliptical galaxies by as much as a factor ≈ 2 only if (i) the flattest IMFs at low stellar masses discussed for the Milky Way are applicable, and simultaneously (ii) a short distance scale ($H_0 \simeq 80\,\mathrm{km\,s^{-1}\,Mpc^{-1}}$) turns out to be correct. For lower values of H_0 and/or

other IMFs the difference is smaller. Together with (4) this makes it likely that elliptical galaxies have indeed nearly maximal M/L_B ratios (minimal halos).

(8) In the models with maximum stellar mass, the dark matter contributes $\sim 10 - 40\%$ of the mass within R_e. The flat CVC models, when extrapolated beyond the range of kinematic data, predict equal interior mass of dark and luminous matter at $\sim 2 - 4R_e$, consistent with results from X-ray analyses.

(9) Even in maximum stellar mass models, the halo core densities and phase-space densities are at least ~ 25 times larger and the halo core radii ~ 4 times smaller than in maximum disk spiral galaxies with the same circular velocity (Persic, Salucci & Stel 1996). Correspondingly, the increase in M/L sets in at ~ 10 times larger acceleration than in spirals. This could imply that elliptical galaxy halos collapsed at high redshifts or perhaps even that some of the dark matter in ellipticals might be baryonic.

References

1. Arnaboldi, M., Freeman, K.C., Hui, X., Capaccioli, M., Ford, H., 1994, The Messenger, 76, 40
2. Bender, R., 1990, A&A, 229, 441
3. Bender, R., Saglia, R.P., Gerhard, O., 1994, MNRAS, 269, 785
4. Cretton, N., Rix, H.-W., de Zeeuw, T., 2000, ApJ, 536, 319
5. Dehnen, W., Gerhard, O.E., 1993, MNRAS, 261, 311
6. Dejonghe, H., De Bruyne, V., Vauterin, P., Zeilinger, W.W., 1996, A&A, 306, 363
7. Douglas, N.G., et al., 2002, PASP, submitted
8. Forbes, D.A., Ponman, T.J., 1999, MNRAS, 309, 623
9. Gebhardt, K., et al. , 2000, AJ, 119, 1157
10. Gerhard, O.E., 1993, MNRAS, 265, 213
11. Gerhard, O.E., Jeske, G., Saglia, R.P., Bender, R., 1998, MNRAS, 295, 197
12. Gerhard, O.E., Kronawitter, A., Saglia, R.P., Bender, R., 2001, AJ, 121, 1936
13. Ikebe, Y., et al., 1996, Nature, 379, 427
14. Kaeppeli, A., 1999, Diploma Thesis, Univ. of Basel
15. Kronawitter, A., Saglia, R.P., Gerhard, O.E., Bender, R., 2000, A&AS, 144, 53
16. Magorrian, J., Ballantyne, D., 2001, MNRAS, 322, 702
17. Matthias, M., Gerhard, O.E., 1999, MNRAS 310, 879
18. Méndez, R.H., et al., 2001, ApJ, 563, 135
19. Persic, M., Salucci, P., Stel, F., 1996, MNRAS, 281, 27
20. Richtler, T., et al., 2001, in 'Extragalactic star clusters', IAU Symp. 207, ed. E.K. Grebel et al., ASP, in press
21. Rix, H.-W., de Zeeuw, P.T., Cretton, N., van der Marel, R., Carollo, C.M., 1997, ApJ, 488, 702
22. Romanowsky, A.J., Kochanek, C.S., 2001, ApJ, 553, 722
23. Saglia, R.P., et al., 1993, ApJ, 403, 567
24. Saglia, R.P., Kronawitter, A., Gerhard, O.E., Bender, R., 2000, AJ, 119, 153
25. Statler, T.S., Dejonghe, H., Smecker-Hane, T., 1999, ApJ, 117, 126

Dust Attenuation and the Stellar Kinematical Evidence for Dark Halos Around Elliptical Galaxies

Herwig Dejonghe and Maarten Baes

Sterrenkundig Observatorium, Universiteit Gent, Krijgslaan 281 S9, B-9000 Gent, Belgium

Abstract. We present a set of dynamical models for elliptical galaxies, in which dust attenuation is included through a Monte Carlo technique. We find that interstellar dust affects the observed kinematics significantly, in the way that it mimics the presence of a dark halo. As a result, we are faced with a new mass-dust degeneracy. Taking dust attenuation into account in dynamical modelling procedures will hence reduce or may even eliminate the need for a dark matter halo at a few effective radii.

1 Introduction

During the past few years, a consensus has developed that elliptical galaxies, like spiral galaxies, contain dark halos. Stellar kinematics are generally considered as the most important tracer for these halos at a few effective radii. Several authors have adopted stellar kinematics to constrain the dark matter distribution in a number of elliptical galaxies ([1],[2],[3]). Meanwhile, it has become well established that elliptical galaxies also contain a surprisingly large amount of interstellar dust, most of it believed to be distributed diffusely over the galaxy ([4]). Because dust grains efficiently absorb and scatter optical photons, they will affect all observable quantities, including the observed kinematics.

We constructed a Monte Carlo code in order to calculate the observed kinematics of elliptical galaxies, including the effects of both absorption and scattering by interstellar dust. We present a simple two-component model, consisting of a stellar and a dust component. The star-dust geometry is chosen such that the dust distribution is shallower than the stellar density, and that it has a total visual optical depth of order unity ([5]). For more details we refer to [6].

2 Results

The effect of dust attenuation on the observed kinematics is illustrated in Figure 1. For the lines of sight that pass near the galaxy center, the kinematics are only slightly affected, in a way comparable with our previous modelling where only absorption was included ([7]). At large projected radii, however, the kinematics are seriously affected: the velocity dispersion profile drops less steeply, and the h_4 parameter is significantly larger compared to the dust-free case. These effects are caused by photons emitted by high-velocity stars in the center of the

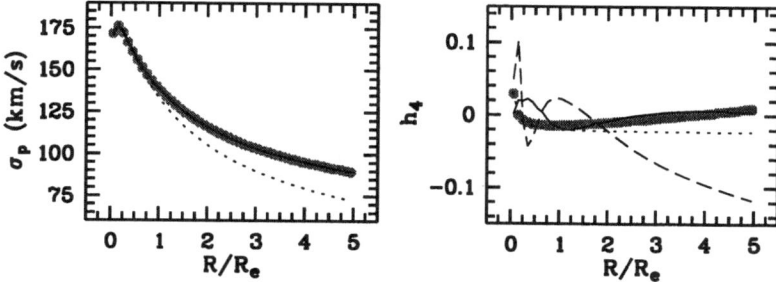

Fig. 1. The projected dispersion profile and the h_4 profile of the models described in the text. The dotted lines and the thick grey lines represent the projected kinematics of the input model, respectively without and with dust attenuation taken into account. The latter results are used as input for the dynamical modelling procedure. The dashed lines represent the best fitting model with constant M/L; the solid line is the best fitting model with a dark matter halo

galaxy, that are scattered in the outskirts of the galaxy into lines of sight with a large projected radii. Curiously, these effects are strikingly similar to the kinematical signature of a dark halo, which is characterized by a slowly decreasing dispersion profile and a positive h_4 profile.

To check this into more detail, we considered our dust-affected kinematics as an observational data set, and modelled it in the usual way, i.e. without taking dust attenuation into account. We found that it was impossible to fit both the photometry and the kinematics with a constant mass-to-light ratio model. A dark matter halo is hence necessary to explain the effects caused by dust attenuation. In the best fitting model with a dark halo, the dark matter contributes roughly a third of the total mass within 1 R_e, and half of the total mass within the last data point.

These results clearly demonstrate that the effects of dust attenuation can mimic the presence of a dark matter halo. In analogy with the mass-anisotropy degeneracy, we are now faced with a new degeneracy, which could be called the mass-dust degeneracy. The new mass-dust degeneracy strongly complicates the use of stellar kinematics as a tracer for the mass distribution in elliptical galaxies: taking dust attenuation into account in dynamical modelling procedures will reduce or may even eliminate the need for a dark matter halo at a few R_e.

References

1. H.-W. Rix et al., ApJ **488**, 702 (1997)
2. O. Gerhard et al., MNRAS **295**, 197 (1998)
3. A. Kronawitter et al., A&AS **144**, 53 (2000)
4. P. Goudfrooij, T. de Jong, A&A **298**, 784 (1995)
5. M. S. Wise, D. R. Silva, ApJ **461**, 155 (1996)
6. M. Baes, H. Dejonghe, ApJ **563**, L19 (2001)
7. M. Baes, H. Dejonghe, MNRAS **313**, 153 (2000)

Galaxy Merging and the Fundamental Plane of Elliptical Galaxies

Carlo Nipoti[1], Pasquale Londrillo[2], and Luca Ciotti[2,3]

[1] Dipartimento di Astronomia dell'Università degli Studi di Bologna,
   via Ranzani 1, 40127 Bologna, Italy
[2] Osservatorio Astronomico di Bologna, via Ranzani 1, 40127 Bologna, Italy
[3] Scuola Normale Superiore, Piazza dei Cavalieri 7, 56126 Pisa, Italy

Abstract. We present preliminary results of numerical simulations of dissipationless merging of stellar systems, aimed at exploring the consequences of merging between gas free, spheroidal systems. In particular, we study the dynamical and structural characteristics of hierarchical merging between equal mass stellar systems, and we compare the properties of the end-products with the most important structural and dynamical scaling relations obeyed by spheroids. In the explored hierarchy of four successive mergings we find that the FP tilt is marginally conserved, but both the Faber-Jackson and Kormendy relations are *not* conserved.

1 Introduction

From a *theoretical* point of view, in the scenario of hierarchical galaxy formation elliptical galaxies (Es) formed by merging of smaller systems [19,24,8]. On the other hand, from *observations* we know that Es satisfy many tight scaling relations: for example the Fundamental Plane [15,16] (FP), the $M_{\mathrm{BH}} - \sigma_0$ [18,25], the $\mathrm{Mg}_2 - \sigma_0$ [6], and the color–magnitude [11] relations. In particular, the FP of Es relates their central velocity dispersion σ_0^2, total luminosity L_B and circularized effective radius $\langle R \rangle_e$, with a 1-sigma scatter of $\simeq 15\%$ in $\langle R \rangle_e$ for fixed L_B and σ_0 [26].

Here we focus on the constraints imposed by the existence of the FP on the role of dissipationless merging in the formation of Es. In other words we want to verify, by using numerical simulations, whether the FP is "closed" with respect to the merging process.

Among the motivations of this exploration is the fact that, in the merging of two galaxies with masses (M_1, M_2) and virial velocity dispersions $(\sigma_{v,1}, \sigma_{v,2})$, the virial velocity dispersion of the merger (in case of no mass loss and negligible initial interaction energy of the galaxy pair when compared to their internal energies) is given by

$$\sigma_{v,1+2}^2 = \frac{M_1 \sigma_{v,1}^2 + M_2 \sigma_{v,2}^2}{M_1 + M_2}. \tag{1}$$

It follows that $\sigma_{v,1+2} \leq \max(\sigma_{v,1}, \sigma_{v,2})$, i.e., *the virial velocity dispersion cannot increase in a merging process of the kind described above.* On the other hand, the Faber-Jackson relation [10] (FJ) indicates that the *projected central velocity dispersion* increases with galaxy luminosity. In addition, since in a purely gas free merging process the stellar mass–to–light ratio cannot increase, *the FP can be maintained only by structural and/or dynamical non–homology.*

2 Results

The details of the adopted (one and two component) numerical models are given in [2]. We describe here the case of a *merging hierarchy*: in other words, we merge together the end–products of previous mergers up to four generations, for a total increase of mass of a factor of 16.

- As expected, in all the merging events $\sigma_{v,1+2}$ does not differ significantly from those of the progenitors (the largest deviation, due to particle escape, is less than 4%).

- On the contrary, the luminosity-weighted projected velocity dispersion inside $\langle R \rangle_e/8$ of the end–products $\sigma_a(\langle R \rangle_e/8)$ (an estimate of the observed σ_0), is larger than in the progenitors, while $\sigma_a(\infty) \simeq \sigma_{v,1+2}$, in accordance with the projected virial theorem[17].

- In general the end–products have σ_0 *lower* and $\langle R \rangle_e$ *larger* than those predicted by the FJ and Kormendy [1] relations, respectively. The effects curiously compensate and the end–products remain near the FP: however, the scatter in $\langle R \rangle_e$ can be as large as 35%, when compared to $\langle R \rangle_e$ derived from the FP relation when using the total luminosity (mass) and central velocity dispersion of the end–products.

We are now running high-resolution numerical simulations analogous to those described here, in order to exclude numerical effects on the above results, and to quantitatively check the effects of massive dark matter halos on the properties of the end-products.

References

1. R. Bender, D. Burstein, S.M. Faber: ApJ, **411**, 153 (1993)
2. R.G. Bower, J.R. Lucey, R.S. Ellis: MNRAS, **254**, 601 (1992)
3. L. Ciotti: Celest. Mech. & Dyn. Astron., **60**, 401 (1994)
4. S. Cole, C.G. Lacey, C.M. Baugh, C.S. Frenk: MNRAS, **319**, 168 (2000)
5. G. Djorgovsky, S. Davis: ApJ, **313**, 59 (1987)
6. A. Dressler, S.M. Faber, D. Burstein, R.L. Davies, D. Lynden-Bell, R.J. Terlevich, G. Wegner: ApJ, **313**, 37 (1987)
7. S.M. Faber, R.E. Jackson: ApJ, **204**, 668 (1976)
8. L. Ferrarese, D. Merritt: ApJ, **539**, L9 (2000)
9. K. Gebhardt, et al.: ApJ, **539**, L13 (2000)
10. I. Jorgensen, M. Franx, P. Kjaergaard: MNRAS, **280**, 167 (1996)
11. G. Kauffmann: MNRAS, **281**, 487 (1996)
12. S.D.M. White, M.J. Rees: MNRAS, **183**, 341 (1978)
13. J. Kormendy: ApJ, **218**, 333 (1977)
14. C. Nipoti, P. Londrillo, L. Ciotti: MNRAS, submitted (2001)

Mass Distributions in Early-Type Galaxy Halos

Aaron J. Romanowsky[1], Nigel G. Douglas[1], Konrad Kuijken[1],
Michael R. Merrifield[2], Magda Arnaboldi[3], Kenneth. C. Freeman[4], and
Keith Taylor[5]

[1] Kapteyn Institute, Postbus 800, 9700 AV Groningen, The Netherlands
[2] School of Physics & Astronomy, University of Nottingham, NG7 2RD, England
[3] Osservatorio Astronomico di Capodimonte, Via Moiariello 16, I-80131 Naples, Italy
[4] RSAA, Mt Stromlo Observatory, Cotter Road, Weston Creek, ACT 2611, Australia
[5] California Institute of Technology, MC: 105-24, Pasadena, CA 91125, USA

Abstract. One of the most promising avenues for determining the distribution of
mass in the outer parts of early-type galaxies is through the kinematics of planetary
nebulae (PNe). We have used new techniques and instrumentation on the WHT and
the VLT to obtain velocities for hundreds of PNe around several nearby galaxies. We
show simple mass models and describe more rigorous orbit modeling methods for the
combined analysis of different dynamical constraints in galaxy halos.

The halo masses of nearby early-type galaxies (ellipticals and S0s) are still
poorly known. X-ray gas emission and stellar absorption line studies indicate that
dark matter is dominant in the outer parts ($\gg R_{\text{eff}}$, an effective radius) of some
ellipticals [1] [2], but strong general constraints on the radial mass distributions
$\rho(r)$ are less forthcoming. For this purpose, kinematical tracers such as globular
clusters (GCs) and planetary nebulae (PNe) are quite promising – PNe especially
so because they directly couple to the stars.

Until now, measuring extragalactic PN velocities in the large numbers (hundreds) needed for dynamical analyses has been prohibitively inefficient. We have
pursued a multi-pronged program using improved techniques, including multi-
fiber spectroscopy with WYFFOS+AUTOFIB2 on the 4.2-m WHT, masked
counter-dispersed imaging with FORS2+MXU on UT2, and counter-dispersed
imaging with the newly-commissioned *Planetary Nebula Spectrograph (PN.S)*
on the WHT. We have so far obtained velocities of 100–200 PNe in each of the
galaxies NGC 821 (E3), NGC 4472 (E1), NGC 4486 (E1), and NGC 7457 (S0),
at distances of 15–25 Mpc, and to radii of 5–6 R_{eff}.

For illustrative purposes, we first model the data using the spherical isotropic
Jeans equations (see Fig. 1). For NGC 4472 and NGC 4486, both the stellar
and GC data indicate that the circular velocity profile $v_c(r) \equiv [GM(r)/r]^{1/2}$ is
constant or rising with radius at 4 R_{eff}. The inferred profile differs depending
on whether the stellar or GC data are used, indicating that the assumption of
isotropy is probably wrong for one or both subsystems. NGC 821 is not yet
modeled, but the projected velocity dispersion profile $\sigma_p(R)$ suggests the presence of a dark halo that is less dominant than in the previous two galaxies;
NGC 4697 is a similar elliptical whose PN dispersion profile also declines with
radius [3]. For NGC 7457, $\sigma_p(R)$ suggests a dominant dark halo.

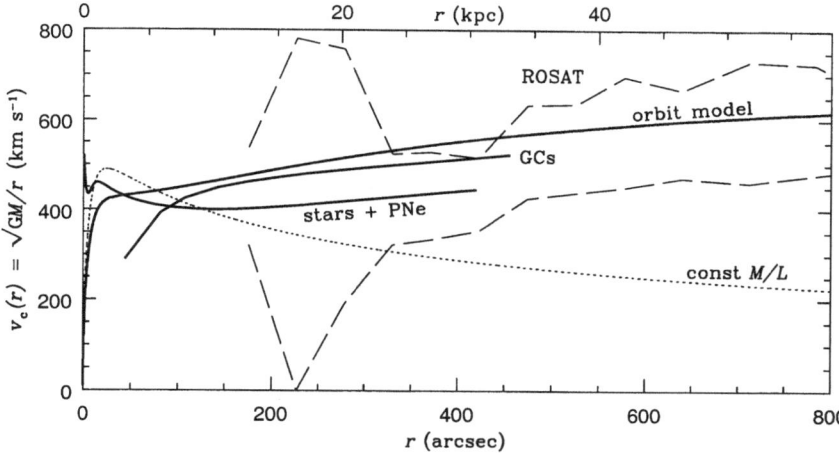

Fig. 1. Modeled circular velocity radial profile for NGC 4486. Shown are the results from Jeans modeling, using the stellar (including PNe) data and globular cluster data; a constant M/L model; confidence limits from ROSAT X-ray constraints [5]; and a best-fit orbit model. The effective radius is $R_{\rm eff} \simeq 100''$

More rigorous determinations of $v_c(r)$ require sophisticated modeling methods that allow nonparametrically for variations in the orbital anisotropies, and that exploit fully the information contained in discrete velocity data. We have developed such a method based on the now established approach of orbit modeling [4]. Upon analysis of the published integrated stellar spectroscopy and GC velocity data for NGC 4486, we have found a density law of $\rho(r) \sim r^{-1.5}$ at 4 $R_{\rm eff}$ (see Fig. 1). This implies that the galaxy is embedded in a core of a massive dark halo associated with the Virgo Cluster itself, as also evidenced by the galaxy's X-ray emission [5]. Similar results have been found for the Fornax Cluster cD galaxy NGC 1399 [6].

Further orbit modeling is now underway in these galaxies, incorporating the constraints from all the available data (integrated stellar light, PNe, GCs, X-rays), and further observations with the PN.S are planned for a dozen nearby ellipticals. This approach should provide strong new limits on the mass distribution in early-type galaxy halos.

References

1. A. Kronawitter, R.P. Saglia, O. Gerhard, R. Bender: A&AS, **144**, 53 (2000)
2. M. Loewenstein, R.E. White, III.: ApJ, **518**, 50 (1999)
3. R.H. Mendéz, et al.: A&A, **astro-ph/0109075** (2001)
4. A.J. Romanowsky, C.S. Kochanek: ApJ, **553**, 722 (2001)
5. P.E.J. Nulsen, H. Böhringer: MNRAS, **274**, 1093 (1995)
6. R.P. Saglia, A. Kronawitter, O. Gerhard, R. Bender: AJ, **119**, 153 (1999)

Studying the Dynamics of Star Forming and IR Luminous Galaxies with Infrared Spectroscopy

Reinhard Genzel, Linda J. Tacconi, Marco Barden, Matthew D. Lehnert, Dieter Lutz, Dimitra Rigopoulou, and Niranjan Thatte

Max-Planck Institut für extraterrestrische Physik, Garching, FRG

Abstract. With the advent of efficient near-IR spectrometers on 10m-class telescopes, exploiting the new generation of low readout noise, large format detectors, OH avoidance and sub-arcsecond seeing, 1-2.4μm spectroscopy can now be exploited for detailed galaxy dynamics and for studies of high-z galaxies. In the following we present the results of three recent IR spectroscopy studies on the dynamics of ULIRG mergers, on super star clusters in the Antennae, and on the properties of the rotation curves of z~1 disk galaxies, carried out with ISAAC on the VLT, and NIRSPEC on the Keck.

1 Ultra-Luminous Infrared Galaxies: Ellipticals and QSOs in Formation?

Recent deep mid-IR [8] and submillimeter [4] surveys have discovered a population of distant, dust enshrouded starbursts that may contribute about half of the cosmic star formation activity at z$\geq$1 [13]. These IR-luminous starbursts may be large bulges/ellipticals in formation. (Ultra)-luminous infrared galaxies ((U)LIRGs: L(1-1000μm)$\geq 10^{11}(\geq 10^{12})L_\odot$, [24]) may be the local-Universe analogues of the high-z population. Almost all ULIRGs are advanced mergers of gas rich disk galaxies. Following the 'ellipticals through mergers scenario' of Toomre and Toomre [33], Kormendy and Sanders [17] proposed that ULIRGs may evolve into ellipticals through merger induced, dissipative collapse. In the process, such mergers may go through a very luminous starburst phase and later evolve into classical QSOs [25]. Studies of local ULIRGs thus may be a key to better understand the properties and evolution of the high-z population.

To test the 'ellipticals in formation' and 'QSOs in formation' scenarios we have recently begun a program of determining the fundamental structural and dynamical properties of the stellar hosts of ULIRGs [14, 31]. If ULIRGs evolve into ellipticals, late stage ULIRG merger remnants should lie on or near the fundamental plane ($\log \sigma - \log r_{eff} - \mu_{eff}$) of early type galaxies. If they also evolve into QSOs, the hosts and central black holes of ULIRGs and QSOs of similar luminosities should have similar properties. ULIRG mergers are not in equilibrium. Recent numerical simulations have shown, however, that because of the rapid action of violent relaxation and tidal torques, the dynamical properties of late stage, compact merger remnants are already fairly close to their final equilibrium values [21, 2]. Our project required infrared imaging and spectroscopy since ULIRGs are highly obscured (A$_V$(screen)~10-40). It required 10m class

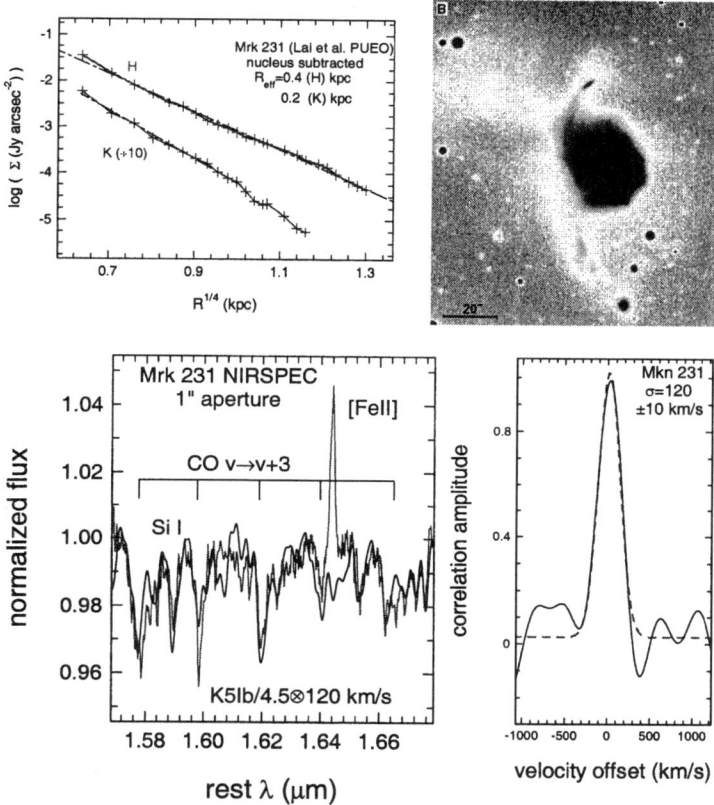

Fig. 1. Observations of the ULIRG/QSO Mkn231 (z=0.04), $L_{IR} = 3 \times 10^{12}$ $L_\odot$). Upper left: H- (red) and K-band (blue) surface brightness distributions ($\sim 0.14''$ resolution) after removal of the nucleus, from the adaptive optics observations of [19]. Upper right: B-band image of the outer parts of Mkn 231, showing the two tidal tails, and the central compact body of the galaxy [5]. Lower left: H-band spectrum of the central region (red, excluding central $\sim 0.8''$), obtained with NIRSPEC on Keck [31]. The black curve shows a late type star template spectrum, convolved with the best fitting Gaussian (right inset) and diluted by a factor of 4.5. The strongest CO and SiI stellar absorption features are marked, as is the [FeII] emission line. Right inset: Line of sight velocity distribution of the stars (continuous black), as derived from a Fourier quotient analysis of the spectrum in the left inset. The best fitting Gaussian (red dashed) has a dispersion of 120 km/s

telescopes because high quality (SNR~100) spectra are necessary to reliably extract velocity dispersions.

We selected our program ULIRGs from the BGS, 2 Jy and 1 Jy IRAS catalogs, culling from these catalogs those sources that have single nuclei, or are compact ($\leq$a few kpc) double nuclei systems on near-IR images, but have definite signatures for a recent merger, such as tidal tails. We picked sources with redshifts $\leq$0.16, where reasonably strong stellar absorption features fall in the J, H, or K-bands. We have presently data on 18 ULIRG merger remnants, 8 of which contain QSO-like active galactic nuclei. As an example, Figure 1 shows our Keck NIRSPEC data of Mkn 231, along with a B-band image from [5], and the nucleus subtracted, H/K surface brightness distribution from [19]. Mkn 231 is the most luminous ULIRG within z=0.05 ($\log L_{IR} = 12.5$) and is also an IR-excess BAL QSO. The two tidal tails and its compact structure with a single bright nucleus suggest it is a late stage merger of near equal mass disk galaxies. The highly AGN diluted central spectrum can be reasonably well fit by a Gaussian velocity dispersion of 120 km/s, which is the lowest of our entire ULIRG sample.

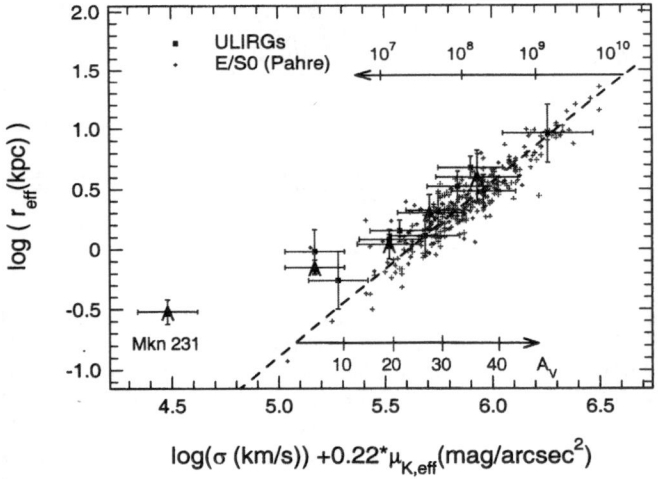

Fig. 2. Distribution of ULIRGs (dark symbols with 1σ error bars, sources dominated by AGNs marked with symbol 'A') and nearby ellipticals ([22], grey crosses) in the fundamental plane, as viewed from a projection perpendicular to the plane (best fit given by dashed line). The direction and magnitudes of various amounts of reddening and different stellar ages (relative to a >10 Gyr population) are shown at the top and bottom of the diagram. Obscuration and younger populations tend to cancel each other

Figure 2 shows an edge-on projection of the fundamental plane of early type galaxies. Small grey crosses denote ellipticals/S0s from the compilation of [22].

Large symbols (with error bars) mark the locations of our ULIRGs, with the AGN dominated sources marked with the letter 'A'. ULIRGs fall remarkably close to the fundamental plane. Their stellar kinematics is largely pressure supported and in most cases stellar rotation, if present, is smaller than the rotation of the gaseous component of the same system. They populate a wide range of the plane, are on average similar to moderate mass, disky ellipticals, but are well offset from giant ellipticals. Their typical effective radii (a few kpc) and velocity dispersions ($\langle \sigma_{ULIRG} \rangle = 186 \pm 15$ km/s) are comparable to the parameters of L* disky ellipticals. In contrast giant ellipticals have effective radii $r_{eff} > 10$ kpc and an average velocity dispersion of $\langle \sigma_{gEs} \rangle = 269 \pm 7$ km/s. Late stage ULIRG mergers also show significant rotation ($\langle v_{rot}/\sigma \rangle \geq 0.4$), and are found in low density environments, again similar to disky ellipticals and unlike giant ellipticals. All 117 ULIRGs of the 1 Jy catalog [16] are found in the field (or possibly small groups), none in clusters. For comparison, of the giant ellipticals in the [1] and [9] samples, 50% reside in clusters of Abell richness class ≥ 0. Our observations thus strongly support the 'ellipticals in formation' scenario but indicate that ULIRG remnants will evolve into moderate mass disky ellipticals, or lenticulars. Giant ellipticals must have a different formation path (see also [9]).

The close proximity of the ULIRG mergers to the fundamental plane is surprising. ULIRGs are heavily obscured and contain a population of young stars. While the σ-coordinate is relatively insensitive to extinction effects, the K-band surface brightness and even the effective radius are affected by absolute and differential extinction and by population effects. The arrows in Figure 2 show how a point in the fundamental plane would move with extinction and aging of the underlying population. In addition the NICMOS images of [27] show that extinction increases toward the nuclei in most ULIRGs, resulting in effective radii decreasing with wavelength (see Figure 1). It is thus important to determine effective radii from near-IR images. Typical ULIRGs have A_V (screen) $\sim 10 - 50$ mag and a stellar population with a characteristic age of a few hundred Myr [31], the combination of which approximately cancel each other in Figure 2. So the good agreement of the ULIRGs with the fundamental plane in the surface brightness coordinate is to some extent the result of a 'cosmic conspiracy'.

In the scenario of [25], the dusty ULIRGs are initially powered by star formation, but as the merger progresses, accretion onto the central black hole(s) increasingly dominates the bolometric luminosity. As the dust and gas is cleared out in the last stages of the merger, the dusty (infrared excess) QSO evolves into a classical, optically bright QSO. If this scenario is correct, late stage ULIRGs (likely AGN dominated) and optically selected QSOs of similar luminosities should have similar host properties. To test this hypothesis, we have compared our ULIRGs to the sample of 33 radio quiet and radio loud QSOs, and radio galaxies studied in detail by Dunlop and coworkers (e.g. [6] and references therein). The Dunlop AGNs sample a similar redshift range and luminosity distribution (Figure 3) as the ULIRGs. Figure 3 shows that these QSOs/radio galaxies occupy the upper right right of the μ_{eff} - r_{eff} projection of the plane, coincident with the locus of giant ellipticals. The Dunlop et al. AGNs, which are

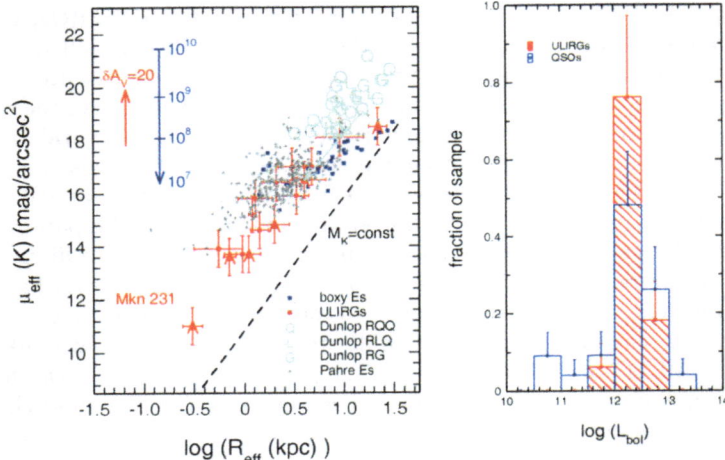

Fig. 3. Left inset: ULIRGs (red symbols, sources dominated by AGNs marked with letter 'A') and ellipticals (grey crosses, [22]) in the $\mu_{eff}(K) - r_{eff}$ projection of the fundamental plane. Boxy giant ellipticals [1, 9] are marked as dark blue filled squares. Radio quiet (letter 'Q'), radio loud (letter 'L') QSOs and radio galaxies (letter 'G') [6] are in light blue. Right inset: distributions of nuclear bolometric luminosity for the Dunlop et al. [6] sample (blue, unfilled histogram) and of the ULIRG luminosities (red hatched histogram). In the case of the Dunlop et al. sources we multiplied the nuclear R-band luminosity be a factor 14, for the ULIRGs we used the infrared luminosity, corrected upwards by 25% to take into account the contribution of the optical/UV bands

fairly representative of optically selected, local Universe QSOs/radio galaxies, have effective radii >10 kpc, and 80% of them reside in clusters. While there are no measurements of velocity dispersions in the Dunlop AGNs yet, [3] have compiled velocity dispersions for 73 nearby radio galaxies. The sample average velocity dispersion of those radio galaxies is 256 km/s, in good agreement with the hypothesis that most local Universe QSOs/radio galaxies are large, massive early type galaxies. Some of the size difference between ULIRGs and QSOs may be attributable to the nuclear concentration of bright young star forming regions in the ULIRGs that will fade as the population ages, and thus lead to an increase of effective radius with time. However, the difference in velocity dispersions between the ULIRGs and the radio galaxies and giant ellipticals should be unaffected by population effects. We thus conclude that the average ULIRG host cannot evolve into a host of a typical QSO or radio galaxy.

After correction for a mean sample K-band extinction of A_K(screen)=0.7 the average ULIRG has an absolute K-magnitude of -25.8 (5 L*), similar to that of the radio quiet QSOs. However, the average total dynamical mass of the ULIRG sample is 1.2×10^{11} $M_\odot$, or about 0.9 m*. Hence the L/m ratio of the average ULIRG host is about 6 times greater than that of an old early type galaxy, corresponding to an effective age of the stellar population of a few hundred Myr. Taking the local black hole mass to bulge mass ratio [10, 12], we find that the average black hole mass in our ULIRGs is 7×10^7 M$_\odot$. In contrast Dunlop et al. estimate an average black hole mass of their sample to be 1.3×10^9 M$_\odot$. Since the nuclear luminosities of ULIRGs and QSOs are similar (Figure 3), ULIRGs must accrete at >50% the Eddington rate, rather than the ~10% efficiency estimated for the Dunlop QSO sample.

In summary, we conclude that ULIRGs have less massive hosts than optically selected, low-z QSOs or radio galaxies. ULIRGs live in lower density environments. Their black holes are more akin to Seyfert galaxies. Nevertheless they can attain QSO-like bolometric and near-IR luminosities because they accrete more (and more efficiently) and have younger and thus brighter hosts. Once the merger induced, enormous influx of matter onto the central black hole(s) ceases, ULIRGs will become inactive, moderately massive field ellipticals or, if their black holes are fed, into objects akin to the hard-X ray luminous, early type galaxies found recently by Chandra. The average ULIRG cannot evolve into a classical optically selected QSO.

2   What Is the Nature of the Super Star Clusters in the Antennae Merger?

During the last years, many interacting and merging galaxies were discovered to contain large numbers of very luminous young star clusters (e.g. [15, 36]. Their overall spectral properties suggest that they may be the progenitors of the globular cluster populations seen in normal nearby ellipticals and spirals (e.g. [37, 11, 26]).

To test the 'globular clusters in formation' scenario and determine cluster masses and lifetimes, we observed several of the brightest young clusters in the Antennae (NGC4038/39) merger system with ISAAC and UVES high resolution spectroscopy [20]. Figure 4 marks the observed clusters on a K-band image. The target clusters were selected to be bright and exhibit large equivalent widths of 2.3μm 0-2 CO stellar absorption, indicative of the presence of late type supergiants. The left upper inset of Figure 4 shows ISAAC spectra of the 0-2 CO overtone bands for the three clusters observed with ISAAC $(\sigma_{instr}(FWHM) = 14km/s)$, along with the best fit, Gaussian broadened stellar templates. In the case of UVES $(\sigma_{instr}(FWHM) = 3.4km/s)$ we targeted the CaT and other metal absorption lines in the 8500-8800 Å range in a total of four clusters, one of them ([W99]2) common between ISAAC and UVES. For all 5 clusters we used I-band HST imaging from Whitmore to derive half power radii from King-model fitting. We find a range of cluster velocity dispersions from 9 to 20 km/s, and half power radii from 3.6 to 6 pc. In the cluster common to UVES and ISAAC, both data sets are in excellent agreement and yield velocity dispersions of 14.3 and 14.0 km/s, respectively. Masses are derived from the Viral Theorem, which should be applicable since the clusters obviously already have survived for 20-50 crossing times. The corresponding dynamical masses range

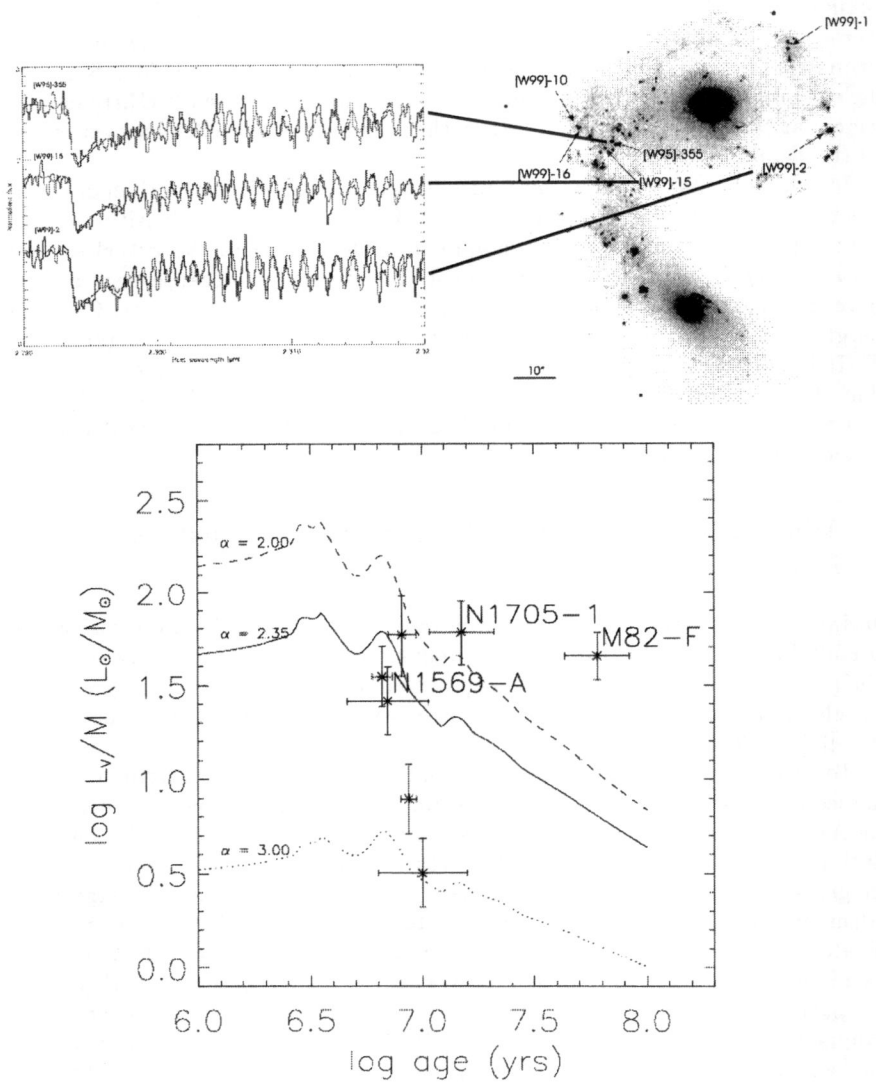

Fig. 4. Observations of the dynamics of super star clusters in NGC 4038/39. Upper right: Ks-band image of the merger system, with clusters observed with ISAAC and UVES marked. Upper left: R=9500 ISAAC spectra of three of the clusters, along with the best fitting stellar template (red). Bottom: V-band luminosity to mass ratio as a function of age, for 4 of the Antennae clusters [20], along with a cluster in M82, and two clusters in the blue dwarfs NGC 1705 and NGC 1569 [29, 30]. The different curves are IMFs with different power law slopes, over a mass range from 0.1 to 100 $M_\odot$

from 6–50 × 10^5 $M_\odot$. Our data thus unambiguously show that the brightest Antennae clusters are indeed massive, with masses well above typical globular clusters (> 10^5 $M_\odot$). Our results are in good agreement with the hypothesis that the Antennae star clusters are globular clusters in formation, given that the observed clusters are at the top of the Antennae cluster luminosity function, that the HST imaging may somewhat overestimate cluster sizes (and thus masses) and that stellar evolution will remove mass from the clusters. Comparison with the N-body simulations of [32] indicates that the Antennae clusters should survive for several Gyrs but lose a significant fraction of their present mass.

In addition to the derived dynamical masses, we determined from our data the ages (8-10 Myr: from CO/CaT and HI Brγ equivalent widths) and the V-band/K-band luminosities of the 5 clusters. Taken together, our measurements constrain the distribution function of stellar masses in the clusters, which should be close to their birth/Initial Mass Function (IMF). The bottom inset in Figure 4 shows the results for four of our clusters in the plane L_V/M vs log(t). For comparison we also plot the positions of a cluster in the starburst galaxy M82 and two clusters in the blue compact dwarfs NGC1705 and NGC1569 [30, 29]. Also plotted are L_V/M(t) ratios for power law IMFs with different slopes α, assuming a mass range from 0.1 to 100 $M_\odot$. A Salpeter IMF (solid line in Figure 4) has α=2.35, and a Salpeter IMF with a mass range of 1-100 $M_\odot$ has an L/M ratio that is a factor of 2.6 greater. At face value the data of the different clusters in Figure 4 require different IMF slopes (see also the analysis of the IMF in the Galactic star cluster NGC 3603 [7]. Most clusters are consistent with a Salpeter IMF, but (three of) the Antennae clusters require a steeper IMF, and the M82 and NGC 3603 clusters require a shallower IMF. To be consistent with a common (Salpeter-like) IMF, all authors would have had to significantly underestimate the errors in the analysis of their data. It is interesting to note that the clusters requiring a steeper IMF are all found in the dusty 'overlap region' between NGC 4038 and NGC 4039. Figure 8 thus provides the tantalizing result that IMFs vary in different environments. Obviously it is of great interest to increase the statistics and confidence of this potentially far-reaching result.

3   First Results of an IR Tully-Fisher Study of Star Forming Disk Galaxies at z∼1

As part of the Ph.D. thesis of one of us (M.B.), we have recently begun a program of near-IR spectroscopy of distant disk galaxies for a determination of Hα rotation curves at z∼1. Our program extends to higher redshift earlier [OII] studies at z∼0.4–0.9 undertaken by the Lick group [34, 35, 18]. We have observed with ISAAC 20 inclined disk galaxies, selected from the CFRS, Hawaii medium deep and Caltech faint galaxy surveys to lie in the redshift range 0.6–1.6. HST images and photometry are available for nearly all of these sources. We gave preference to those systems with significant [OII] emission in the existing optical spectra. So far we have been able to extract Hα rotation curves for 16 galaxies. The top

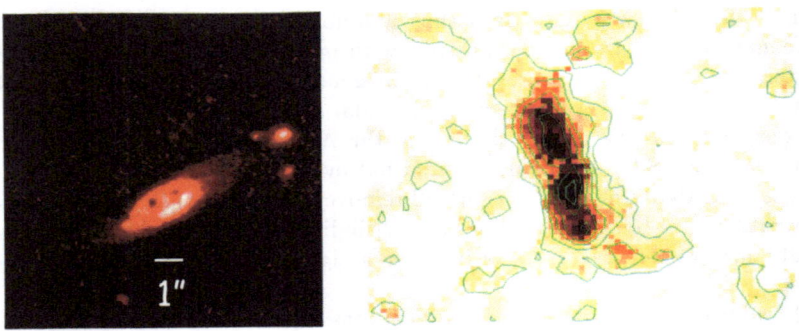

Tully-Fisher Relation at z=1

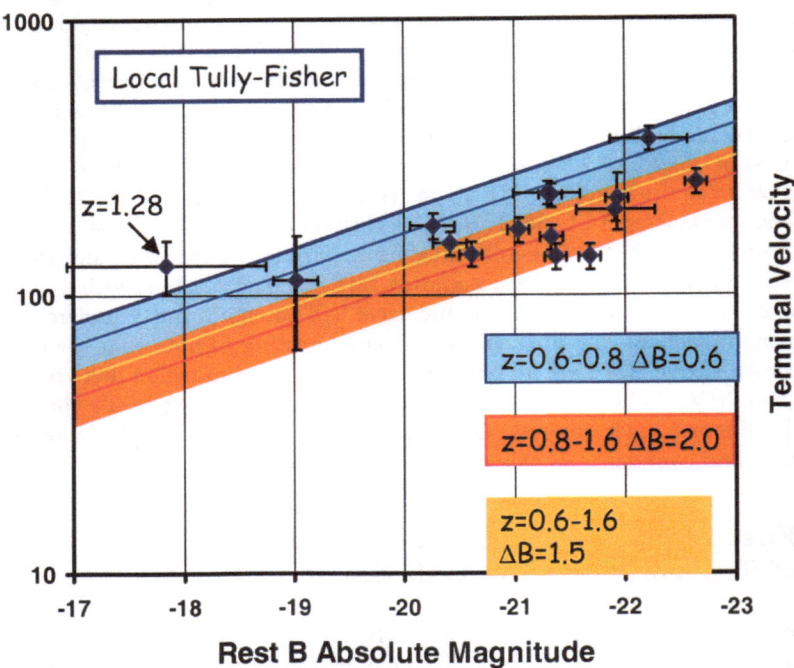

Fig. 5. Tully-Fisher diagram for 15 disk galaxies between z=0.6 and 1.6, derived from ISAAC Hα rotation curves. The local Tully-Fisher curve is denoted by a thick blue line. The z~1 data are consistent with the same slope as that of the local curve, but appear to require a brightening of 1.5 mag in the rest-B band

insets of Figure 5 show a typical (HST) image and a position-velocity diagram for one of our galaxies. In that case, the turnover of the observed rotation curve to the flat part appears to be well sampled. In other cases, the derivation of the true rotation curve requires careful modeling with an input rotation curve and taking into account the spatial and spectral convolution of our finite resolution data (see [34,35] for a discussion). A first cut Tully-Fisher curve is shown in the lower part of Figure 5 (for a $\Omega_m = 0.3, \Omega_A = 0.7, h=0.7$ cosmology). Our data are consistent with the same slope as the local Tully-Fisher relationship but indicate a 1.5 mag brightening in the rest-wavelength B-band. For comparison [35] did not find any significant evolution of the Tully-Fisher relation at z~0.5 ($\Delta B \leq 0.6$ mag), while [23] and [28] find a brightening of ~1.5-2 mag at z~0.2-0.3 for blue, compact emission line galaxies. Our sample contains mainly large disk galaxies (disk scale lengths of 5±3 kpc), similar to [35], perhaps indicating a more significant evolution at the higher redshifts we are sampling (see also Weiner, this symposium). We are presently investigating in more detail the possible influence of the selection criteria of our sample on the results.

References

1. R. Bender, D. Burstein, S.M. Faber: ApJ **399**, 462 (1992)
2. G.J. Bendo, J. Barnes: MNRAS **316**, 315 (2000)
3. D. Bettoni, R. Falomo, G. Fasano, F. Gavoni, M. Salvo, R. Scarpa: A&A **380**, 471 (2001)
4. A.W. Blain, I. Smail, R.J. Ivison, J.-P. Kneib, D.T. Frayer: Phys.Rep. **in press**, (astro-ph 0202228) (2002)
5. G. Canalizo, A. Stockton: AJ **120**, 1750 (2000)
6. J.S. Dunlop, R.J. McLure, M.J. Kukula, S.A. Baum, C.P. O'Dea, D.H. Hughes: MNRAS **in press**, (astro-ph 0108397)(2002)
7. F. Eisenhauer, A. Quirrenbach, H. Zinnecker, R. Genzel: ApJ **498**, 278 (1998)
8. D. Elbaz, C.J. Cesarsky, P. Chanial, H. Aussel, A. Franceschini, D. Fadda, R.R. Chary: A&A **in press**, (astro-ph 0201328) (2002)
9. S.M. Faber, et al.: AJ **114**, 1771 (1997)
10. L. Ferrarese, D. Merritt: ApJ **539**, L9 (2000)
11. U. Fritze-v.Alvensleben: A&A **342**, L25 (1999)
12. K. Gebhardt, et al.: ApJ **539**, L13 (2000)
13. R. Genzel, C. Cesarsky: ARAA **38**, 761 (2000)
14. R. Genzel, L.J. Tacconi, D. Rigopoulou, D. Lutz, M. Tecza: ApJ **563**, 527 (2001)
15. J.A. Holtzmann, et al.: AJ **103**, 691 (1992)
16. D.C. Kim, D.B. Sanders: ApJS **119**, 41 (1998)
17. J. Kormendy, D.B. Sanders: ApJ **390**, L53 (1992)
18. D.C. Koo: these proceedings, (Springer, Heidelberg 2002) (astro-ph 0112552)
19. O. Lai, D. Rouan, F. Rigaut, E. Arsenault, E. Gendron: A&A **334**, 783 (1998)
20. S. Mengel, M. Lehnert, N. Thatte, R. Genzel: A&A **in press**, (astro-ph 0111560) (2002)
21. J.C. Mihos: Ap&SS **266**, 195 (1999)
22. M. Pahre: ApJS **124**, 12 (1999)
23. H.-W. Rix, et al.: MNRAS **285**, 779 (1997)
24. D.B. Sanders, I.F. Mirabel: ARAA **34**, 749 (1996)

25. D.B. Sanders, B.T. Soifer, J.H. Elias, B.F. Madore, K. Matthews, G. Neugebauer, N.Z. Scoville: ApJ **325**, 74 (1988)
26. F. Schweizer: In IAU Symposium 207, "Extragalactic Star Clusters", eds. E. Grebel, D. Geisler, D. Minniti (ASP, San Francisco 2002) (astro-ph 0106345)
27. N.Z. Scoville, et al.: ApJ **119**, 991 (2000)
28. L. Simard, et al.: ApJ **519**, 563 (1999)
29. L.J. Smith, J.S. Gallagher: MNRAS **326**, 1027 (2001)
30. A. Sternberg: ApJ **506**, 721 (1998)
31. L.J. Tacconi, R. Genzel, D. Lutz, D. Rigopoulou, A. Baker, C. Iserlohe, M. Tecza: ApJ **submitted** (2002)
32. K. Takahashi, S.F. Portegies-Zwart: ApJ **535**, 759 (2000)
33. A. Toomre, J. Toomre: ApJ **178**, 623 (1972)
34. N.P. Vogt, et al.: ApJ **465**, L15 (1996)
35. N.P. Vogt, et al.: ApJ **479**, L121 (1997)
36. B.C. Whitmore, Q. Zhang, C. Leitherer, S.M. Fall, F. Schweizer, B.W. Miller: AJ **118**, 1551 (1999)
37. S.E. Zepf, K.M. Ashman: MNRAS **264**, 611 (1993)

Luminosities of Radio-Loud Active Galaxies

Aldo Treves[1], Nicoletta Carangelo[1], and Renato Falomo[2]

[1] Università dell'Insubria, Via Valleggio 11, Como, Italy
[2] Osservatorio Astronomico di Padova, Vicolo dell'Osservatorio 5, Padova, Italy

Abstract. We investigate the host galaxy luminosities of three different samples of radio loud active nuclei at z<0.5 including BL Lac Objects (BLLs), Radio Loud Quasars (RLQs) and Radio Galaxies (FRs). The data have been extracted from images obtained with HST (for BLLs and RLQs) and with the ESO-2.2m and NOT telescopes (for RGs) All sources are well resolved and their host galaxies are luminous bulge dominated systems. After a homogeneous treatment of the data we find that the host galaxy average absolute magnitude R are: $< M_R >_{RLQ} = -24.3$, $< M_R >_{BLL} = -23.7$, $< M_R >_{FRI} = -24.1$, $< M_R >_{FRII} = -23.6$. Host of RLQ are ~ 0.5 mag brighter than those of BLLs which are intermediate between RG of FR I and FR II type. These galaxy luminosities are used to evaluate the central black hole masses from the M_{BH} -L_{bulge} relation found for nearby spheroidal galaxies. Finally we construct the Host Galaxy Luminosity Function for the different classes of AGN and compare its shape with that of elliptical galaxies.

1 Introduction

There is nowadays a large body of evidence that radio-galaxies, radio-loud quasars and BL Lac objects are all hosted in massive ellipticals (eg. [24], [7], [12]). These galaxies are difficult to characterize because of the distance and of the presence the bright nucleus. A problem that becomes more relevant with the increasing of the redshift. Nevertheless in the last decade a number of samples of radio loud active galaxies have been investigated either with large ground based telescopes with optimal seeing conditions or using HST images.

In this paper we perform a comparative study of the galaxies hosting different kinds of AGN using ground based data for FRs and observations by HST for RLQs and BLLs. For this study we consider only objects at z<0.5 in order to better constraint their host properties. A homogeneous treatment of the data is applied to the different samples in order to construct a uniform dataset for each class of objects. In this analysis $H_0=50$ Km s^{-1} Mpc^{-1} and $\Omega_0=0$ were used.

2 Luminosity Distributions of Galaxies Hosting RLQs, BL Lacs and FRs

All magnitude are converted into the R (Cousins) band. Color corrections for objects observed in other filters were derived using the expected colors of galaxies given by [10], K-correction was performed following [13], and galactic reddening as in [21]. Note that the lack of the latter correction has generated some confusion in the literature.

2.1 RLQs

The most comprehensive paper on the subject is that of [7], who report HST data on N=10 RLQs with 0.1< z <0.3 (see also McLure and Dunlop at this Conference). As shown in [22] the [7] sample is fully compatible with that of [1] (N=8), and that of [2] (N=5). We therefore have considered the combination of the three samples (N=18) taking averages values for objects observed twice. From the reported apparent magnitudes we calculated the absolute magnitudes as specified above, and the resulting distributions are given in Fig. 1 and Table 1.

2.2 BL Lacs

The HST snapshot image survey of BL Lacs [24], [21] has provided a homogeneous set of 110 short exposures high resolution images through the F702 filter. From the data set we have considered all resolved objects at z<0.5, yielding N=57 sources (see Fig. 1 and Table 1).

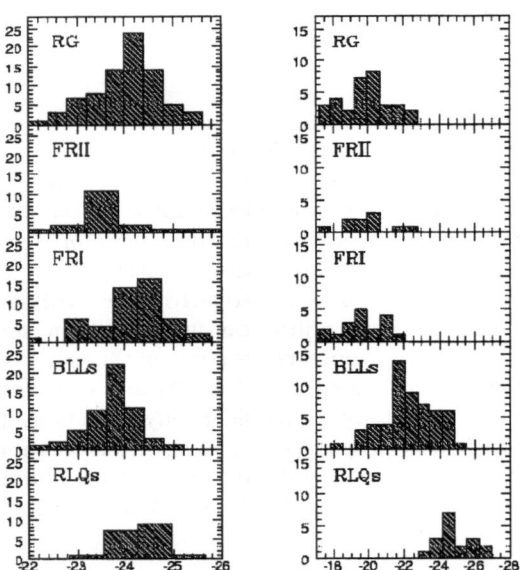

Fig. 1. Host galaxies absolute R magnitude distribution (left panels) and nuclear absolute B magnitude distribution (right panel) of RLQs, BL Lacs, FRI and FRII

Table 1. Properties of the datasets: Column (1) dataset; (2) references: (a) [24], (b) [22], (c) [12]; (3) number of objects; (4) average redshift; (5) average B nuclear magnitude; (6) average R host magnitude; (7) average black hole mass

Dataset	Ref.	N_{obj}	$<z>$	$<M_B(nuc)>$	$<M_R(host)>$	$<(M_{bh}/M_\odot)>$
(1)	(2)	(3)	(4)	(5)	(6)	(7)
BLLs	(a)	57	0.20±0.11	-22.3±1.5	-23.7±0.5	$(7.7\pm0.6)\times10^8$
RLQs	(b)	18	0.25±0.07	-24.8±1.0	-24.3±0.4	$(1.4\pm0.7)\times10^9$
FRIs	(c)	50	0.05±0.03	-19.5±1.4	-24.1±0.6	$(1.2\pm0.5)\times10^9$
FRIIs	(c)	19	0.06±0.02	-20.0±1.3	-23.6±0.7	$(7\pm0.5)\times10^8$

2.3 FRI and FRII

For radio-galaxies we refer to the ground observed sample of [12] of 50 FRI and 19 FRII, all objects at z<0.12. For some HST observations see [7], [5] and [3]. The results are given in Fig. 1 and Table 1.

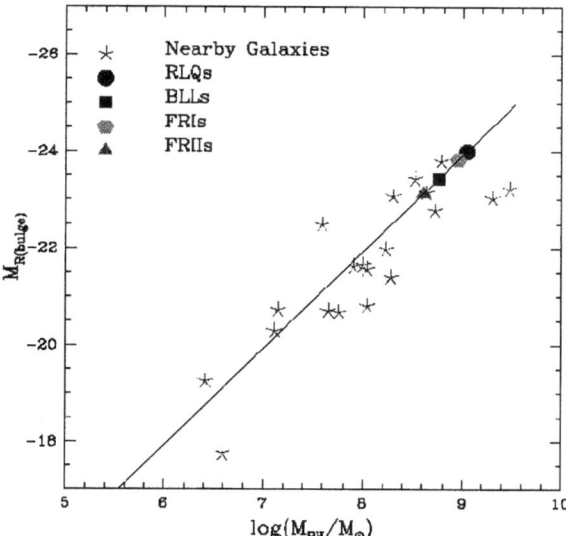

Fig. 2. Absolute R-band bulge magnitude versus dynamical black hole mass estimate. The solid line is the best fitting linear relation obtained for the sample of 20-inactive E-type galaxies [15]

For all three classes the host galaxies are very luminous: they are ~1-2 mag more luminous than the typical galaxy luminosity (M*(R)≃-22.75). RLQ hosts are somewhat (~0.7 mag) more luminous than BL Lac hosts. The luminosity of BLL is closer to that of FRII than to that of FRI.

3   Mass of the Central Black Hole

Dynamical studies of nearby elliptical galaxies have shown that there is a correlation between the galaxy luminosity and the mass of the central black hole ([15] and references therein). We use the form of the correlation reported by [18] (see also the contributions of authors at the meeting). Assuming that the dependence can be used to calculate the masses of the central black hole for active Ellipticals, we obtain the average values reported in Table 1. The dispersion refers to that of the luminosity distributions.

In figure 2, adapted from [7], the $M_{BH} - M_R$ is reproduced together with the average values of BH masses for our samples of radio-loud AGNs: the best fit including the extinction correction is $\log(M_{bh}/M_\odot)=-0.45(\pm0.06)M_R-1.78(\pm1.33)$.

The similarity of the galaxy luminosities reflects in that of the hole masses. Note that total luminosities of RLQs are larger than those of BL Lacs (and FRs) by at least 2 orders of magnitude. This implies a drastic difference of the

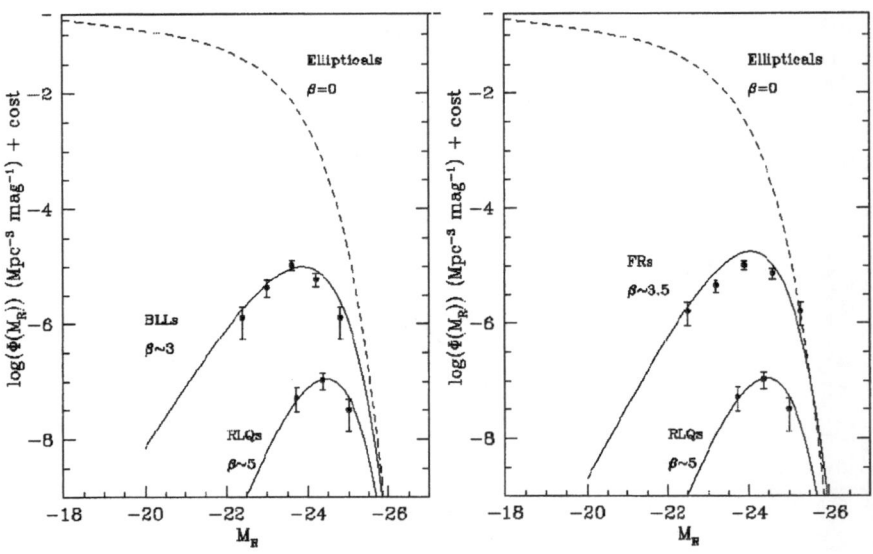

Fig. 3. The HGLF of RLQs, BL Lacs and FRs with arbitrary normalization. The *dashed curve* is the elliptical galaxy luminosity function of [19] and the *solid lines* are the fits with a modified Schechter function Φ

Eddington ratio $\xi_E = L/L_E$ where $L_E=1.25\times10^{38}\times(M_{bh}/M_\odot)$ erg s^{-1} (on this point see [20] and [22]).

The central black hole mass of Ellipticals can be obtained also through the mass-velocity dispersion relation ([9]; [11]). For some preliminary results on the mass of black holes in BL Lacs, see Falomo et al. at this Conference.

4   The Host Galaxy Luminosity Functions (HGLF)

The three samples that we have considered were essentially selected on the basis of nuclear properties. Therefore on a first approximation they are unbiased with regard to the host galaxy luminosity. Even if the samples are small, we have constructed a rough luminosity function for each class (see Fig.3, and see also [4]) and we compared it in shape with that of Elliptical Galaxies.

We fit the luminosity function of the host galaxies with a modified Schechter function $\Phi=K \times \Phi_S \times (L/L^*)^\beta$, where Φ_S is the Schechter function for elliptical galaxies ([19]): $\Phi_S=\Phi^* \times (L/L^*)^\alpha \times \exp(-L/L^*)$, with $\Phi^* = 8.5 \times 10^{-2}$ Mpc^{-3}, $\alpha=-1.2$ and L$^*=2.25 \times 10^{44}$ erg s^{-1}. The best fit has been estimated minimizing χ^2 for the function Φ. We find $\beta \sim 3$ for BLLs, $\beta \sim 5$ for RLQ hosts and $\beta \sim 3.5$ for FRs.. The derived HGLFs are given in Fig.3.

The HGLFs of radio-loud AGNs are remarkably different in shape from that of inactive Ellipticals indicating that these AGNs are preferentially drawn from the bright tail of elliptical galaxy luminosity function. The HGLFs of RLQs, BL Lacs and RGs have similar shape but with a trend for brighter galaxies in RLQs. This is quantified by comparison of the β parameter: $\beta_{RLQs} \sim 5 > \beta_{FRs} \sim 3.5 > \beta_{BLL} \sim 3$.

References

1. J.N. Bahcall, S. Kirhakos, D.H. Saxe & D. P. Schneider: ApJ **479**, 642 (1997)
2. P.J. Boyce, M.J. Disney, j.C.Blades, A. Boksenberg, P. Crane, J.M. Deharveng, F.D. Macchetto, C.D. Mackay, W.B. Sparks: MNRAS 298, 121 (1998)
3. A. Capetti, H.R. de Ruiter, R. Fanti, R. Morganti, P. Parma and M.-H. Ulrich: A&A **362**, 871 (2000)
4. N. Carangelo, A. Treves, R. Falomo: astro-ph 0110301 (2001)
5. M. Chiaberge, A. Capetti and A. Celotti: A&A **355**, 873 (2000)
6. M.J. Disney, P.J. Boyce, J.C. Blades, A. Boksenberg, P. Cane, J.M. Deharveng, F. Macchetto, C.D. Mackay, W.B. Sparks, S. Phillipps: Nature **376**, 150 (1995)
7. J.S. Dunlop, R.J. McLure, S.A. Baum, C.P. O'Dea & D.H. Hughes: astro-ph 0108397 (2001)
8. R. Falomo, J. Kotilainen, A. Treves: ApJ **547**, 124 (2001)
9. L. Ferrarese, D. Merritt: ApJ **539L**, 9F (2000)
10. M. Fukugita, K. Shimasaku, T. Ichikawa: PASP **107**, 945 (1995)
11. K. Gebhardt, R. Bender, G. Bower, A. Dressler, S.M. Faber, A.V. Filippenko, R. Green, C. Grillmair, L.C. Ho, J. Kormendy, T.R. Lauer, J. Magorrian, J. Pinkney, D. Richstone, S. Tremaine: ApJ **539L**, 13G (2000)
12. F. Govoni, R. Falomo, G. Fasano, R. Scarpa: A&AS **143**, 369 (2000)

13. B. M. Poggianti: A&AS **122**, 399 (1997)
14. K.I. Kellermann, R. Sramek, M. Scamidt, D.B. Shaffer, R. Green: AJ **98**, 1195 (1989)
15. J. Kormendy & K. Gebhardt: astro-ph/0105230 (2001)
16. J. M. Kukula, J.S. Dunlop, R.J. McLure, L. Miller, W.J. Percival, S.A., Baum, C.P. O'Dea: MNRAS.**326**, 1533K (2001)
17. R.J. McLure et al.: MNRAS **308**, 377 (1999)
18. R.J. McLure & J.S. Dunlop: astro-ph/0108417 (2001)
19. N. Metcalfe, A. Ratcliffe, T: Shanks and R. Fong: MNRAS **294**, 147 (1998)
20. M. O'Dowd, C.M. Urry, R. Scarpa, R. Falomo, J.E. Pesce, A. Treves: "The nucleus-host galaxy connection in radio-loud AGN" to appear in the proceedings of *QSO Hosts and their Environments, Granada, January 10-12, 2001*
21. R. Scarpa, C.M. Urry, R. Falomo, J. Pesce & A. Treves: ApJ **532**, 740 (2000)
22. A. Treves, N. Carangelo, R. Falomo: astro-ph 0107129 (2001)
23. C.M. Urry, P. Padovani: PASP **107**, 803 (1995)
24. C. M. Urry, R. Scarpa, M. O'Dowd, R. Falomo, J.E. Pesce, A. Treves: ApJ **532**, 816 (2000)

The Evolution of Galaxy Mass in Hierarchical Models

C.M. Baugh[1], A.J. Benson[2], S. Cole[1], C.S. Frenk[1], and C. Lacey[1]

[1] Department of Physics, Durham University, South Road, Durham, DH1 3LE.
[2] Caltech, MC105-24, 1200 E. California Blvd., Pasadena, CA 91125

Abstract. Advances in extragalactic astronomy have prompted the development of increasingly realistic models which aim to describe the formation and evolution of galaxies. We review the philosophy behind one such technique, called semi-analytic modelling, and explain the relation between this approach and direct simulations of gas dynamics. Finally, we present model predictions for the evolution of the stellar mass of galaxies in a universe in which structure formation is hierarchical.

1 Modelling the Formation and Evolution of Galaxies

An incredibly wide range of physical processes are believed to be influential in the formation of galaxies. Some of these processes are well understood, for example, the build up of dark matter haloes through mergers or the accretion of smaller units; the formation of haloes has been studied extensively using N-body simulations and can be described analytically with a reasonable degree of success (e.g. Lacey & Cole 1993, 1994). On the other hand, we are still some distance away from being able to simulate the formation of stars. An impressive initial step in this direction has been taken by Abel *et al.* (2002) with a simulation that leads up to the formation of the first star in the universe. However, the conditions in this calculation are much simpler than would be typical for the formation of the bulk of the stars in the universe and the simulation is stopped once additional physics not currently included in the calculation, such as radiative transfer, become important.

The absence of a complete theory of star formation need not be an obstacle to the development of a theory of galaxy formation. One can take a phenomenological approach in which a physically motivated recipe is adopted to describe star formation within a galaxy. The recipe will inevitably contain one or more uncertain parameters but these can be fixed by comparing model predictions with observational data.

Adopting this pragmatic approach, two techniques have been developed to model the formation and evolution of galaxies. The first of these is the direct simulation of gravitational instability and gas dynamics. The second class of technique is semi-analytic modelling (Kauffmann *et al.* 1993; Cole *et al.* 1994). In such models, the merger trees of dark matter haloes can either be grown using a Monte-Carlo algorithm or they can be extracted from an N-body simulation. The gas physics, namely shock heating, radiative cooling, star formation and supernova feedback (along with galaxy mergers), is followed using approximations and simple rules.

The two techniques have complementary pros and cons. Direct simulations do not require the specialised assumptions that are necessary in the semi-analytic models, e.g. the imposition of spherical symmetry in the calculation of the gas cooling time. On the other hand, semi-analytic models are fast and flexible, allowing a wide range of parameter space to be explored. The modular structure of the semi-analytic models means that new prescriptions for processes such as star formation can be readily evaluated.

In certain respects, the two techniques are actually very similar. The direct simulation approach necessarily breaks down at some level because of the finite resolution that is attainable. It is not possible to achieve the sub-parsec resolution needed to simulate star formation in a cosmologically representative volume. Coupled with the lack of knowledge of the relevant micro-physics, this means that recipes like those used in the semi-analytic models have to be deployed in order to produce a fully specified model.

The first comparisons of the two techniques have recently been carried out (Benson et al. 2001a; Helly et al. 2002; Yoshida et al. 2002). These studies considered the rate at which gas cools in SPH simulations and in "stripped-down" semi-analytic models in which star formation and feedback have been switched off. The two approaches are in remarkably good agreement, which inspires confidence in the cooling model adopted in the semi-analytic schemes.

2   Constructing a Model

In the phenomenological approach to galaxy formation, the values of the parameters in the recipes that describe processes such as star formation and feedback have to be specified to produce model predictions. This task is performed by comparing the model predictions to a subset of the available observational data. Different groups of modellers have different priorities when attempting to reproduce the data. The Munich group, for example, has attached most importance to matching the slope and zero point of Tully & Fisher's (1977) correlation between the luminosity and rotation speed of disk dominated galaxies. The Durham group instead try hardest to match the form of the present day galaxy luminosity function. The luminosity function is the most basic description of the galaxy population and is now known to a high level of accuracy in the optical from the 2dFGRS (Norberg et al. 2002) and SDSS (Blanton et al. 2000) and in the near-infrared from 2MASS photometry (Cole et al. 2001; Kochanek et al. 2001). The predictions of the fiducial model of Cole et al. (2000) are compared with these recent estimates of the local luminosity function in Fig. 1.

Although most weight is given to reproducing the luminosity function when setting model parameters, matching other datasets, such as the Tully-Fisher relation, the distribution of disk scale lengths, the metallicity of gas in spiral disks and of stars in ellipticals, and the gas fraction in spiral disks, is also important. This greatly restricts the viable range of parameter space of the models.

One criticism levelled at semi-analytic models that has entered into popular folklore is the inability of the models to match the zeropoint of the Tully-Fisher

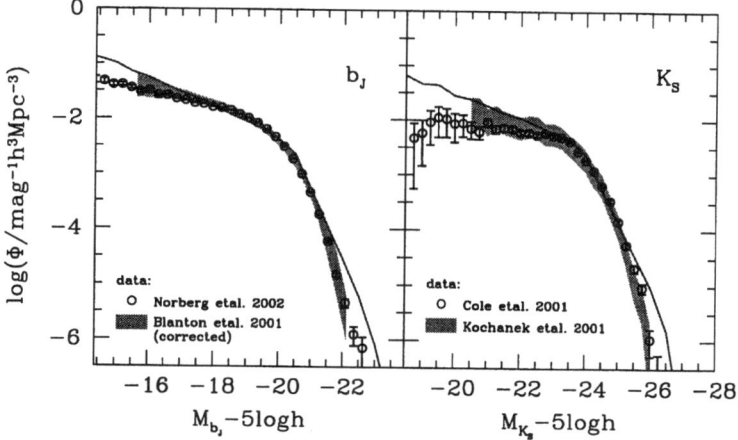

Fig. 1. The local galaxy luminosity function, in the b_J- and K_S- bands. The predictions of the fiducial model from Cole *et al.* (2000) are shown by the solid line in each panel. In the left panel, the symbols shows an estimate of the luminosity function from the 2dFGRS (Norberg *et al.* 2002). The shaded region shows an estimate based on the analysis of SDSS data in Blanton *et al.* (2000) (see Norberg *et al.* 2002 for full details). In the right panel, a combination of 2dFGRS redshifts and 2MASS photometry was used to estimate the near infrared luminosity function (Cole *et al.* 2001). The shaded region shows another observational estimate which also uses 2MASS photometry (Kochanek *et al.* 2001)

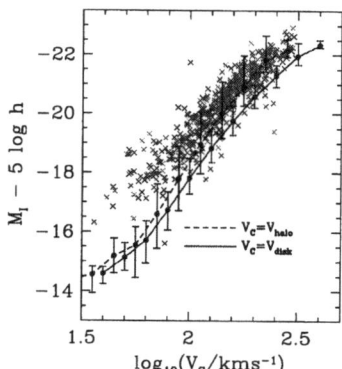

Fig. 2. The Tully-Fisher relation for star forming disk galaxies. The crosses show data from the sample of Mathewson, Ford & Buchhorn (1992). The dashed line shows the model prediction for the Tully-Fisher relation when the rotation speed of the halo at the virial radius is plotted. The solid line shows the predictions when the rotation speed at the half mass radius is plotted instead

relation at the same time as reproducing the break in the luminosity function at L_*. The Tully-Fisher relation of the fiducial model of Cole *et al.* is compared with the observed relation in Fig. 2. The solid line shows the model prediction when the rotation speed at the half-mass radius of the disk is plotted; the dashed line shows how the zeropoint shifts when the rotation speed of the halo at the virial radius is plotted instead, which is much closer to the observed zeropoint. The shift is around 20% - 30%, which is comparable to the accuracy one might expect in the calculation of the rotation speed at the half-mass radius. This depends upon several effects, such as the self gravity of the baryons and their gravitational influence on the halo dark matter.

3 Model Predictions – An Example

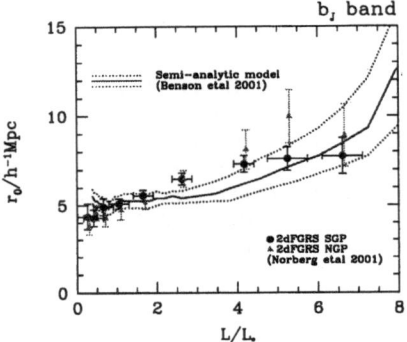

Fig. 3. The correlation length in real space, obtained by fitting a power law to the measured correlation function, $\xi(r) = (r_0/r)^\gamma$, plotted as a function of luminosity. The solid line shows the model predictions taken from Benson *et al.* (2001b). The dotted lines show the Poisson errors derived from the pair counts. The symbols show the subsequent measurements made from the 2dFGRS (Norberg *et al.* 2001). In this case, the errors are estimated from mock 2dFGRS catalogues constructed from N-body simulations and include sample variance

Now that we have arrived at a fully specified model by comparing the output against a subset of the observations to fix the model parameters, we can make predictions for other quantities. Benson *et al.* (2000a,b; 2001b) populated a high resolution N-body simulation with galaxies using the semi-analytic model of Cole *et al.*. The simulation gives the spatial distribution of galaxies and allows their clustering to be measured. Remarkably, without any further adjustment to the model parameters, Benson *et al.* found that the fiducial ΛCDM model of Cole *et al.* predicts a correlation function that is in extremely good agreement with that measured for APM galaxies (Baugh 1996). This is particularly noteworthy as the galaxy correlation function is close to a power law, whereas the correlation function of the dark matter shows considerable curvature.

Benson *et al.* (2000b; 2001) presented predictions for the dependence of clustering strength on luminosity in the same model and found an approximately linear dependence of correlation length on luminosity; galaxies six times more luminous than L_* have a correlation length 50% longer than that predicted for L_* galaxies. At the time, the picture emerging from the data was unclear. This has now been resolved by measurements from the 2dFGRS (Norberg *et al.* 2001) and SDSS (Zehavi *et al.* 2002), which are in reasonable agreement with the trend predicted by the semi-analytic models.

4   The Evolution of the Stellar Mass of Galaxies

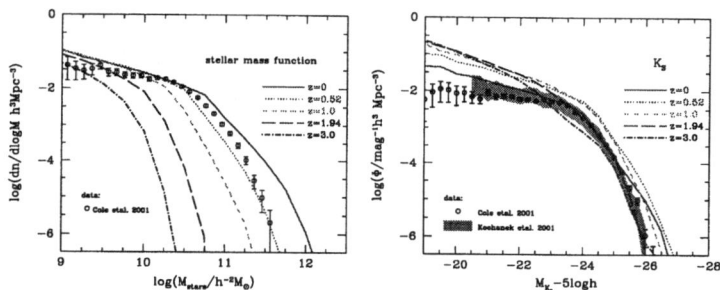

Fig. 4. *Left:* The evolution of the stellar mass function with redshift. The lines show the model predictions at different redshifts, as indicated by the key. The datapoints show the present day stellar mass function inferred from the K_S-band luminosity function by Cole *et al.* (2001). *Right:* The evolution with redshift of the *observer* frame K_S-band luminosity function. The symbols and shaded region show the present day K_S-band luminosity function estimated with 2MASS photometry

Advances in detectors that operate in the near infrared have led to a huge increase in the size of K selected samples over the past decade. The first direct estimate of the K-band luminosity function from a K-selected sample used ~ 500 galaxies (Gardner *et al.* 1997); the estimate of the K_S-band luminosity function by Cole *et al.* (2001), using 2MASS photometry and 2dFGRS redshifts was made from over 17,000 galaxies. The K-band luminosity of a galaxy gives a reasonable indication of its stellar mass. The output from the semi-analytic model suggests that the scatter in the stellar mass$--K$-band magnitude relation is a factor of ~ 2, showing the relative insensitivity to star formation history.

It is important to make a fair comparison between observational estimates and theoretical predictions for stellar mass. The stellar mass inferred from the K-band light is sensitive to the choice of IMF. Also, one needs to be clear whether recycling of gas is included i.e. whether the quantity under consideration is the mass locked up in stars or the total mass that had been turned into stars (some of which is subsequently expelled in stellar winds and supernovae). Cole *et al.* (2001) estimated the stellar mass function from the K_S-band luminosity function (shown by the symbols in Fig. 4), and found that only a small fraction of the baryons in the universe, perhaps as little as 5%, is actually locked up in stars. (Similar results were obtained by Kochanek *et al.* 2001.)

We plot the evolution of the stellar mass function in Fig. 4. There is a steady increase in the typical stellar mass with time; the value of M_* increases by a factor of ~ 2 between $z = 1$ and the present. The observable counterpart to the stellar mass function, the *observer* frame $K-$band luminosity function shows more complex evolution (Fig. 5). This is due to band shifting.

5   Summary

We have given an outline of the semi-analytic approach to modelling galaxy formation. This technique is complementary to direct simulation of the relevant gas dynamic processes. In fact, both methods rely upon physically motivated recipes to deal with star formation and feedback. The model predicts strong evolution in the mass of stellar systems, with more than an order of magnitude increase in the abundance of $10^{11}h^{-2}M_\odot$ systems between $z = 1$ and the present day. Constraints on these predictions are now beginning to emerge, with the advent of the first results from deep, near-infrared photometry (see, for example, Drory *et al.* 2001, and the contributions by Drory and by Papovich *et al.* to this volume).

References

1. Abel, T., *et al.*, Science, **295**, 93. (2002).
2. Baugh, C.M., MNRAS, **280**, 267. (1996).
3. Benson, A., *et al.*, MNRAS, **311**, 793. (2000).
4. Benson, A., *et al.*, MNRAS, **316**, 107. (2000).
5. Benson, A., *et al.*, MNRAS, **320**, 261. (2001a).
6. Benson, A., *et al.*, MNRAS, **327**, 1041. (2001b).
7. Blanton, M., *et al.* (the SDSS team), AJ, **121**, 2358. (2001).
8. Cole, S., *et al.*, MNRAS, **271**, 781. (1994).
9. Cole, S., *et al.*, MNRAS, **319**, 168. (2000).
10. Cole, S., *et al.* (the 2dFGRS team), MNRAS, **326**, 255. (2001).
11. Drory, N., *et al.*, ApJ, **562**, L111. (2001).
12. Gardner *et al.*, ApJ, **480**, L99. (1997).
13. Kauffmann, G., *et al.*, MNRAS, **264**, 201. (1993).
14. Helly, J., *et al.*, MNRAS submitted, astro-ph/0202485. (2002).
15. Kochanek, C., *et al.*, ApJ, **560**, 566. (2001).
16. Lacey, C. & Cole, S., MNRAS, **262**, 627. (1993).
17. Lacey, C. & Cole, S., MNRAS, **271**, 676. (1994).
18. Mathewson *et al.*, ApJS, **81**, 413. (1992).
19. Norberg, P., *et al.* (the 2dFGRS team), MNRAS, **328**, 64. (2001).
20. Norberg, P., *et al.* (the 2dFGRS team), submitted, astro-ph/0111011. (2002).
21. Tully, R.B., & Fisher, J.R., A&A, **54**, 661. (1977).
22. Yoshida, N., *et al.*, MNRAS submitted, astro-ph/0202341. (2002).
23. Zehavi, I., *et al.* (the SDSS team), ApJ in press, astro-ph/0106476. (2002).

Growth of Massive Bulges
by Mergers of Dense Satellites

Marc Balcells[1], J. Alfonso L. Aguerri[1,2], and Reynier F. Peletier[3]

[1] Instituto de Astrofísica de Canarias, 38200 La Laguna, Tenerife, Spain
[2] Astronomisches Institut der Universitat Basel, 4102 Binningen, Switzerland,
[3] School of Physics and Astronomy, University of Nottingham, NG7 2RD, UK

Abstract. We present merger models suggesting that bulge growth via accretion of dense satellites generates two of the observed trends of galaxies along the Hubble sequence, namely the increase of Sersic index toward more massive bulges, and the appearance of a thick disk with higher scale length than that of the thin disk. We argue that bulges with exponential surface brightness profiles cannot have grown significantly from collisionless satellite accretion.

1 The Accretion Hypothesis for Bulge Growth

The surface brightness profiles of bulges are well described with a Sersic profile [9]: Sersic's concentration index n scales with galaxy morphology, with late types showing $n \leq 1$ and early types $n \sim 4 - 6$ [2][5] (see Fig. 1). We have studied the contribution of collisionless satellite accretion to setting the observed trend [1]. We model the merger of dense satellites onto disk-bulge-halo galaxies using collisionless N-body simulations. Before the merger, the bulges have an exponential profile. After the accretion, the surface density profile of bulges is recovered via a disk-plus-Sersic fit. We describe the evolution of the bulges' concentration index n with growth vectors in the n vs. B/D plane. Results are shown in Fig. 1.

We find that, when growing by mergers, bulge surface brightness profiles easily evolve away from the exponential shape. Roughly a 25% growth in B/D is associated to an increase $\Delta(n) = 1$, and accreting a satellite as massive as the bulge itself results in an $r^{1/4}$ final bulge. The increase of n scales with the satellite mass, and does not strongly depend on other merger parameters such as orbital energy or angular momentum. The evolution drawn by the mergers in broad terms reproduces the distribution of observed bulges, although it is steeper in the simulations. This may be due in part to the high density of our satellites. Also, we have not modeled the lower-end of the observed B/D distribution.

The evolution is due both to mass deposition (satellite plus bar-driven disk infall) and to heating of the pre-existing bulge. The final n is similar for mergers with King and Hernquist satellites of the same mass, emphasizing that the level of incompleteness of the relaxation determines the final profile shape, rather than the structural details of the satellite and bulge. If the satellite density is low and the satellite disrupts before reaching the bulge (e.g. for dwarf galaxies), no evolution occurs.

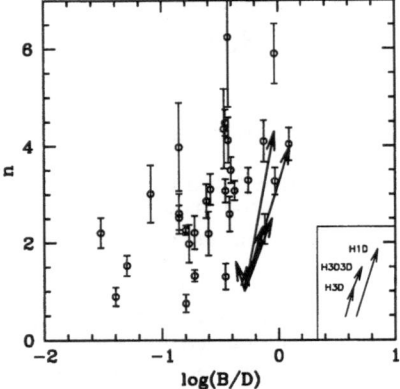

Fig. 1. Merger-driven growth vectors in the n vs B/D plane, and the data from [2]. **Inset:** The cumulative effect: two 1/3-mass mergers (H3D) yield 2/3 as much evolution as a 1:1 merger (H1D)

As expected, the merger heats up the disk. In addition to the scale height, the scale length of the final disk is higher than that typical of thin disks. Because this is a common property of thick disks [4][6][7], we argue that the disks in our merger remnants may be related to present-day thick disks. The hypothesis of accretion-driven bulge growth then predicts a higher prevalence of thick disks toward the early-types of the Hubble sequence, something supported by observations [6]. A match to real galaxies presupposes the formation of a new thin disk out of the gas left over in the merger [8], so, despite the pre-existing disk, this scheme is one of bulge-before-the-disk. The stellar populations of such bulges are expected to be as old as those of the precursor galaxies, perhaps with a younger component from a merger-driven star formation burst. The fragility of exponential profiles suggests that the bulges of late-type galaxies, which are $n \sim 1$, have not formed by dissipationless mergers.

References

1. J. A. L. Aguerri, M. Balcells, R. F. Peletier: A&A, 367, 428 (2001)
2. Y. C. Andredakis, R. F. Peletier, M. Balcells: MNRAS, 275, 874 (1995)
3. M. Balcells, R. F. Peletier: AJ, 107, 135 (1994)
4. D. Burstein: ApJ, 234, 829 (1979)
5. A. Graham: AJ, 121, 820 (2001)
6. R. de Grijs, R. F. Peletier: A&A, 320, L21 (1997)
7. J. A. Larsen, R. M. Humphreys: ApJ, 468, L99 (1996)
8. G. Kauffmann, B. Guiderdoni, S. D. M. White: MNRAS, 267, 981 (1994)
9. J. L. Sersic: *Atlas de Galaxias Australes* (Observatorio Astronómico, Córdoba 1968)

Spheroidal Galaxies/QSOs Connection

Luigi Danese[1], Gian Luigi Granato[2], Laura Silva[3], Manuela Magliocchetti[1], and Gianfranco De Zotti[2]

[1] SISSA/ISAS, Via Beirut 2–4, I-34014 Trieste, Italy
[2] Osservatorio Astronomico di Padova, Vicolo dell'Osservatorio 5, I-35122 Padova, Italy
[3] Osservatorio Astronomico di Trieste, Via Tiepolo 11, I-34131 Trieste, Italy

Abstract. In view of the extensive evidence of a tight inter-relationship between spheroidal galaxies (and galactic bulges) and massive black holes hosted at their centers, a consistent model must deal jointly with the evolution of the two components. We describe one viable model, which successfully accounts for the local luminosity function of spheroidal galaxies, their photometric and chemical properties, deep galaxy counts in different wavebands, including those in the (sub)-mm region which proved to be critical for current semi-analytic models stemming from the standard hierarchical clustering picture, clustering properties of SCUBA galaxies, of EROs, and of LBGs, as well as for the local mass function of massive black holes and for quasar evolution. Predictions that can be tested by surveys carried out by SIRTF are presented.

1 Introduction

The hierarchical clustering model with a scale invariant spectrum of density perturbations in a Cold Dark Matter (CDM) dominated universe has proven to be remarkably successful in matching the observed large-scale structure as well as a broad variety of properties of galaxies of different morphological types (e.g. [6,18]). However, serious shortcomings of this scenario have also become evident in recent years. The critical point can be traced back to the relatively large amount of power on small scales predicted by this model which would imply far more dwarf galaxies or substructure clumps within galactic and cluster mass halos than are observed (the so-called "small-scale crisis" ([20,54,24,37]), unless star formation in small objects is strongly suppressed (or the small scale power is reduced by modifying the standard model).

At the other extreme of the galaxy mass function we have another strong discrepancy with model predictions, that we might call "the massive galaxy crisis": even the best semi-analytic models ([18,9]) hinging upon the standard picture for structure formation in the framework of the hierarchical clustering paradigm, fall short by a substantial factor (up to about 10) to account for the (sub)-mm (SCUBA and MAMBO) counts of galaxies, most of which are probably massive objects undergoing a very intense star-burst (with star formation rates $\sim 1000 \, M_\odot \, yr^{-1}$) at $z > 2$ (see, e.g. [10]). Recent optical data confirm that most massive ellipticals were already in place and (almost) passively evolving up to $z \simeq 1$–1.5, implying that they were fully assembled by $z \sim 2.5$, although the issue is still somewhat controversial ([47,13,8,30,48,5,23,31]). These data are more consistent with the traditional "monolithic" approach whereby giant ellipticals formed most of their stars in a single gigantic starburst at substantial redshifts, and underwent essentially passive evolution thereafter.

On the contrary, in the canonical hierarchical clustering paradigm the smallest objects collapse first and most star formation occurs, at relatively low rates, within relatively small proto-galaxies, that later merged to form larger galaxies. Thus, the expected number of galaxies with very intense star formation is far less than detected in SCUBA and MAMBO surveys and the surface density of massive evolved ellipticals at $z \gtrsim 1$ is also smaller than observed. The "monolithic" approach, however, is inadequate to the extent that it cannot be fitted in a consistent scenario for structure formation from primordial density fluctuations.

2 Relationships Between Quasar and Galaxy Evolution

The above difficulties, affecting even the best current recipes, may indicate that new ingredients need to be taken into account. A key new ingredient may be the mutual feedback between formation and evolution of spheroidal galaxies and of active nuclei residing at their centers ([25,29,21,59,14–16,33–35]). In this framework, [19] elaborated the following scheme (see also [11,53,36]):

- Feed-back effects, from supernova explosions and from active nuclei (note that supernova feedback alone falls short of solving the dearth of dwarf galaxies, [27], but photo-ionization by the UV background re-ionizing the inter-galactic medium (IGM) could do the job [54]) delay the collapse of baryons in smaller clumps while large ellipticals form their stars as soon as their potential wells are in place; *the canonical hierarchical CDM scheme – small clumps collapse first – is therefore reversed for baryons.*
- Large spheroidal galaxies therefore undergo a phase of high (sub)-mm luminosity.
- At the same time, the central black-hole (BH) grows by accretion and the quasar luminosity increases; when it reaches a high enough value, its action (ionization and heating of the gas) stops the star formation and eventually expels the residual gas. This explains the observed correlation between BH and host spheroidal masses (see [53,12]). The same mechanism distributes in the IGM a substantial fraction of metals and may pre-heat the IGM. The onset of quasar activity (and the corresponding squelching of star formation) occurs earlier for more massive objects. The duration of the star-burst increases with decreasing mass from ~ 0.5 to $\sim 2\,\mathrm{Gyr}$.
- This implies that the star-formation activity of the most massive galaxies quickly declines for $z \lesssim 3$, i.e. the redshift distribution of SCUBA/MAMBO galaxies should peak at $z \gtrsim 3$, just before quasars reach their maximum luminosity (at $z \simeq 2.5$). This:
    - explains why very luminous quasars are more easily detected at (sub)-mm wavelengths for $z \gtrsim 2.5$ [39]. The latter authors argue that a large

fraction of the observed (sub)-mm emission is powered by a starburst, as expected from this model

- implies an extremely steep (essentially exponential) decline of (sub)-mm counts at bright fluxes, as indicated by recent data.

- A "quasar phase" follows, lasting 10^7–10^8 yrs.
- A long phase of passive evolution of galaxies ensues, with their colors becoming rapidly very red [Extremely Red Object (ERO) phase].
- Intermediate- and low-mass spheroids have lower Star Formation Rates and less extreme optical depths. They show up as Lyman-Break Galaxies (LBGs).
- Therefore, in this scenario, large ellipticals evolve essentially as in the "monolithic" scenario, yet in the framework of the standard hierarchical clustering picture.

The various aspects and implications of this compound scheme have been addressed by our group in a series of papers. [51] estimated the mass function of quiescent BHs at the centers of local galaxies. They found a dichotomy between large and small BHs: the former are hosted in elliptical galaxies and tend to be not obscured, while the latter, found in the bulges of spiral galaxies, can be reactivated by interactions and are frequently obscured. Their results are consistent with the high luminosity quasar activity occurring as a single short-lived event at relatively high redshift with luminosity close to the Eddington limit. Later, lower luminosity, nuclear activity may be due to re-activation of the central BH, due, e.g., to interactions. [36] analyzed the evolution of the quasar luminosity function in the framework of a model in which the spheroidal galaxies (and bulges of later-type galaxies) and the BHs at their centers form and evolve in parallel. Adopting the standard [46] formation rate of dark matter halos, they found that consistency with the data is achieved if, and only if, BHs in bigger galactic halos form earlier. In other words, the interval between the onset of star formation and the peak of quasar luminosity is shorter for the more massive objects. Some of the other, most recent, results are briefly described in the next section.

3   Some Recent Results and Predictions

3.1   Counts at (sub)-mm Wavelengths

The (sub)-mm counts are expected to be very steep because of the combined effect of the strong cosmological evolution of dust emission in spheroidal galaxies and of the strongly negative K-correction (the dust emission spectrum steeply rises with increasing frequency). The model by [19] has extreme properties in this respect: above several mJy its 850 μm counts reflect the high-mass exponential decline of the mass function of dark halos. In this model, SCUBA/MAMBO galaxies correspond to the phase when massive spheroids formed most of their stars at $z \gtrsim 2.5$; such objects essentially disappear at lower redshifts. On the contrary, the counts predicted by alternative models (which are essentially phenomenological [3,55,50,41,56]), while steep, still have a power law shape, and the redshift distribution has an extensive low-z tail.

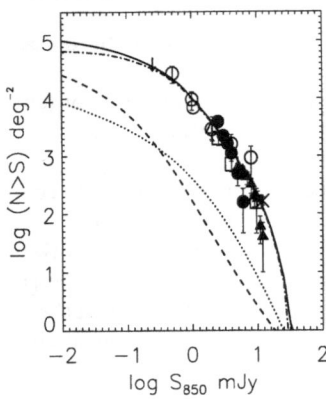 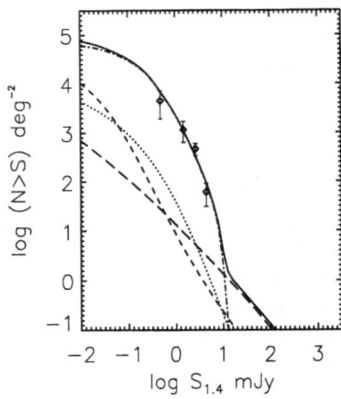

Fig. 1. Integral source counts at 850 μm (left panel) and at 1.4 mm (right panel) predicted by the model by [19] compared with observations. The dotted, dashed and dot-dashed lines show the contributions of starburst, spiral, and forming elliptical galaxies, respectively. The long-dashed line (shown only at 1.4 mm) gives the counts of radio sources, after [57]. The solid line shows the total counts. References for the data points can be found in [44]

As illustrated by Fig. 1, the recent relatively large area surveys [52,4] are indeed suggestive of an exponential decline of the 850 μm counts above several mJy. Further evidence in this direction comes from MAMBO surveys at 1.2 mm [2]; see [44]).

3.2 Lensing Effects on (sub)-mm Counts

A direct consequence of the extreme steepness of the (sub)-mm counts predicted by this model is that their bright tail is strongly affected by gravitational lensing [43,44]. In fact, although the probability of strong lensing is very small, it has a power-law tail ($p(A) \propto A^{-3}$) extending up to large values of the magnification A. Thus, if counts are steep enough, the fraction of lensed sources at bright fluxes may be large.

[43] and [44] find that, according to the model by [19], essentially all proto-spheroidal galaxies brighter than $S_{850\mu m} \simeq 60$–70 mJy are gravitationally lensed. Allowing for the other populations of sources contributing to the bright mm/sub-mm counts, they find that the fraction of gravitationally lensed sources may be $\simeq 40\%$ at fluxes slightly below $S_{850\mu m} = 100$ mJy. If so, large area surveys such as those to be carried out by PLANCK/HFI or by forthcoming balloon experiments like BLAST and ELISA will probe the large scale distribution of the peaks of the primordial density field. For comparison, the maximum fraction of lensed sources predicted by current phenomenological models is $\lesssim 5\%$.

3.3   Clustering of SCUBA/MAMBO Galaxies

Since, in the above scheme, SCUBA galaxies correspond to the rare, massive density peaks at high-z, they are expected to be highly biased tracers of the dark matter distribution, and therefore to be strongly clustered. The recent analyses by [28] and [44] have shown that the implied angular correlation function is indeed consistent with the results by [17] on the clustering of LBGs, and by [52] and [40] on clustering of SCUBA galaxies. The estimated $w(\theta)$ for bright SCUBA galaxies [52] and for EROs [7] indicate dark halo masses $10^{13}\,M_\odot$ for these sources; the $w(\theta)$ for LBGs [17] indicate typical masses at least 10 times lower (see also [38]).

3.4   Chemical and Photometric Evolution

The problem has been addressed by [49]. The short duration of the star-burst in spheroidal galaxies, and its occurrence at $z \gtrsim 3$, assumed by the model, are consistent with the supra-solar Mg/Fe ratio in ellipticals, with the tightness of their fundamental plane and of their color–σ relation. But the model also compares successfully with a broad variety of detailed observational data: the correlation between the abundances of Fe and those of "enhanced" elements (C, N, O, Ne, Mg, Si, S), the correlation between Fe and Mg_2, the template Spectral Energy Distribution of local ellipticals, the color $(V - K)$-magnitude (M_V) and the $(J - K)$–$(V - K)$ relations. The key ingredients allowing to match the observational data, and in particular the relationships among abundances of the various elements are, on one side, the decrease of the duration of the star-burst with increasing galactic mass, and, on the other side, the fact that the effect of stellar feedback is higher in less massive objects, so that the amount of material enriched by SNII that can be retained is lower. The main difference with the majority of previous chemical evolution models is that star formation is stopped not by supernova-driven winds (which also fall short of providing enough preheating of the proto-cluster medium, [58,1,26]), but by the energy injected by the active nucleus.

Note that this model also accounts for the roughly solar, or supra-solar, metallicity of the quasar broad-line regions and of supra-solar N/C and Fe/α ratios in the more luminous objects observed out to $z > 4$ [22] and even up to $z \sim 6$ [42]. This implies that the medium must have been chemically enriched before the quasars became observable and on short timescales (≤ 1 Gyr, at least at the highest redshifts).

Another interesting implication of the model, also discussed by [49], is the dependence of the ratio $M_{\text{darkmatter}}/M_{\text{stars}}$ on M_{stars}. This ratio has a minimum value $\simeq 20$ for $M_{\text{stars}} \simeq 10^{10}$–$10^{11}\,M_\odot$, in good agreement with the results by [32], and increases at smaller masses (in keeping with the findings by [45]) due to the increasing fraction of gas expelled by supernova feedback, as well as at higher masses, due to the increase of the cooling time of the gas in the outer regions of the galactic halo.

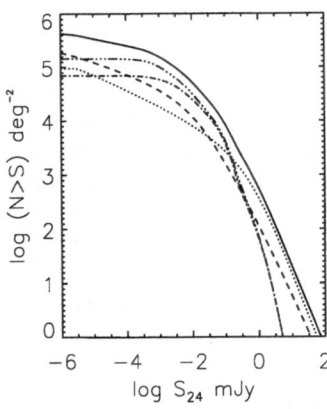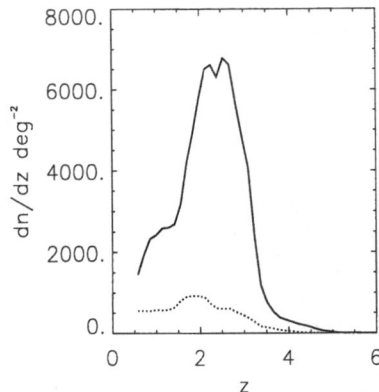

Fig. 2. Integral source counts (left-hand panel) and redshift distribution (right-hand panel) for a flux limit of 30μJy (right panel) at 24 μm predicted by the model by [19]. In the left-hand panel, the dotted, dashed and dot-dashed lines show the contributions of starburst, spiral, and forming elliptical galaxies, respectively, while the three-dots/dashed line shows the total counts of ellipticals, including also those where the star-formation has ended. In the right-hand panel, the solid and the dotted lines show the redshift distributions of ellipticals during and after the star-formation phase, respectively

3.5   Predictions for SIRTF Surveys

SIRTF surveys have the potential of providing further tests of the model. In particular, the GOODS (http://www.stsci.edu/science/goods) 24 μm survey should reach the confusion limit at 30–100 μJy. According to the model, about 50% of the detected galaxies should be spheroidal galaxies forming their stars at $z \gtrsim 2$. About 400–600 such objects are expected over an area of 0.1 square degree (see Fig. 2). Their redshift distribution is predicted to peak at z slightly above 2, with a significant tail extending up to $z \gtrsim 3$.

Acknowledgements. We benefited from many helpful exchanges with C. Baccigalupi, F. Matteucci, F. Perrotta, D. Romano. Work supported in part by ASI and MIUR.

References

1. M.L. Balogh, F. Pearce, R.G. Bower, S. Kay: MNRAS **326**, 1228 (2001)
2. F. Bertoldi, K.M. Menten, E. Kreysa, C.L. Carilli, F. Owen: in *Cold Gas and Dust at High Redshift*, ed. by D.J. Wilner, Highlights of Astronomy, Vol. 12 (2000)
3. A.W. Blain, A. Jameson et al.: MNRAS **309**, 715 (1999)
4. C. Borys, S. Chapman, M. Halpern, D. Scott D.: MNRAS, submitted (2001)
5. J.G. Cohen: AJ **121**, 2895 (2001)
6. S. Cole, C.G. Lacey, C.M. Baugh, M. Carlton, C.S. Frenk: MNRAS **319**, 168 (2000)
7. E. Daddi, A. Cimatti, et al.: A&A **361**, 535 (2000)

8. E. Daddi, A. Cimatti, A. Renzini: A&A **362**, L45 (2000)
9. J.E.G. Devriendt, B. Guiderdoni: A&A **363**, 851 (2000)
10. J.S. Dunlop: New Astr. Rev. **45**, 609 (2001)
11. S.A. Eales, M.G. Edmunds: MNRAS **280**, 1167 (1996)
12. A.C. Fabian: MNRAS **308**, L39 (1999)
13. H.C. Ferguson, M. Dickinson, R. Williams: 2000, ARA&A **38**, 667 (2000)
14. L. Ferrarese, D. Merritt: ApJ **539**, L9 (2000)
15. K. Gebhardt, R. Bender et al.: ApJ **539**, L13 (2000)
16. K. Gebhardt, J. Kormendy et al.: ApJ **543**, L5 (2000)
17. M. Giavalisco, C.C. Steidel et al.: ApJ **503**, 543 (1998)
18. G.L. Granato, C.G. Lacey et al.: ApJ **542**, 710 (2000)
19. G.L. Granato, L. Silva et al.: MNRAS **324**, 757 (2001)
20. Z. Haiman, R. Barkana, J.P. Ostriker: in *Proc. 20th Texas Symp.* (2001)
21. P.B. Hall, R.F. Green: ApJ **507**, 558 (1998)
22. F. Hamann, G. Ferland: ARA&A **37**, 487 (1999)
23. M. Im, S.M. Faber et al.: AJ, **122**, 750 (2001)
24. M. Kamionkowski, A.R. Liddle: Phys. Rev. Lett. **84**, 4525 (2000)
25. J. Kormendy, D. Richstone: ARA&A **33**, 581 (1995)
26. A.V. Kravtsov, G. Yepes: MNRAS **318**, 227 (2000)
27. M.-M. MacLow, A. Ferrara: ApJ **513**, 142 (1999)
28. M. Magliocchetti, L. Moscardini at al.: MNRAS, **325**, 1553 (2001)
29. J. Magorrian, S. Tremaine et al.: AJ **115**, 2295 (1998)
30. P. Martini: AJ **121**, 2301 (2001)
31. P.J. McCarthy, R.G. Carlberg et al.: ApJ **560**, L131 (2001)
32. T.A. McKay, E.S. Sheldon, et al.: ApJ, submitted, astro-ph/0108013 (2001)
33. R.J. McLure, J.S. Dunlop: MNRAS **327**, 199 (2001)
34. D. Merritt, L. Ferrarese: MNRAS **320**, L30 (2001)
35. D. Merritt, L. Ferrarese: ApJ **547**, 140 (2001)
36. P. Monaco, P. Salucci, L. Danese: MNRAS **311** 279 (2001)
37. B. Moore, S. Ghigna et al.: ApJ, **524**, L19 (1999)
38. L.A. Moustakas, R.S. Somerville: ApJ, submitted, astro-ph/0110584 (2001)
39. A. Omont, P. Cox et al.: &A **374**, 371 (2001)
40. J.A. Peacock, M. Rowan-Robinson et al.: MNRAS **318**, 535 (2000)
41. C.P. Pearson, H. Matsuhara et al.: MNRAS **324**, 999 (2001)
42. L. Pentericci, X. Fan et al.: AJ, submitted, astro-ph/0112075 (2001)
43. F. Perrotta, C. Baccigalupi et al.: MNRAS **329**, 445 (2002)
44. F. Perrotta, M. Magliocchetti et al.: MNRAS, submitted, astro-ph/0111239 (2001)
45. M. Persic, P. Salucci, F. Stel: MNRAS **281**, 27 (1996)
46. W.H. Press, P. Schechter: ApJ **187**, 425 (1974)
47. A. Renzini, A. Cimatti: 1999, in *The Hy-Redshift Universe: Galaxy Formation and Evolution at High Redshift*, ASP conf. ser. **193**, p. 312 (1999)
48. G. Rodighiero, A. Franceschini, G. Fasano: MNRAS **324**, 491 (2001)
49. D. Romano, L. Silva, F. Matteucci, L. Danese: preprint (2001)
50. M. Rowan-Robinson: ApJ **549** 745 (2001)
51. P. Salucci, E. Szuszkiewicz, P. Monaco, L. Danese, L.: MNRAS **307** 637 (1999)
52. S.E. Scott, M. Fox et al.: MNRAS, submitted, astro-ph/0107446 (2001)
53. J. Silk, M.J. Rees: A&A **331**, L1 (1998)
54. R.S. Somerville: ApJL, submitted, astro-ph/0107507 (2001)
55. J.C. Tan, J. Silk, C. Balland: ApJ **522**, 579 (1999)
56. T.T. Takeuchi, R. Kawabe et al.: PASP **113**, 586 (2001)
57. L. Toffolatti, F. Argüeso Gómez et al.: MNRAS **297**, 117 (1998)
58. P. Valageas, J. Silk: A&A **350**, 725 (1999)
59. R.P. van der Marel: AJ **117**, 744 (1999)

What Does the Local Black Hole Mass Distribution Tell Us About the Evolution of the Quasar Luminosity Function?

Luca Ciotti[1,2,3], Zoltan Haiman[3], and Jeremiah P. Ostriker[4,3]

[1] Osservatorio Astronomico di Bologna, via Ranzani 1, 40127 Bologna, Italy
[2] Scuola Normale Superiore di Pisa, Piazza dei Cavalieri 7, 56126 Pisa, Italy
[3] Princeton University Observatory, Peyton Hall, 08544 NJ, USA
[4] Institute of Astronomy, Cambridge University, Madingley Road, CB3 0HA, Cambridge, UK

Abstract. We present a robust method to derive the duty cycle of QSO activity based on the empirical QSO luminosity function and on the present-day linear relation between the masses of supermassive black holes and those of their spheroidal host stellar systems. It is found that the duty cycle is substantially less than unity, with characteristic values in the range $3 - 6 \times 10^{-3}$. Finally, we tested the expectation that the QSO luminosity evolution and the star formation history should be roughly parallel, as a consequence of the above–mentioned relation between BH and galaxy masses.

1 Introduction

The discovery of remarkable correlations between the masses of supermassive BHs hosted at the centers of galaxies and the global properties of the parent galaxies themselves [18,25,8] leads to a natural link between the cosmological evolution of QSOs and the formation history of galaxies [11]. The investigation of such interesting correlations looks promising not only to better understand how and when galaxies formed, but also to obtain information on the QSO population itself [19]. Here we focus on two specific points raised by the general remarks above: 1) The use of the "Magorrian relation" to determine the QSO duty cycle at redshift $z = 0$; 2) The expected relation between the cosmological evolution of the total luminosity emitted by star–forming galaxies and that of the total luminosity emitted by QSOs. As we will see, an interesting consequence of this last point is the possible existence of a physical process limiting gas accretion onto BHs at high redshifts. The *observational* inputs of our analysis are the Magorrian relation [8], the galaxy mass–to–light ratio (from the Fundamental Plane) [15,16], the present-day luminosity function of spheroids [2], the present-day and the integrated QSO cosmological (light) evolution [1], and finally the star formation history [24]. A possible alternative to the use of the mass-to-light ratio is the use of the Faber-Jackson [10] relation coupled with the so-called $M_{\mathrm{BH}} - \sigma$ relation [18,25]. The technical details will be given elsewhere [19].

2   Results

We start our analysis by deriving the QSO's mean accretion efficiency associated with the BH's growth (see, e.g., [3]) as $\epsilon \equiv E_Q^T/M_{BH}^T c^2 \simeq 0.06$, where M_{BH}^T is the present-day total mass of BHs at the center of stellar spheroids, and E_Q^T is the total energy emitted by all QSOs over the entire life of the Universe. We then derive the QSO (mean) duty cycle f_Q in two ways:

$$< f_Q >_x \equiv \frac{\epsilon c^2 M_{BH}(x)}{\int_0^\infty L_Q(x,t)dt}, \quad < f_Q >_x \equiv \frac{N_{BH}^>[M_{BH}(x)]}{N_Q^>[L_Q(x,0)]}, \tag{1}$$

where $L_Q(x,t)$ is such that QSOs brighter than $L_Q(x,t)$ (whose number is $N_Q^>[L_Q(x,t)]$) emit the fraction x of $L_Q^T(t)$, and $M_{BH}(x)$ is such that all BHs heavier than $M_{BH}(x)$ (whose number is $N_{BH}^>[M_{BH}(x)]$) sum up to $x M_{BH}^T$ (at z=0). We found that the two ways give consistently $< f_Q >_{0.1} \simeq 0.003$ and $< f_Q >_{0.9} \simeq 0.006$, in good agreement with theoretical results [6]. Our result is similar in spirit but different in detail from that derived in [26], who obtained a *QSO lifetime* of $\simeq 10^7$ yr at a Hubble epoch of $\simeq 10^9$ yr or, in our terms, $f_Q \simeq 10^{-2}$. Finally, an interesting expectation based on the Magorrian relation would be a parallel evolution of the QSO accretion and star formation histories. As is well known, at $z \lesssim 2$ the QSO total luminosity and the UV luminosity associated to the star formation history are indeed roughly parallel but at $z \gtrsim 2$ they have clearly divergent slopes. A fit shows that the QSO luminosity evolution is reasonably well fitted under the assumptions that BHs are accreting at *one tenth* of the Eddington luminosity, as computed under the standard assumption of pure electron scattering. A global picture could then be that at low redshift the BH accretion (and star formation) are limited by the available amount of gas, while at high redshift some extra source of opacity (as for example bremsstrahlung opacity [17]), able to reduce the Eddington luminosity of one order of magnitude (with respect to the pure electron scattering case), is at work [19].

References

1. L. Ciotti, J.P. Ostriker: ApJ, **551**, 131 (2001)
2. L. Ciotti, T.S. van Albada: ApJ, **552**, L13 (2001)
3. L. Ciotti, B. Draine, J.P. Ostriker: in preparation
4. L. Ciotti, Z. Haiman, J.P. Ostriker: in preparation
5. G. Djorgovski, M. Davis: ApJ, **313**, 59 (1987)
6. A. Dressler et al.: ApJ, **313**, 42 (1987)
7. S.M. Faber, R.E. Jackson: ApJ, **204**, 668 (1976)
8. L. Ferrarese, D. Merritt: ApJ, **539**, L9 (2000)

9. K. Gebhardt, K., et al.: ApJ, **539**, L13 (2000)
10. M.G. Haehnelt, P. Natarajan, M.J. Rees: MNRAS **300**, 827 (1998)
11. P. Madau, L. Pozzetti: MNRAS, **312**, 9 (2000)
12. J. Magorrian, et al.: AJ, **115**, 2285 (1998)
13. Y.C. Pei: ApJ, **438**, 623 (1995)
14. P. Salucci et al.: MNRAS, **307**, 637 (1999)
15. A. Soltan: MNRAS, **200**, 115 (1982)

The Black Hole Mass of BL Lacs from the Stellar Velocity Dispersion of the Host Galaxy

Renato Falomo[1], Jari Kotilainen[2], and Aldo Treves[3]

[1] Osservatorio Astronomico di Padova, Vicolo dell'Osservatorio 5, Padova, Italy
[2] Tuorla Observatory, University of Turku, Väisäläntie 20, Piikkiö, Finland
[3] Università dell'Insubria, Via Valleggio 11, Como, Italy

1 Introduction

One of the most important quantities in theoretical models of AGNs is the black hole mass (M_{BH}) that, together with the total luminosity, defines the fraction of the Eddington luminosity at which the AGN is emitting. Determination of M_{BH} in AGN is difficult mainly because of the bright emission from the nucleus and their large distance. The main method that has proved to be successful in AGN is reverberation mapping, which is extremely time consuming and gives results on M_{BH} that depend on the assumed geometry of the accretion disk. Therefore, only for a few well studied quasars and Seyfert galaxies M_{BH} is known (see e.g. [7], [11] and references therein). This method cannot obviously be employed for BL Lac objects because they lack prominent emission lines. Therefore other methods need to be applied to infer M_{BH} for BL Lacs. The discovery of a relation between M_{BH} and the luminosity of the bulge in nearby early-type galaxies offers now a new tool for estimating the mass of the central BH (see e.g. review [10]). This has been done for two samples of nearby quasars [9] and BL Lacs [12].

Recently, a tighter correlation was found relating M_{BH} with the central stellar velocity dispersion σ of the spheroidal component in nearby galaxies [5,2], that can also be used to estimate M_{BH} in AGN. The relationship appears to predict more accurately [10] M_{BH}, but requires the measurement of σ in the host galaxies of AGN that is difficult to obtain, in particular for objects at moderately high redshift and with very luminous nuclei. On the other hand, for BL Lacs that have relatively fainter nuclei than quasars, this measurement (at least for low redshift objects) can be secured with observations at medium-sized telescopes.

We present here the first estimates of stellar velocity dispersion of BL Lacs from our ongoing program aimed specifically at deriving M_{BH} from the $M_{BH} - \sigma$ correlation. We selected a sample of nearby ($z<0.2$) BL Lacs for which high quality images were obtained either from the ground using the Nordic Optical Telescope (NOT) [3] or with HST+WFPC2 [13,4]. From the images a characterization of the host galaxies and of the nuclear luminosity are obtained. This allows us to compare M_{BH} with the mass (and the luminosity) of the host galaxy and also to evaluate the Eddington ratio, provided that the nuclear emitted power is corrected for the beaming factor. Moreover, a comparison of M_{BH} for BL Lacs with different jet/ disk luminosities can be used to test the hypothesis (see e.g. [8]) that the accretion rate changes from largely sub-Eddington, for low luminosity weak-lined sources, to near-Eddington for high luminosity, strong-lined sources. If the accretion rate in terms of Eddington ratio were the same in both classes, the BH masses should differ almost by three orders of magnitude.

2 Observations, Data Analysis and First Results

We secured medium resolution (R ~3000) optical spectra of a sample of the BL Lacs using the NOT 2.5m and the ESO 3.6m telescopes equipped with ALFOSC and EFOSC2, respectively. The chosen grisms combined with a 1 arcsec slit yield a spectral resolution for velocity dispersion measurement of ~60-80 km/s. The used spectral range includes the absorption lines of Ca II (3933-68 Å), Mg I (5175 Å), E-band (5269 Å) and Na I (5892 Å) from the host galaxies. The measurement of σ was done with the Fourier Quotient method using template spectra of late-type (G and K) stars.

Here we report the first results on three BL Lacs: Mkn 501, Mkn 180 and PKS 2201+04. From the measurement of σ we have evaluated the black hole masses assuming that the $M_{BH} - \sigma$ relationship [2] is valid for the BL Lac hosts. We found similar BH masses ($\log(M_{BH}) = 8.5$ and 8.7) for Mkn 501 and Mkn 180, and a significantly lower value ($\log(M_{BH}) = 7.5$) for PKS 2201+04.

From the values of σ and the effective radius R_e we have also estimated the mass of the host galaxy (using the relation $M(host) = 5\ R_e\sigma^2/G$; [1] and compared it with that of the BH. It turns out that $M_{BH}/M(host)$ for the BL Lacs is in the range of 0.03 - 0.1%.

References

1. Bender, R., Burstein,D., Faber, S.M. 1992, ApJ 399, 462
2. Ferrarese L., Merritt D, 2000, ApJ 539, L9
3. Falomo, R., Kotilainen, J., 1999, A&A 352, 85
4. Falomo et al 2001, ApJ 547, 124
5. Gebhardt et al. 2000, ApJ 539, L13
6. Haehnelt M.G. and Kauffmann G.,2000, MNRAS 318, L35
7. Kaspi et al 2000, ApJ 533, 631
8. Maraschi, L. & Tavecchio F., 2000, (astro-ph/0102295)
9. McLure & Dunlop 2001 (astro-ph/0108417)
10. Merritt D & Ferrarese L., 2001, astro-ph/0107134
11. Nelson C.H., 2000, ApJ 544, L91
12. Treves, A., et al, ASP Conf Series, in press (astro-ph/0107129)
13. Urry C.M., et al, 2000, ApJ 532, 816

The Central Mass Distribution in Dwarf and LSB Galaxies

R.A. Swaters

Dept. of Physics and Astronomy, Johns Hopkins University, 3701 San Martin Drive, Baltimore, MD 21218, and STScI, 3700 San Martin Drive, Baltimore, MD 21218, USA

1 Introduction

It has proved remarkably hard to establish the inner slope of the dark matter distribution observationally. It is usually constrained by fitting the observed rotation curves with mass models that include the dark and luminous components. However, because the mass-to-light ratio (M/L) of the stellar disk is unknown, the observed rotation curves can be explained by a range in mass models, often including ones in which the stellar disk dominates in the inner parts and with a core dominated halo, and by ones in which the stellar disk is less important and the mass distribution is dominated by a centrally concentrated halo.

This degeneracy can be avoided by studying dwarf and low surface brightness (LSB) galaxies. For reasonable stellar M/Ls, these galaxies are dominated by dark matter at all radii. Early studies found that these galaxies have slowly rising rotation curves and that their halos can be well described by pseudo-isothermal spheres with constant density cores (Swaters 1999 and references therein). Because of their shallow slopes, these rotation curves were found to be inconsistent with the steep halos predicted by cold dark matter (e.g., Moore 1994, Flores & Primack 1994).

Recent studies raised the concern that some of these HI rotation curves may have been affected by beam smearing due to the relatively poor angular resolution of the HI observations, and found that the rotation curves of dwarf and LSB galaxies rise more steeply when beam smearing is taken into account (Swaters 1999, Blais-Ouellette et al. 1999, Swaters et al. 2000). Mass models fitted to HI rotation curves that have been corrected for beam smearing indicated that dwarf and LSB galaxies are consistent both with constant density cores and cusps with $\alpha = 1$. Slopes with $\alpha = 1.5$ appear difficult to reconcile with the observations (van den Bosch et al. 2000, Swaters 2000a, van den Bosch & Swaters 2001).

HI observations, with typical resolutions of $15''$, may not be best suited to determine the inner slopes. The large range in mass models consistent with the observations might simply be a reflection of the uncertainties in the HI observations. High angular resolution observations, e.g., based on Hα long slit or Fabry-Perot, seem more suited to measure the inner slopes. Yet even among studies based on high resolution observations a controversy remains. Some studies find that dwarf and LSB galaxies may have steep inner slopes (Swaters 2000b), and others that find that these galaxies are not consistent with steep inner slopes (Blais-Ouellette, Amram, & Carignan 2001, de Blok et al. 2001).

Here, we investigate this controversy further, based on $H\alpha$ long slit spectroscopy for a sample of 15 dwarf and LSB galaxies. The data for the LSB galaxies has been published in Swaters et al. (2000). We have measured the inner slopes of their mass distributions, and determined the range of mass models consistent with the observations, taking into account the systematic effects that may have been introduced by the observations. A more detailed discussion of this work will be presented in Swaters et al. (2002, submitted).

2 Inner Slopes

In principle one can derive the density distribution of a galaxy by inverting the observed rotation curve (see e.g., de Blok et al. 2001). For disk-like mass distributions the density distributions derived from such an inversion are uncertain, because in these cases the shape of the rotation curves depends on the density distribution at all radii. For a spherical mass distribution the rotation velocity at a given radius depends solely on the mass enclosed within that radius, making the inversion a well-defined procedure. Because dwarf and LSB galaxies are probably dominated by their dark halos, they are particularly suited for determining the density distributions by inversion of their rotation curves. From $\nabla^2\Phi = 4\pi G\rho$ and $\Phi = -GM/r$, and assuming that the mass distribution is spherical, the density distribution is given by

$$\rho(r) = \frac{1}{4\pi G}\left(2\frac{v}{r}\frac{\partial v}{\partial r} + \frac{v^2}{r^2}\right). \tag{1}$$

To determine the inner slope of the density distribution, we first inverted the derived rotation curve using Eq. 1. Following the same procedure as de Blok et al. (2001), we identified a break radius, and next we determined the inner slope by making a weighted least-squares fit to the inner parts.

2.1 Model Observations

Even though $H\alpha$ observations have a higher angular resolution than HI observations, rotation curves derived from these observations may still be affected by systematic effects, introduced by for example the seeing, slitwidth and errors in positioning of the slit. These effects may be particularly important for galaxies at large distances and for galaxies with steeply rising rotation curves. To understand how these systematic effects affect the derived rotation curves we have constructed model observations for which we varied slit width, seeing, and the offset between the slit and the major axis of the model galaxy.

To construct the model observations, we have assumed the model galaxies to be uniformly filled with $H\alpha$ emission. The model galaxies were adopted to have an inclination of $45°$, and a distance of either 10 Mpc or 60 Mpc. The model rotation curves were calculated under the assumption that the dark matter is the only contributor to the total mass. The models were convolved with the seeing and the instrumental velocity dispersion. Next, we extracted from these

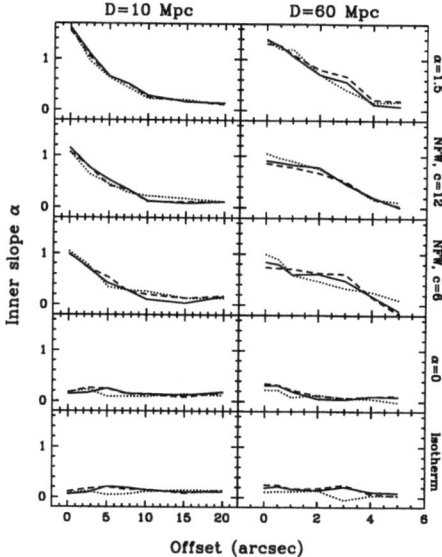

Fig. 1. Derived inner slopes as function of offset for slit width/seeing combinations of 0.5″ (dotted), 1″ (full) and 1.5″ (dashed lines)

observations a spectrum for a given slit width and a given offset from the model galaxy center, and derived a rotation curve from these model observations using the same procedure as for the real observations. Because the effects of seeing and slit width are similar, we have focussed on models in which both the seeing and slitwidth are 0.5″, 1″ or 1.5″.

To investigate the systematic effects on the derived inner slopes of the mass distribution, we used five different models for the dark halo: a halo with $\alpha = 1.5$ and $c = 1$, an NFW halo with $c = 12$, and one with $c = 6$, a halo with $\alpha = 0$ and $c = 32$, and finally a pseudo-isothermal halo with $r_c = 1$ kpc and $\rho_0 = 0.11$ M$_\odot$ pc^{-3}. We determined the inner slopes following the procedure outlined above, for models with a slit width and seeing of 0.5″, 1″ or 1.5″, both at 10 Mpc and 60 Mpc, and a range in offsets.

The derived inner slopes for the three combinations of seeing and slit width are shown in Fig. 1, plotted versus the offset between the model slit and the galaxy major axis. From Fig. 1 it can be seen that if the slit is aligned with the major axis, i.e., the offset is 0″, the inner slope can be accurately retrieved for the galaxies at 10 Mpc, and the effects of seeing and slit width are negligible. At large distances, however, the seeing and the slit width encompass a significant velocity gradient, and as a result the inner slope will be smeared out. This leads to an underestimate of the inner slope, even if the slit is aligned with the major axis of the galaxy. A much larger effect on the measured inner slope is a possible

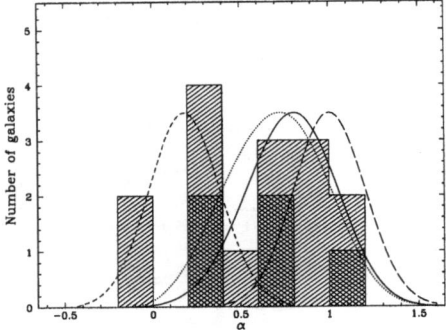

Fig. 2. Histogram of measured inner slopes compared to the expected distributions for different alignment errors and values of the intrinsic inner slopes. See text

offset between the galaxy nucleus and the position of the slit. As can be seen in Fig. 1, an offset of a few arcseconds can lead to a substantial underestimate in the derived inner slope. Note that the flat inner slope of an isothermal halo is not affected by slit offsets.

2.2   Measured Inner Slopes

We determined the inner slopes of the mass distributions from the rotation curves of the galaxies in our sample. In Fig. 2 a histogram of the derived inner slopes is presented. The values for α_m span a range from 0 to 1, with a large clump with 9 galaxies with α_m between 0.5 and 1, and 6 galaxies with lower values of α_m.

We have used the results from the models presented in Section 2.1 to calculate the expected distribution of α_m for a given model and a given Gaussian error distribution in the placement of the slit with dispersion σ_{off}. We estimate that for the LSB galaxies the slit alignment was accurate to $1''$, and to $2.5''$ for the dwarf galaxies. The long dashed line in Fig. 2 gives the expected distribution of α_m if the slit is perfectly aligned with the major axis (i.e, $\sigma_{\text{off}} = 0''$), for a dark matter profile with $\alpha = 1$ and a model distance of 10 Mpc. The solid line is the same, except with an offset error $\sigma_{\text{off}} = 2.5''$, and the dotted line represents $\sigma_{\text{off}} = 5''$. The short dashed line is the expected distribution for dark halos with $\alpha = 0$. For the models with $\alpha = 0$, the expected distribution does not depend much on the alignment error. In all cases we assumed a random measurement error $\sigma_\alpha = 0.25$. This is the error we found from the modeling, and it is similar to the typical error we found on the measurement of α_m. The model offsets have been calculated for an inclination of $45°$, which is average inclination of the galaxies in our sample.

For the LSB galaxies, which are at distances of around 60 Mpc, the expected distributions are similar as the plotted ones, but for offsets that are a factor of 3 smaller than those for the model galaxies at 10 Mpc,(i.e., $\sigma_{\text{off}} = 0'', 1'', 2''$ for the long-dashed line, the solid line and the dotted line, respectively).

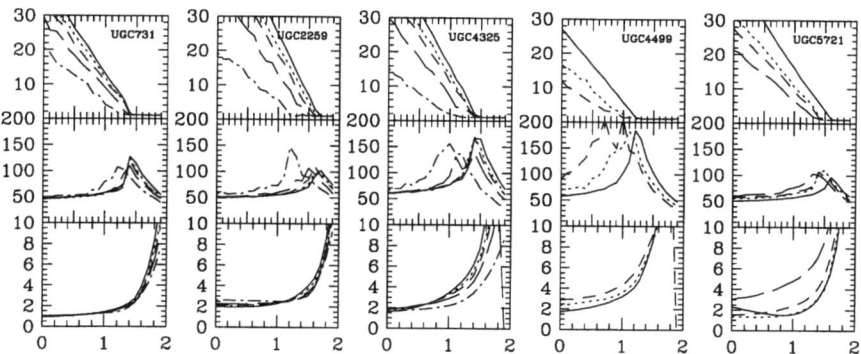

Fig. 3. Best fit mass models based on Eq. 2 for 5 out of the 15 galaxies in our sample. The x-axis gives α, the y-axis v_{rot}. The top panel plots c, the middle v_{200}, the bottom χ_r^2, for M/Ls of 0 (full), 0.5 (dotted), 1 (short dashed), 2 (long dashed), 4 (dot dashed)

The inner slopes for galaxies at larger distances may be affected in two ways. On the one hand, the width of the list, the seeing and a possible offset may lead to an underestimate of the inner slopes. On the other, as was also pointed out by de Blok et al. (2001), they may overestimated because part of the steeper, outer parts of the profile may be included in the fit. To investigate this, we have coded the dwarf and the LSB galaxies in different ways in Fig. 2. Although based on only a few galaxies, there is no indication that the LSB galaxies cause the high values of α_m. The distribution of α_m is not uniquely consistent with any of the models overplotted here. We will discuss this below.

2.3   Mass Models

Measurements of the inner slope as derived above provide an upper limit to the intrinsic slope of the dark matter distribution, because the contribution of the baryons was ignored. Using mass modeling, we have explored what the effect of the baryons on the derived inner slope of the dark matter distribution is.

We have fitted mass models based on the generic dark halo given in Eq. 2 to the observed rotation curves. We have imposed

$$\rho(r) = \frac{\rho_0}{(r/r_s)^\alpha (1 + r/r_s)^{3-\alpha}}, \qquad (2)$$

which reduced to the NFW profile (Navarro et al. 1997) for $\alpha = 1$ (see e.g., van den Bosch et al. 2001). Adiabatic contraction has been taken into account. We have made mass models for a range in M/Ls from 0 to 4, unless the maximum disk solution has a lower M/L. We have not left α as a free parameter because α is poorly constrained from the mass modeling because of the cusp-core degeneracy (van den Bosch & Swaters 2001). Instead, we have determined the best fitting combination of c and v_{200} for a given combination of α and Υ_*. As an example of the fitting results, we show the best fitting parameters and the corresponding χ_r^2 for 5 out of the 15 galaxies in our sample in Fig. 3.

From Fig. 3 and the 10 other galaxies that are not shown here, a number of general trends appear. The halo concentration c of the best fitting model decreases with increasing α, because the enclosed mass for the different models has to be similar. The value of v_{200} reaches a maximum for $c = 1$. We imposed $c > 1$ in these fits, because Eq. 2 does not give an appropriate description of the dark matter distribution for $c < 1$. For most galaxies and M/Ls, the mass models are equally well fitted by dark halos with $0 < \alpha < 1$. For α slightly larger than 1, the quality of the fits rapidly decrease, even for M/Ls of 0. If the M/Ls are close to the maximum disk values (not shown here), the rotation curves can only be fitted by core dominated dark halos.

3 Discussion

Both the measurement of the inner slopes obtained by inverting the observed rotation curves and the fitting of mass models show that more than two thirds of the galaxies in the sample presented here are consistent with having cuspy halos with $\alpha = 1$. At the same time, *all* galaxies in this sample can be well explained by an isothermal halo with a constant density core. At first sight, this may suggest that isothermal halos provide better fits to the observed rotation curves, but, as has been shown here, systematic effects, such as seeing, slit width and slit offsets may affect the data and lead to an underestimate of the inner slope. These effects mostly affect the derived inner slopes of galaxies with steep intrinsic slopes, and hardly affect those with shallow slopes, and hence they may explain part of the measured shallow inner slopes. Still, there are 4 galaxies that have inner slopes that cannot be explained by even a very large offset. Interestingly, three out of these four have strong bars. Clearly, to resolve the controversy on the inner slopes of dwarf and LSB galaxies, high resolution integral field spectroscopy is needed to map the non-circular motions in the central parts.

References

1. Blais-Ouellette, S., Amram, P., & Carignan, C. 2001, AJ, 121, 1952
2. de Blok, W. J. G., McGaugh, S. S., Bosma, A., & Rubin, V. C. 2001, ApJ, 552, L23
3. Flores, R. A., Primack, J. R. 1994, ApJ, 427, L1
4. Moore, B. 1994, Nature, 370, 629
5. Swaters, R. A., 1999, PhD thesis, Rijksuniversiteit Groningen
6. Swaters, R. A., 2000a, in 'Gas and Galaxy Evolution', eds. J.E. Hibbard, M. Rupen, J. H. van Gorkom, ASP Conf. series 240
7. Swaters, R.A., 2000b, in: Galaxy Disks and Disk Galaxies,eds. J.G. Funes S.J., E.M. Corsini, ASP conf. series (astro-ph/0009370)
8. Swaters, R. A., Madore, B. F., & Trewhella, M. 2000, ApJ, 531, L107
9. van den Bosch, F. C., Swaters, R. A., 2001, MNRAS, in press
10. van den Bosch, F. C., Robertson, B. E., Dalcanton, J. J., & de Blok, W. J. G. 2000, AJ, 119, 1579

Masses of Nearby Galaxies
from WIYN IFU Spectroscopy†‡

Marianne Takamiya[1], Mark Chun[1], Inger Jørgensen[1], and Lancelot Kao[2]

[1] Gemini Observatory, Hilo HI 96720, USA
[2] City College of San Francisco, CA 94112, USA

Abstract. The dynamical masses of NGC 6052 and I Zw 207 are estimated from WIYN IFU data to be of order $10^{10} M_\odot$. Both galaxies have large star formation rates and appear to be interacting systems. They could be the local counterpart of the *irregular* galaxies found in large numbers with the Hubble Space Telescope (HST).

We present results from IFU spectroscopy of two galaxies: NGC 6052 and I Zw 207. We obtained high resolution spectra with DensePak [1] on WIYN in queue mode. DensePak is a 91 red-optimized fiber array forming a rectangle $43'' \times 27''$ with fiber diameters of $3''$. The fiber cable feeds the Bench Spectrograph which was configured in echelle mode (order 8) with the 316 lines/mm grating at an angle of 63.454°.

NGC 6052 and I Zw 207 are galaxies with large Hα equivalent widths [2] and morphologies that suggest they are the result of galaxy interactions. Both galaxies appear to be highly disturbed systems and are interesting in the context of galaxy evolution because they could be the local counterparts of the irregular galaxies and the so-called chain galaxies found in large numbers by the HST [6]. NGC 6052 is probably the result of a merger of two disk galaxies with similar masses. We can detect at least two kinematically distinct components with similar shapes in the reconstructed Hα image. Figure 1 shows the Hα image and the positive and negative velocity fields which show two systems with major axis in the north-south direction. The Hα image of I Zw 207 (not presented here) shows two kinematic components, however, of different shapes: a nucleus and a tail.

NGC 6052 and I Zw 207 provide a good laboratory to estimate their masses. Assuming a simple model of circular orbit of equal mass galaxies, we estimate the masses from: $m = v^2 r / G$, where v is the relative speed between the galaxies, r, the projected separation, and G, the gravitational constant. Since the systems are unlikely to be relaxed, these results are only approximate. The crossing time (t_c) can also be estimated from r/v which sets a limit on the age of the system. In Table 1 we present the data determined for both galaxies. Using this approach,

† Supported by the Gemini Observatory, which is operated by the Association of Universities for Research in Astronomy, Inc., on behalf of the international Gemini partnership of Argentina, Australia, Brazil, Canada, Chile, the United Kingdom and the United States of America.

‡ The WIYN Observatory is a joint facility of the University of Wisconsin-Madison, Indiana University, Yale University, and the National Optical Astronomy Observatories.

the masses of high-z galaxies can be estimated using 8-30m telescopes with either carefully positioned slits or, more interestingly, with multiple deployable IFUs.

Table 1. Galaxy Physical Parameters

Galaxy	Mass $M_\odot$	z	r kpc	v km s^{-1}	t_c yr
NGC 6052	6.3×10^{10}	0.0152	2.3	343	$\sim 6 \times 10^6$
I Zw 207	7.7×10^9	0.0183	9.4	59	$\sim 1.5 \times 10^8$

References

1. Barden, S., Sawyer, D., Honeycutt, R. 1998, Proc. SPIE 3355, 892
2. Gallagher, J. S., Bushouse, H., and Hunter, D. A. 1989, AJ, 97, 700
3. Cowie, L.L., Hu, E.M., and Songaila, A. 1995, AJ, 110, 1576

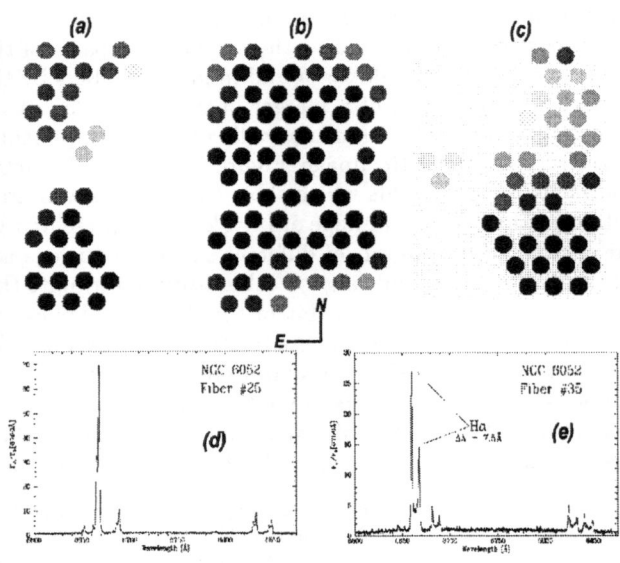

Fig. 1. IFU spectroscopy of NGC 6052: positive (a) and negative (c) velocity fields; b) reconstructed image in Hα; spectra of fibers 35 (d) and 25 (e)

The Origin of the Correlation Between the Spin Parameter and the Baryon Fraction of Galactic Disks

Andreas Burkert[1], Frank C. van den Bosch[2], and Rob A. Swaters[3]

[1] Max-Planck-Institut für Astronomie, Heidelberg, Germany
[2] Max-Planck-Institut für Astrophysik, Garching, Germany
[3] Carnegie Institution of Washington, Washington DC 20015, USA

Abstract. The puzzling correlation between the spin parameter λ of galactic disks and the disk-to-halo mass fraction f_{disk} is investigated. We show that such a correlation arises naturally from uncertainties in determining the virial masses of dark matter halos. This result leads to the conclusion that the halo properties derived from fits to observed rotation curves are still very uncertain which might explain part of the disagreements between cosmological models and observations. We analyse λ and f_{disk} as function of the adopted halo virial mass. Reasonable halo concentrations require $f_{disk} \approx 0.01 - 0.07$ which is significantly smaller than the universal baryon fraction. Most of the available gas either never settled into the galactic disks or was ejected subsequently. In both cases it is not very surprising that the specific angular momentum distribution of galactic disks does not agree with the cosmological predictions which neglect these effects.

1 Introduction

Within the framework of hierarchical cosmological structure formation galactic disks form from gas that falls into dark matter halos, where it cools and settles into the equatorial plane. The disk scale lengths and their rotation curves are determined by the gravitational potential and by their specific angular momentum distribution which has been acquired from cosmological torques (Hoyle 1953; Peebles 1969) and the random merging of subunits (Maller, Dekel & Somerville 2001) with additional modification during the dissipative protogalactic collapse phase. Cosmological simulations (Van den Bosch et al. 2002) have shown that the initial angular momentum distribution of the baryonic and dark matter component is similar. This initial condition could explain the observed scale lengths and various other properties of galactic disks, provided that the disk material retained its initial specific angular momentum when settling into the galactic plane (e.g. Fall & Efstathiou 1980; Mo, Mao & White 1998; Firmani & Avila-Reese 2000; van den Bosch 2001; Buchalter, Jimenez & Kamionkowski 2001).

In the past couple of years, high-resolution cosmological NBody/SPH simulations have however uncovered problems with this scenario. Baryons tend to lose a large fraction of their angular momentum to the dark matter while settling into a disk component. As a result, simulated galactic disks are an order of

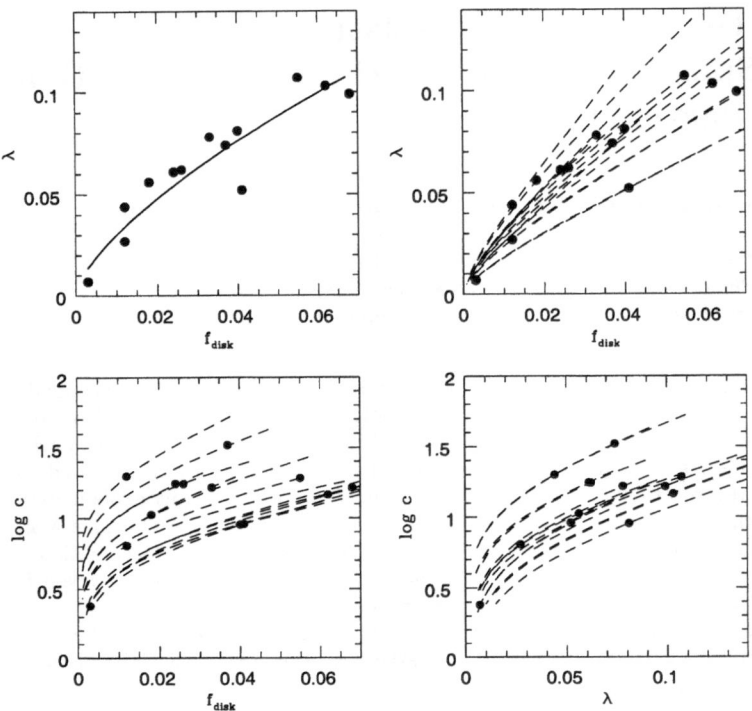

Fig. 1. Correlations between the disk spin parameter, the disk mass fraction and the halo concentration for the Swaters galaxy sample. Data points in the upper left panel show the best fit values with no constraints on the virial parameters, adopting a stellar mass-to-light ratio $\Upsilon_R = 1.0(M/L)_\odot$. The solid line in the upper left panel shows the correlation that would result from errors in determining the virial radii of dark matter halos. The dashed curves in the other panels show for each galaxy how the data points shift if one determines the best fitting rotation curve for different values of the virial radius

magnitude smaller than observed (Navarro & Benz 1991; Navarro & Steinmetz 1997). Even if the angular momentum would be conserved, the observed disk angular momentum distribution does not agree with theoretical predictions. This was shown by van den Bosch, Burkert & Swaters (2001), who measured in detail the angular momentum distribution for a sample of dwarf disk galaxies by fitting a NFW profile (Navarro, Frenk & White 1997) to the observed rotation curves, taking into account the disk stars and the HI gas and considering adiabatic contraction and beam smearing. They confirmed that the mean specific angular momentum of the disk material is of the same order as expected if angular momentum is conserved during the protogalactic collapse phase. Their case by case studies however revealed a mismatch of the specific angular momentum profiles of galactic disks, compared with the predicted universal dark halo angular momentum distribution of Bullock et al. (2001): the cosmologically predicted mass fraction with low angular momentum is much larger than observed.

The problem of angular momentum loss during gas infall might partly be solved by energetic feedback processes. Thacker & Couchman (2001), for example, showed that stellar heating could decouple the dynamical evolution of the protogalactic gas with respect to the dark halo, leading to galactic disks with a specific angular momentum that is within 10% of the observed value (see also Sommer-Larsen et al. 1999, Navarro & Steinmetz 2000, Maller & Dekel 2002). Still, as demonstrated by van den Bosch et al. (2002), a large fraction of the baryonic component would have very low or even negative specific angular momentum, in contrast with the observed disk angular momentum distribution. It has also been suggested that this low-angular momentum gas could form large galactic bulges instead of disks (Thacker & Couchman 2002, van den Bosch et al. 2002). These bulges are however not observed in the LSB galaxies, studied by Van den Bosch, Burkert & Swaters (2001).

In addition, van den Bosch, Burkert & Swaters (2001) detected a strong correlation between the disk spin parameter and the disk mass fraction for the Swaters sample. A similar correlation for a much larger sample of LSB and HSB galaxies has been found by Jimenez, Verde & Oh (2002). This result is puzzling. It is not clear why the fraction of baryonic material that forms the observed galactic disks should correlate with the disk spin parameter.

2 The Origin of the Correlation Between Spin and Disk Mass Fraction

The upper left panel of figure 1 shows the observed correlation between the disk spin parameter

$$\lambda = \gamma \frac{j_{tot}}{\sqrt{2} R_{vir} V_{vir}} \tag{1}$$

and the disk mass fraction $f_{disk} = \frac{M_{disk}}{M_{vir}}$, with M_{disk} the total disk mass and $M_{vir} = V_{vir}^2 R_{vir}/G$ the virial mass of the dark halo. Here γ is a geometrical factor which depends on the dark matter density distribution, j_{tot} is the observed total disk angular momentum and R_{vir} and V_{vir} are the virial radius and virial mass of the dark halo, respectively. These values represent the best fit to the rotation curves if no constraints are imposed on R_{vir} and V_{vir}. A dependence of disk rotation on the disk mass fraction might provide interesting new insight into the evolution of disk galaxies. However it also could emerge from uncertainties in determining the dark halo properties. All the information about the structure of the dark matter halos is gained through disk rotation curves which are restricted to the inner halo regions. The outer halo regions and especially their virial masses or virial radii are poorly constrained. In addition, tests show that the fits to the observed rotation curves are almost equally good for a large range of virial values. Both, λ and f_{disk} depend on R_{vir}. As γ does not vary strongly with halo mass and with $V_{vir} \sim R_{vir}$ we find $\lambda \sim R_{vir}^{-2}$ and $f_{disk} \sim R_{vir}^{-3}$. Any error in R_{vir}

will therefore shift the data points along a curve $\lambda \sim f_{disk}^{2/3}$ which is shown by the solid curve in the upper left panel of figure 1. The good agreement of the distribution of the data points with this relationship indicates indeed that the correlation results from errors in determining R_{vir}. This problem is shown in more details and for each galaxy separately in the right upper panel of figure 1, where the dashed curves show the best fit values of λ and f_{disk} for all galaxies, adopting different values of R_{vir}.

3   Conclusion

The correlation between λ and f_{disk} that has been found by van den Bosch et al. (2001) or Jimenez et al. (2002) can be explained as a result of uncertainties in determining the dark halo virial radii or masses. Cosmological models predict that most of the protogalactic gas with a cosmological baryon fraction (for LCDM) of $f_{bar} = \Omega_{bar}/\Omega_0 \approx 0.13$ loses 90% of its angular momentum and settles into the equatorial plane, leading to typical values of $\lambda \approx 0.005$ and $f_{disk} \approx 0.13$. Even if the virial radius is unknown and treated as a free parameter, the upper right panel of figure 1 clearly shows that these values can be ruled out.

Dark matter halos have typical concentrations of order $c \approx 12 - 15$ which should not be affected strongly by the dynamical evolution of the baryonic component in dark matter dominated LSB galaxies. The lower panels of figure 1 show that $f_{disk} \approx 0.01 - 0.07 < f_{bar}$ for these halo concentrations. The galaxies either lost a substantial fraction of their baryons in a galactic wind or accreted only a small fraction of the gas that has been available initially. In both cases, there exists no reason why the specific angular momentum distribution of the disk component should match the dark halo angular momentum distribution as assumed e.g. by Mo, Mao & White (1998). The cosmological angular momentum problem of disk galaxies might therefore be connected directly with the origin of their low baryon fractions.

References

1. A. Buchalter, R. Jimenez & M. Kamionkowski: MNRAS **322**, 43 (2001)
2. J.S. Bullock, A. Dekel, T.S. Kolatt, A.V. Kravtsov, A.A. Klypin, C. Porciani & J.R. Primack: ApJ **555**, 240 (2001)
3. S.M. Fall & G. Efstathiou: MNRAS **193**, 189 (1980)
4. C. Firmani & V. Avila-Reese: MNRAS **315**, 457 (2000)
5. F. Hoyle: ApJ **118**, 513 (1953)
6. R. Jimenez, L. Verde & S.P. Oh: astro-ph/0201352 (2002)
7. A.H. Maller, A. Dekel & R.S. Somerville: MNRAS **329**, 423 (2001)
8. A.H. Maller & A. Dekel: astro-ph/0201187 (2002)
9. H.J. Mo, S. Mao & S.D.M. White: MNRAS **295**, 319 (1998)
10. J.F. Navarro & W. Benz: ApJ **380**, 320 (1991)
11. J.F. Navarro & M. Steinmetz: ApJ **478**, 13 (1997)
12. J.F. Navarro, C.S. Frenk & S.D.M White: ApJ **490**, 493 (1997)
13. J.F. Navarro & M. Steinmetz: ApJ **538**, 477 (2000)

14. P.J.E. Peebles: ApJ **155**, 393 (1969)
15. J. Sommer-Larsen, S. Gelato & H. Vedel: ApJ **519**, 501 (1999)
16. R.J. Thacker & H.M.P. Couchman: astro-ph/0106060 (2001)
17. F.C. van den Bosch, T. Abel, R.A.C. Croft, L. Hernquist & S.D.M. White: astro-ph/0201095 (2002)
18. F.C. van den Bosch: MNRAS **327**, 1334 (2001)
19. F.C. van den Bosch, A. Burkert & R.A. Swaters: MNRAS **326**, 1205 (2001)

Probing the Evolution of Massive Galaxies with the K20 Survey*

Andrea Cimatti

Osservatorio Astrofisico di Arcetri, Largo E. Fermi 5, I-50125, Firenze, Italy

Abstract. The motivations and the status of the K20 survey are presented. The first results on the evolution of massive galaxies and the comparison with the predictions of the currently competing scenarios of galaxy formation are also discussed.

1 Introduction

Understanding the evolution of massive galaxies (e.g. $M_{stellar} \gtrsim 10^{11}$ $M_\odot$) is important to constrain the different scenarios of structure and galaxy formation. In particular, the question on the formation of the present-day massive spheroidals is still one of the most debated issues of galaxy evolution (see [1] for a review). In one scenario, massive spheroidals are formed at early cosmological epochs (e.g. $z > 3$) through a short and intense episode of star formation (with $SFR \sim 100 - 1000$ $M_\odot \mathrm{yr}^{-1}$), followed by a passive evolution (or pure luminosity evolution, PLE) of the stellar population to nowadays. In marked contrast, the hierarchical scenarios predict that massive spheroidals are the product of rather recent merging of pre-existing disk galaxies taking place mostly at lower redshifts and with moderate star formation rates [2,3]. In hierarchical merging scenarios, fully assembled massive field spheroidals with $M_{stellar} \gtrsim 10^{11}$ $M_\odot$ at $z \gtrsim 1$ are very rare objects [4] (see also Baugh, this volume).

From an observational point of view, a direct way to test the above scenarios is to study the evolution of massive galaxies by means of spectroscopic surveys of field galaxies selected in the K-band [5–8]. Since the near-IR light is a good tracer of the galaxy mass[9,4], K-band imaging allows to select massive galaxies at high-z. A galaxy with a stellar mass of about 10^{11} $M_\odot$ is expected to have $18 < K < 20$ for $1 < z < 2$ [4], thus implying that moderately deep K-band surveys can efficiently select massive galaxies in that redshift range. Deep spectroscopy with 8-10m class telescopes can then be used to search for massive systems and to constrain their redshift distribution.

---

* The collaborators in the K20 survey include T. Broadhurst (HUJ), S. Cristiani (ECF & Trieste), S. D'Odorico (ESO), E. Daddi (Firenze & ESO), A. Fontana (Roma), E. Giallongo (Roma), R. Gilmozzi (ESO), N. Menci (Roma), M. Mignoli (Bologna), F. Poli (Roma), L. Pozzetti (Bologna), S. Randich (Arcetri), A. Renzini (ESO), P. Saracco (Milano), J. Vernet (Arcetri), G. Zamorani (Bologna)

2   The K20 Survey

2.1   The Observations and the Database

In order to investigate the evolution of massive galaxies and to constrain the currently competing galaxy formation scenarios, we started in 1999 a project that was called "K20 survey". For such a project, 17 nights were allocated to our team in the context of an ESO VLT Large Program distributed over a period of two years (1999-2000) (see also http://www.arcetri.astro.it/~k20/).

The prime aim of such a survey is to derive the redshift distribution of about 550 K-selected objects with $Ks \leq 20$. The targets were selected from a 32.2 arcmin^2 area of the Chandra Deep Field South (CDFS; [10]) using the images from the ESO Imaging Survey public database (EIS; http://www.eso.org/science/eis/; the raw Ks-band images were reduced and calibrated by our group), and from a 19.8 arcmin^2 field centered at 0055-269 using NTT+SOFI Ks-band imaging (Fontana et al. in preparation).

Optical multi-object spectroscopy was made with the ESO VLT UT1 and UT2 equipped with FORS1 (October-November 1999) and FORS2 (November 2000) during 0.5″-1.5″ seeing conditions and with 0.7″-1.2″ wide slits depending on the seeing. The grisms 150I, 200I, 300I were used with typical integration times of 1-3 hours. Dithering of the targets along the slits between two fixed positions was made for most observations in order to efficiently remove the CCD fringing and the strong OH sky lines at $\lambda_{obs} > 7000$ Å. The spectra were calibrated using standard spectrophotometric stars, dereddened for atmospheric extinction, corrected for telluric absorptions and scaled to the total R-band magnitudes. A small fraction of the K20 sample was observed with near-IR spectroscopy using the VLT UT1+ISAAC in order to derive the redshifts of the galaxies which were too faint for optical spectroscopy and/or expected to be in a redshift range for which no strong features fall in the observed optical spectral region (e.g. $1.5 < z < 2.0$). However, due to the lack of a multi-object spectroscopy mode in ISAAC, it turned out very hard and inefficient to obtain redshifts of faint galaxies in this manner, which was successful only for a few galaxies at $1.3 < z < 1.9$ with strong Hα in emission.

In addition to spectroscopy, $UBVRIzJKs$ imaging is also available for both fields, thus providing the possibility to estimate photometric redshifts for all the objects in the K20 sample, to "calibrate" them through a comparison with the spectroscopic redshifts and to assign reliable photometric redshifts to the objects for which it was not possible to derive the spectroscopic z (see Fontana et al., these proceedings).

The spectroscopic observations were completed in December 2000. The spectral analysis was done by means of automatic software (IRAF: rvidlines and xcsao) and through visual inspection of the 1D and 2D spectra. Because of four nights lost due to the bad weather, only 94% of the sample with $Ks <$ 20 could be observed. The efficiencies in deriving the spectroscopic redshifts for the observed targets was high: $N_{identified}/N_{observed}$=95%, 93%, 91% for $Ks < 19.0, 19.5, 20.0$ respectively. The overall spectroscopic redshift completeness is still rather high, with $N_{identified}/N_{total}$=93%, 91%, 85% for $Ks <$ 19.0, 19.5, 20.0 respectively (where $N_{total} = N_{observed} + N_{unobserved}$). The size of the sample, the spectroscopic redshift completeness, and the availability of tested and reliable photometric redshifts make the K20 sample one of the largest and most complete database to study the evolution of K-selected galaxies available to date.

2.2   The Scientific Aims

Kauffmann & Charlot (1998) estimated that $\sim$ 60% and $\sim$ 10% of the galaxies in a $K <$ 20 sample are expected to be at $z > 1$, respectively in a PLE and in a standard CDM hierarchical merging model (cf. their Fig. 4). Such a large difference was in fact one of the main motivations of our original proposal to undertake a redshift survey for all objects down to $K <$ 20. However, more recent models consistently show that the difference between the predictions is less extreme than in the KC98 realization (Menci et al., Pozzetti et al., Somerville et al., in preparation). Part of the effect is due to the now favored ΛCDM cosmology which pushes most of the merging activity at earlier times compared to τCDM and SCDM models, and therefore get closer to the PLE case. Moreover, a different tuning of the star-formation algorithms (to accommodate for more star formation at high z) also reduces the differences between the two scenarios. Our database is currently being used to perform a stringent comparison between the observed redshift distribution and the ones predicted by the most recent models of galaxy formation (Cimatti et al., in preparation).

Besides the main goal described here above, our unique database is also being used to address other important questions on galaxy evolution: *(1)* the evolution of the Luminosity and Mass Functions (see Pozzetti et al., this volume), *(2)* the evolution of ellipticals, *(3)* the fraction of dusty starbursts and high-z ellipticals in the ERO population, *(4)* the evolution of galaxy clustering (see Daddi et al., this volume), *(5)* the spectral properties of a large number of galaxies and their evolution as a function of redshift, *(6)* the volume star formation density using different indicators, *(7)* the fraction of AGN in K-selected samples, *(8)* the brown dwarf population at high Galactic latitude.

3   First Results on Extremely Red Objects

A fraction of the galaxies selected in the near-infrared show very red colors (e.g. $R - K > 5$). Such galaxies are known as Extremely Red Objects (EROs), and the most recent surveys demonstrated that they form a substantial field population [11–13]. Since their colors are consistent with being either old passively evolving galaxies or dusty starbursts or AGN, it is therefore of prime importance to determine their nature in order to exploit the stringent constraints that EROs can place on the galaxy formation scenarios. In this section we adopt a cosmology with $H_0 = 70$ km s^{-1} Mpc^{-1}, $\Omega_m = 0.3$ and $\Omega_\Lambda = 0.7$.

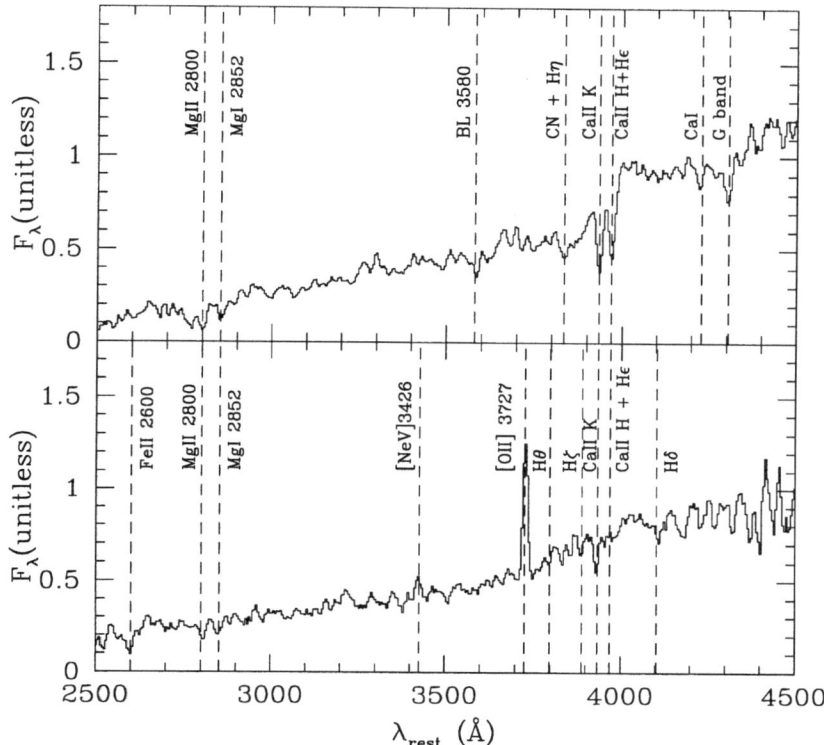

Fig. 1. The average rest-frame spectra (smoothed with a 3 pixel boxcar) of old passively evolving (top; $z_{mean} = 1.000$) and dusty star-forming EROs (bottom; $z_{mean} = 1.096$) with $Ks \leq 20$

From our total $Ks < 20$ sample, 78 EROs with $R - Ks \geq 5.0$ were extracted. About 70% of the EROs with $R - Ks \geq 5.0$ and $Ks \leq 19.2$ was spectroscopically identified with *old* and *dusty star-forming* galaxies at $0.7 < z < 1.5$ [14]. The two classes are about equally populated and for each of them we derived the average spectrum (Fig. 1).

Old EROs have an average spectrum consistent with being old passively evolving ellipticals, and it can be well reproduced with Bruzual & Charlot (2000) spectral synthesis models with no dust extinction and ages $\gtrsim 3$ Gyr if $Z = Z_\odot$, thus implying an average formation redshift $z_f \gtrsim 2.4$ for such a metallicity.

The average spectrum of star-forming EROs can be reproduced only if a substantial dust extinction is introduced, typically in the range of $0.5 < E(B - V) < 1$. Their star formation rates, corrected for the average reddening, suggest a significant contribution (>20%) of EROs to the cosmic star-formation density at $z \sim 1$. However, the detection of [NeV]$\lambda 3426$ emission suggests that a fraction of star-forming EROs also host some AGN activity obscured by dust.

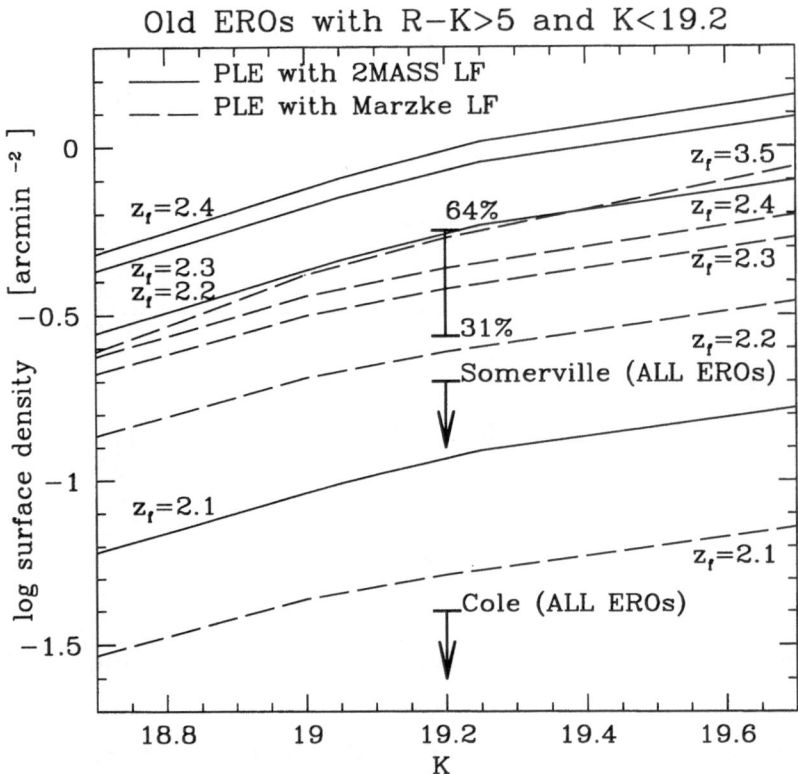

Fig. 2. Comparison between the observed density of old EROs with the predictions of PLE models (see [14] for more details). The range of observed density of old EROs with $Ks < 19.2$ is shown with a vertical bar ranging from the minimum observed density (32% of the total density of EROs observed in the K20 survey) and the maximum possible density (64% of the total density of EROs assuming that all the spectroscopically unidentified EROs are old passive systems). The dashed and solid lines show the predictions of PLE models adopting respectively the local luminosity function of ellipticals [17] and of early-type galaxies (2MASS, [18]), and using Bruzual & Charlot spectral synthesis models with solar metallicity, Salpeter IMF, e-folding time of the star formation $\tau = 0.3$ Gyr, and a set of formation redshifts (z_f). See also [4] for more details on PLE models. The arrows indicate the predicted densities of EROs in two recent hierarchical merging models (the Somerville et al. model shown in Firth et al. 2001, and the Cole et al. 2000 model presented in Smith et al. 2001). Since such models predict the total density of *all* EROs (i.e. old + dusty star-forming) they actually represent upper limits to the predicted density of old passive systems

Since old EROs have spectra consistent with being passively evolving ellipticals, we compared their density with different model predictions (see Fig. 2). The main result is that the density of old EROs observed in our survey is strongly underpredicted by the current hierarchical merging models (a factor of $\sim$4–10 for the models presented by [15] and [16] respectively). On the other hand, PLE models with a reasonable choice of input parameters predict surface and comoving densities in agreement with the observations and imply that massive spheroidals formed at $z_f > 2 - 3$ (see [14] for more details).

Since the luminosities and the stellar masses of the observed old EROs are in the range of 0.5-4L^* and 1-6$\times 10^{11}$ M$_\odot$, this means that fully assembled massive systems were already in place up to $z \sim 1.3$.

The existence of such galaxies with a comoving density of $\gtrsim 2 \times 10^{-4}$ Mpc^{-3} at $z \sim 1$ ([14]) is in contrast with the ΛCDM hierarchical model of [20], where the comoving density of *all galaxy types* with M$> 10^{11}$ M$_\odot$ is about one order of magnitude lower, whereas the difference is less dramatic with the model of [8] (see Fig. 1 of [22]).

References

1. Renzini A., Cimatti A. 2000, in "The Hy-Redshift Universe: Galaxy Formation and Evolution at High Redshift", ed. A.J. Bunker & W.J.M. van Breugel, A.S.P. Conf. Series Vol. 193, in press (astro-ph/9910162)
2. Kauffmann G. 1996, MNRAS, 281, 487
3. Baugh C.M., Cole S., Frenk C.S. 1996, MNRAS, 283, 1361
4. Kauffmann G., Charlot S. 1998, MNRAS, 297, L23
5. Broadhurst T.J., Ellis R.S., Glazebrook K. 1992, Nature, 355,55
6. Cowie L.L. et al. 1996, AJ, 112, 839
7. Cohen J.G. et al. 1999, ApJ, 512, 30
8. Stern D. et al., in ESO/ECF workshop on "Deep Fields", 9-12 October 2000, Garching (astro-ph/0012146)
9. Gavazzi G., Pierini D., Boselli A. 1996, A&A, 312, 397
10. Giacconi R. et al. 2001, ApJ,551,624
11. Thompson D. et al. 1999, ApJ, 523, 100
12. Daddi E. et al. 2000, A&A, 361, 535
13. McCarthy P.J. et al. 2001, ApJL, 560, L131
14. Cimatti A. et al. 2002, A&A Letters, in press (astro-ph/0111527)
15. Firth A.E. et al. 2001, MNRAS, submitted (astro-ph/0108182)
16. Smith G.P. et al. 2001, MNRAS, in press (astro-ph/0109465)
17. Marzke R.O. et al. 1994, AJ, 108, 437
18. Kochanek C.S. et al. 2001, ApJ, 560, 566
19. Daddi E., Cimatti A., Renzini A. 2000, A&A, 362, L45
20. Cole S. et al. 2000, MNRAS, 319, 168
21. Kauffmann G. et al. 1999, MNRAS, 303, 188
22. Benson A.J. et al. 2001, MNRAS submitted (astro-ph/0110387)

Constraining the Mass Assembly in Ellipticals from ERO Clustering[*]

Emanuele Daddi[1,2]

[1] Università di Firenze, Largo E. Fermi 2, I-50125 Firenze, Italy
[2] European Southern Observatory, Karl-Schwarzschild-Str. 2, D-85748 Garching, Germany

Abstract. The redshift evolution of clustering is a powerful tool to constrain the models of galaxy assembly. In particular, the strong clustering of EROs, if mainly due to early type galaxies, suggests a nearly constant comoving correlation length for this population of galaxies from $z = 0$ to $z \sim 1$, in apparent agreement with hierarchical scenarios. However, our recent spectroscopic identifications of a large ERO sample has shown a large fraction of EROs to be dust-reddened star-forming galaxies. We present here a real space clustering analysis of both EROs species that confirm the early-type EROs to be the source of ERO clustering, with dusty star-forming EROs being relatively uncorrelated. Despite the success on accounting for the clustering evolution, we discuss how hierarchical models still fail in reproducing the high comoving density and old ages of early-type galaxies.

1 Clustering and the Assembly of Mass in Galaxies

The evolution of the galaxy two-point correlation function provides important clues to the process driving galaxy formation and evolution The shape and normalization of this function depends on both the cosmic growth of mass structures and on the details of how galaxies trace mass at different epochs - the bias evolution. Competitive models are still under debate for explaining the assembly of mass galaxies: the *ab-initio* hierarchical models predict that massive galaxies are being build during a long-lasting process involving merging of smaller blocks. On the other hand the old fashioned (but still viable) monolithic models postulate an early assembly ($z > 2$) for massive galaxies, as a direct explanation for the old ages and overall homogeneous properties of local early-type galaxies (see e.g. Renzini 1999). The two competitive models imply rather different evolution to high redshift for the bias of galaxies. In the monolithic framework the number density of massive galaxies is conserved over time (at least after the completion of the formation process, e.g. at $z < 2$) and as a result the bias of galaxies increases in an almost linear way (Tegmark & Peebles 1998). In the hierarchical scenario, instead, galaxy merging acts to reduce the number density of massive

---

[*] The collaborators in the K20 survey include T. Broadhurst (HUJ), A. Cimatti (Arcetri), S. Cristiani (ECF & Trieste), S. D'Odorico (ESO), A. Fontana (Roma), E. Giallongo (Roma), R. Gilmozzi (ESO), N. Menci (Roma), M. Mignoli (Bologna), F. Poli (Roma), L. Pozzetti (Bologna), S. Randich (Arcetri), A. Renzini (ESO), P. Saracco (Milano), J. Vernet (Arcetri), G. Zamorani (Bologna)

galaxies, and require the bias to increases rapidly with z, approximately with a square dependence on z (e.g. Moscardini et al. 1998). As a result, high-z massive galaxies are predicted to be more strongly clustered in the hierarchical models, with respect to the monolithic scenario.

2   The Role of EROs

Extremely Red Objects ($R - K > 5$, EROs hereafter) have the colors expected for distant ($z > 0.8$) passively evolving elliptical galaxies and can be used to trace the evolution of elliptical galaxies from $z \sim 1$ to nowadays. The number counts of EROs have been shown to favor the monolithic scenario, as the hierarchical models fails to match them by a factor of at least 5 (Daddi et al. 2000a, Smith et al. 2001, Firth et al. 2001). Recently, we have completed a relatively large deep survey of very red galaxies covering 700 arcmin^2 (Daddi et al. 2000b, D00 hereafter), concluding that EROs are strongly clustered in projection, by an order of magnitude more than general field galaxies at the same limits of $K \leq 18$–19.2. With careful attention to the measurement uncertainty inherent in narrow field data, Daddi et al. (2001, D01 hereafter) showed that the angular clustering of EROs implies a spatial correlation length of $r_0 = 12 \pm 3 \ h^{-1}$ comoving Mpc, with the assumption that the ERO population is dominated by elliptical galaxies. This large clustering amplitude is in agreement with the hierarchical paradigm predictions (D01).

However, in our recent large K20 redshift survey of a flux selected sample of ~ 500 galaxies, limited to $K \leq 20$ over 52 arcmin^2 (Cimatti et al. 2002, C02 hereafter), we obtained redshifts for a sub-sample of 35 EROs. For red objects with $R - K > 5$, we are 70% complete to $K \leq 19.2$ (the same magnitude limit of the D00 clustering measurements based on 2D data), finding approximately equal numbers of old systems (consistent with being passively evolving elliptical galaxies) and dusty starbursts galaxies (C02). While the derived fraction of galaxies with early-type spectra, $50 \pm 20\%$, is consistent with previously derived estimates (e.g. Moriondo et al. 2000), C02 showed that the dusty star-forming objects do contribute significantly to the ERO population at faint magnitudes, thus complicating the interpretation of both ERO surface density and the clustering measured in earlier analyses. In particular, given the strong theoretical interest in the clustering amplitude of early-type galaxies it is important to evaluate the role of star-forming objects to understand if and how much they are clustered, and hence the implications for the clustering of $z \sim 1$ ellipticals once this contaminant is subtracted.

3   Spatial Clustering of Old and Dusty-SF EROs

Table 1 shows the redshifts of the EROs identified in the K20 survey (C02) and classified as old passively evolving or dusty-SF galaxies, sorted with increasing redshift and divided between the two survey fields (32.2 arcmin^2 from CDFS

Table 1. The redshifts of the EROs in the K20 survey. All but four EROs have $K \leq 19.2$. The redshift measurement errors are preliminarily estimated to have a mean dispersion of $\sigma \sim 100\text{--}200$ km/s in the K20 survey.

CDFS		0055-27	
Old	Dusty-SF	Old	Dusty-SF
0.726	0.796	0.790	0.820
1.019[1]	0.863	0.864	0.996
1.039	0.891	0.896	1.210
1.096	0.974	0.896	1.240
1.215	0.996[1]	0.935	1.300[1]
1.222	1.030	1.050	1.419
1.222	1.094	1.104	
	1.109[1]	1.166	
	1.149		
	1.221		
	1.294		
	1.327		

[1] Objects with $19.2 < K \leq 20$

and 19.8 arcmin^2 from 0055-27). We refer to C02 and future papers for details regarding the photometry and the spectral analysis.

Despite being the largest by far sample of EROs with identified redshift, standard methods for evaluating the full two point correlation function cannot still be applied because of the small number of objects. Nevertheless the clustering properties of the old and dusty-SF samples can be investigated by studying the frequency of close pairs, that relates to the integral under the correlation function on small scales, where most of the amplitude lies.

From Table 1, it can be noted that the sample of old EROs contains two pairs that, within the observational redshift accuracy, have the same redshift ($z = 0.896$ in the 0055-27 field and $z = 1.222$ in the CDFS), with an additional object close to the second pair at $z = 1.215$. On the other hand, the sample of dusty-SF EROs contains no really close pair, the closest pair having a relatively large redshift separation $\Delta z = 0.015$ ($z = 1.094$ and $z = 1.109$ in the CDFS). The two old ERO pairs with the same redshift have also quite small angular separations ($\lesssim 1$ arcmin), implying spatial separations of 0.51 and 0.82 h^{-1} Mpc, while the two closest dusty-SF pairs are separated by 24 and 40 h^{-1} Mpc, respectively. The number of independent pairs in the samples is 81 for the dusty-SF EROs and 49 for the old EROs, thus suggesting a relatively higher intrinsic clustering amplitude for the old EROs.

To assess the significance of observed pair counts we first generate random samples. The selection functions are constructed from the observed redshift distributions of the two ERO populations. Simulated samples were built by assigning at random a redshift (rounded to $\Delta z = 0.001$ to match the data redshift measurements) extracted from the appropriate selection function, with sky positions within boundaries matching the area of each of our fields, and number of objects as in the relative observations (Table 1). The resulting probability of finding by chance ≥ 2 pairs of old EROs within a separation $\leq 0.82\ h^{-1}$ Mpc is about $5\ 10^{-5}$, a clear evidence of clustering among the sample of old EROs. On the other hand, the random simulations showed that the probability of finding the closest dusty-SF ERO pair at $\leq 24\ h^{-1}$ Mpc and the two closest pairs at $\leq 40\ h^{-1}$ Mpc are both $\sim 97\%$, consistent with purely random chance.

We can proceed a step further and generate simulated samples incorporating a known 2-point clustering amplitude, in order to derive informations on the clustering of the two classes, and to obtain meaningful estimates of the variance inherent in the pair statistics in the sample. We follow the recipe described in D01, allowing us to generate many samples with a given value of r_0 over very large volumes. We adopt the canonical parameterisation $\xi(r) \propto r^{-\gamma}$ with a slope of $\gamma = 1.8$ (justified by the observed angular slope $\delta = 0.8$, D00) for the 2-point correlation function and allow the amplitude to vary.

The probability to find ≥ 2 pairs within $\leq 0.82\ h^{-1}$ Mpc increases strongly with r_0 and at the 1σ level the observed close pairs statistics requires the correlation length of the present sample of old EROs to lie in the broad range $5.5 \lesssim r_0/(h^{-1}\ \text{Mpc}) \lesssim 16$. On the other hand, for the dusty-SF EROs, the observation of the two closest pairs within $40\ h^{-1}$ Mpc constrains $r_0 < 2.5 h^{-1}$ Mpc at the 3σ confidence level. Fig. 1 summarises concisely the comparison between the fraction of observed pairs below a given scale compared with the random $(r_0 = 0)$ and clustered $(r_0 = 10\ h^{-1}\ \text{Mpc})$ expectations for a range of scales.

3.1   Spatial and Angular Clustering of $K < 19.2$ EROs

If we assume $r_0 < 2.5\ h^{-1}$ Mpc for the observed sample of dusty-SF EROs, this results in an angular clustering amplitude $A(1^o) \lesssim 0.002$ at $K \sim 19$. We recall that EROs as a whole (undivided into old and dusty-SF objects) have a factor of 10 larger angular amplitude than this (D00). A solid result of this analysis is therefore that the dusty-SF EROs cannot be the cause of the strong angular clustering of EROs reported by D00, in agreement with the considerations of D01. It seems clear from our redshift survey that a significant fraction of EROs are weakly clustered dusty-SF galaxies which therefore dilutes the true angular clustering amplitude of the dominant early-type galaxies population responsible for the majority of the clustering signal. A detailed estimate of the amplitude of this dilution effect would need a better knowledge of the relative fractions of both classes and a measure of the cross correlation between the two ERO species. In fact, 3 cross pairs are observed (all from CDFS, Table 1) with $\Delta z \leq 0.002$ and distances within $3.2\ h^{-1}$ Mpc, with a probability of only 2% to happen by chance, suggesting the cross-correlation between old and dusty-SF EROs not to be negligible (as must be expected at some level given that they are found in a similar redshift range). In any case, we note that although our spatial clustering amplitude of $12 \pm 3\ h^{-1}$ Mpc, derived in D01, is more secure based on a relatively large sample and consistent with the present analysis, we must revise

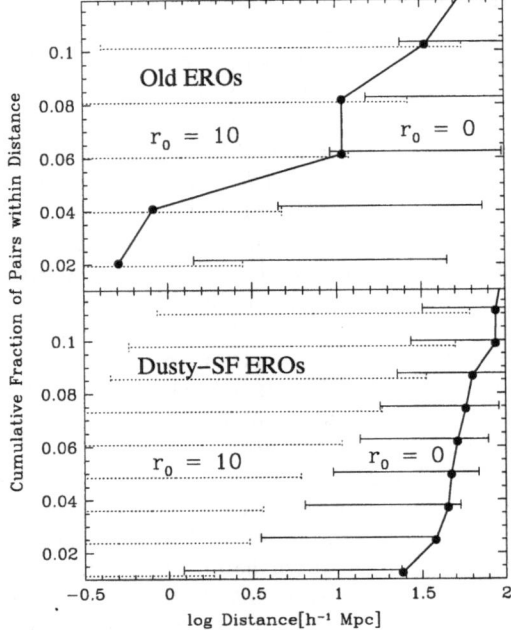

Fig. 1. Top panel: the cumulative distribution of pair separations observed for the old EROs (heavy line with filled circles). The horizontal error bars show the 2σ range recovered from our simulations with random (solid lines) and clustered (dotted lines, $r_0 = 10\ h^{-1}$ Mpc) realizations. Bottom panel: the same but for the dusty-SF EROs. This comparison shows that while the error on an estimate of the correlation length of either sample is quite broad, it is clear that the dusty-SF EROs as a class are completely inconsistent with a correlation length of order $10\ h^{-1}$ Mpc, estimated from projected samples of EROs (D01, Firth et al. 2001)

this amplitude upward in light of the findings presented here, thus confirming further the agreement with the hierarchical merging models.

4   Discussion: The Clustering of Old EROs

The ERO observations provide all the way 3 key properties for bright $L \gtrsim L_*$, $z \sim 1$, field early-type galaxies: (1) a space density consistent with the local value plus pure luminosity evolution (PLE) (C02); (2) an age $\gtrsim 3$ Gyr for their stellar populations (C02); and (3) a comoving correlation length $r_0 \gtrsim 10\ h^{-1}$ Mpc (this paper and D01).

A large correlation length, $r_0 \gtrsim 10\ h^{-1}$ Mpc, is anticipated for the hierarchical merging paradigm of rapid bias growth for massive galaxies to $z \sim 1$ (e.g. Mo & White 1996, Moscardini et al. 1998), while it would not be obvious to be accounted for within a PLE scenario (D01). However, current semianalytical renditions of the hierarchical models seem to be at odds with the first two evidences. In fact, the Cole et al. (2000) model predicts a comoving density (Fig. 1 of Benson et al. 2001) of *all* the $z \sim 1$ galaxies with $10^{11} M_\odot$ (consistent with our $K \leq 19.2$ selection) which is a full order of magnitude below the density of just the old EROs observed by C02. Similarly, the Kauffmann et al. (1999) model [1] predicts a comoving density of $z \sim 1$ EROs ($R - K \geq 5$, $K \leq 19.2$) that is 3(6) times lower the value observed by C02 for old(all) EROs. In addition, in these models $z \sim 1$ galaxies qualified as field early-types appear to have experienced recent star-formation, while the present sample of old EROs is dominated by an old stellar population. We conclude that to our knowledge no semianalytical rendition of the hierarchical merging models can yet account for all the 3 key observed properties of $z \sim 1$ field early type galaxies.

References

1. Benson A.J., Ellis R.S. & Menanteau F., 2001, MNRAS, submitted (astro-ph/0110387)
2. Cimatti A., Daddi E., Mignoli M., et al., 2002, A&A in press (astro-ph/0111527) (C02)
3. Cole S., Lacey C., Baugh C. & Frenk C., 2000, MNRAS, 319, 168
4. Daddi E., Cimatti A. & Renzini A., 2000a, A&A 362, L45
5. Daddi E., Cimatti A., Pozzetti L., et al., 2000b, A&A 361, 535 (D00)
6. Daddi E., Broadhurst T., Zamorani G., et al., 2001, A&A, 376, 825 (D01)
7. Firth A.E., Somerville R.S., McMahon R.G. et al. 2001, MNRAS, submitted (astro-ph/0108182)
8. Kauffmann G., Colberg J.M., Diaferio A., White S.D.M., 1999, MNRAS 307, 529
9. Mo H. & White S.D.M., 1996, MNRAS, 282, 347
10. Moscardini L., Coles P., Lucchin F. & Matarrese S., 1998, MNRAS 299, 95
11. Moriondo G., Cimatti A. & Daddi E., 2000, A&A 364, 26
12. Renzini A., 1999, in "When and How do Bulges Form and Evolve?", ed. by C.M. Carollo, H.C. Ferguson & R.F.G. Wyse (Cambridge University Press) (astro-ph/9902108)
13. Smith G.P., Smail I., Kneib J.-P., et al. 2001, MNRAS, in press (astro-ph/0109465)
14. Tegmark M. & Peebles P.J.E., 1998, ApJ 500, L79

---

[1] http://www.mpa-garching.mpg.de/GIF/

Clustering at High Redshift

S. Cristiani[1,2], S. Arnouts[3], A. Fontana[4], P. Saracco[5], and E. Vanzella[3,6]

[1] ST European Coordinating Facility, European Southern Observatory,
K.-Schwarzschild-Strasse 2, D-85748 Garching bei München, Germany
[2] Osservatorio Astronomico di Trieste, via Tiepolo 11, I-34131 Trieste, Italy
[3] European Southern Observatory, K.-Schwarzschild-Strasse 2,
D-85748 Garching bei München, Germany
[4] Osservatorio Astronomico di Roma, via dell'Osservatorio 2, Monteporzio, Italy
[5] Osservatorio Astronomico di Brera, via E. Bianchi 46, Merate, Italy
[6] Dipartimento di Astronomia, Università di Padova, vicolo dell'Osservatorio 2,
I-35122 Padova, Italy

1   Near IR Imaging of the HDF-S with VLT/ISAAC

The addition of deep near infrared images to the database provided by the HDF WFPC2 is essential to monitor the SEDs of the objects on a wide baseline and address a number of key issues including the total stellar content of baryonic mass, the effects of dust extinction, the dependence of morphology on the rest frame wavelength, the photometric redshifts, the detection and nature of extremely red objects (EROs). For these reasons deep near infrared images were obtained with the ISAAC instrument at the ESO VLT in the Js, H and Ks bands reaching, respectively, 23.5, 22.0, 22.0 limiting Vega-magnitude (5σ in an aperture of diameter $1".2 \equiv 2 \times$ FWHM, [3,4]).

2   A Multi-Color Catalog of the HDF-S

A multi-color (F300, F450, F606, F814, Js, H, Ks) photometric catalog of the HDF-S has been produced [4] developing specific procedures to match HST and VLT data. Having in mind the generation of photometric redshifts we have chosen a conservative approach in the object detection, leading to a list of 1611 sources. After correcting for the incompleteness of the source counts, the object surface density at $I_{AB} \leq 27.5$ is estimated to be 220 arcmin^{-2}. The comparison between the median V-I colour in the HDF-North and South shows a significant difference around $I_{AB} \sim 26$, possibly due to the presence of large scale structure at $z \sim 1$ in the HDF-N.

Using for the object detection the Ks-band image we have selected down to $K_{AB} < 24$ a sample of 15 EROs, defined as sources with $(I - K)_{AB} > 2.7$, corresponding to the colour of passively evolving elliptical galaxies at $z > 1$. The EROs surface density turns out to be 3.2 ± 0.9 arcmin^{-2}, and their distribution, at least from the angular point of view, is remarkably nonuniform: 10 EROs out of 15 are inside the upper WFPC2 chip (0.6 square arc-minute). One of the EROs is a powerful radio-galaxy.

3   Photometric and Spectroscopic Redshifts

Photometric redshifts have been produced both fitting templates to the observed SEDs [1,2] and with neural network techniques [5]. Using colour-colour diagrams, 90 U-band dropouts have been selected down to $I_{AB} = 27$ (19 are brighter than 25 mag). Spectroscopic observations of the 9 candidates with $I_{AB} < 24.5$ have been carried out, confirming all of them to be galaxies with $2 < z < 3.5$. Similarly, 17 B-band dropouts have been selected down to $I_{AB} = 27$ (all with $I_{AB} > 26$). A comparison of the 38 spectroscopic redshifts available so far with the photometric predictions, provides an estimate of the redshift accuracy σ_z of 0.16, 0.13 for the template fitting and neural network technique, respectively.

4   The Redshift Evolution of the Galaxy Clustering

The photometric redshifts for all the galaxies brighter than $I_{AB} < 27.5$ have been used to study the evolution of galaxy clustering in the interval $0 < z < 4.5$ [1]. The clustering signal is obtained in different redshift bins using two different approaches: a standard one, which uses the best redshift estimate of each object, and a second one, which takes into account the redshift probability function of each object and improves the information in the redshift intervals where the contamination from objects with insecure redshifts is important. With both methods, we find that the clustering strength up to $z \sim 3.5$ in the HDF-S is consistent with the previous results in the HDF-N. While at redshift lower than $z \sim 1$ the HDF galaxy population is un/anti-biased ($b < 1$) with respect to the underlying dark matter, at high redshift the bias increases up to $b \sim 2 - 3$, depending on the cosmological model. These results support previous claims that, at high redshift, galaxies are preferentially located in massive haloes, as predicted by the biased galaxy formation scenario. The impact of cosmic errors on the analysis has been quantified, showing that errors in the clustering measurements in the HDF surveys are indeed dominated by shot-noise in most regimes. Future observations with instruments like the ACS on HST will improve the S/N by at least a factor of two and more detailed analyses of the errors will be required. In fact, pure shot-noise will give a smaller contribution with respect to other sources of errors, such as finite volume effects or non-Poissonian discreteness effects.

References

1. Arnouts, S., Moscardini, L., Vanzella, E., Colombi, S., Cristiani, S., Fontana, A., Giallongo, E., Matarrese, S., Saracco, P., 2002, MNRAS, 329, 355
2. Fontana, A., et al. 2002 in preparation
3. Saracco, P., Giallongo E., Cristiani, S., D'Odorico S., Fontana A., Iovino A., Poli F., Vanzella E., 2001, A&A 375, 1
4. Vanzella, E., Cristiani, S., Saracco, P., Arnouts S., Bianchi, S., D'Odorico, S., Fontana, A., Giallongo, E., Grazian, A., 2001, AJ 122, 2190
5. Vanzella, E. et al. 2002 in preparation

The Evolution of the Near-IR Luminosity Function to $z \simeq 1$ in the K20 Redshift Survey

Lucia Pozzetti[1] and the K20 collaboration[2]

[1] Osservatorio Astronomico di Bologna, Via Ranzani 1, 40127 Bologna, Italy
[2] Zamorani G. (Bologna), Cimatti A. (Arcetri), Mignoli M. (Bologna), Fontana A. (Roma), Poli F. (Roma), Renzini A. (ESO), Broadhurst T. (ESO), Cristiani S. (ESO), Daddi E. (Arcetri), D'Odorico S. (ESO), Giallongo E. (Roma), Gilmozzi R. (ESO), Menci N. (Roma), Saracco P. (Milano)

Abstract. We have estimated the evolution of the near-IR Luminosity Function (LF) up to $z \simeq 1.3$ from a deep K-band selected spectroscopic survey (K20 redshift survey). We have derived the LF's in the rest-frame J and K bands in two redshift bins ($z_{mean} \simeq$ 0.5 and $z_{mean} \simeq 1.0$) and compared them to the Local Luminosity Function. The evolution with z of near-IR LF could infer information on the evolution of the Galaxy Stellar-Mass Function. Preliminary analysis shows *a mild luminosity evolution both in J and K Luminosity Function to $z \simeq 1$*, suggesting therefore a slow or no evolution in the Galaxy Stellar Mass Function. On the contrary pure density evolution can not reproduce the observed LF at $z \simeq 1$. Moreover we found that *red and early-type galaxies dominate the bright-end of the LF*, suggesting that early massive galaxies were already in place at $z \simeq 1$.

1 The Near-IR Luminosity Function

The near-IR galaxy Luminosity Function is an important characteristic of the galaxy population as the rest-frame near-IR luminosities are good tracers of evolved stars and hence it provides a direct and quantitative estimate of the Galaxy Stellar-Mass Function. We have derived its evolution up to $z \simeq 1$ from a deep VLT spectroscopic survey selected in the K-band (*K20 redshift survey:* see Cimatti et al., these proceedings) to place constraints to the models of galaxy formation. From a complete sample of about 500 galaxies with K<20, covering an area of about 52 square arcmin, about 93% and 85% of the galaxies at K<19.5 and K<20, respectively, have been spectroscopically identified.

We have computed the Luminosity Function in the rest-frame J and K-bands using a $1/V_{max}$ formalism. After assigning a spectral type to each galaxy, based on their $R - K$ colors, we have derived absolute rest-frame magnitudes M_J and M_K using k-corrections computed with Bruzual & Charlot (2000) models. An advantage of the K-band selection is that the k-corrections are relatively invariant to galaxy type and relatively small also at high redshift (Cowie et al. 1996). Moreover at $z \simeq 0.8$ the observed K-band corresponds to rest-frame J-band. We have used photometric redshifts derived from multi-band photometry (see Fontana et al., these proceedings) for most of the faint z-unidentified objects. We have computed the LF's in two redshift bins: 1) $0.2 < z < 0.65$ ($z_{mean} \simeq 0.5$) and 2) $0.75 < z < 1.3$ ($z_{mean} \simeq 1.0$); the redshift bin 0.65<z<0.75 has been excluded because dominated by two clusters. Thanks to the higher statistical significance and completeness of our sample, it is possible to investigate the evolution of the near-IR LF with a higher confidence level than in previous surveys (Cowie et al. 1996, Cohen et al. 1999). We have compared the LF's derived at high redshift to the Local Luminosity Function (LLF) from 2MASS+2dFGRS (Cole et al. 2001) in J and K bands (Figure 1), and explored two possible scenarios: luminosity and/or density evolution. Preliminary analysis shows that the data are consistent with a mild evolution from $z = 0$ to $z \simeq 1$ both in J and K bands. We have estimated a luminosity evolution at $z \simeq 1$ of the order of $\Delta M_J \simeq -0.6$ ($L_J \propto (1 + z)^{0.8}$) and $\Delta M_K \simeq -0.4$ ($L_K \propto (1 + z)^{0.5}$), rather consistent with a scenario of Passive Luminosity Evolution, while a pure density evolution could not reproduce the observed LF's. According to mass-to-light ratios from spectral models, assigned on the basis of $R - K$ colors, we *suggest therefore a slow or no evolution in the Galaxy Stellar Mass Function* up to $z \sim 1$. Moreover dividing the galaxy population in two subsamples according to their $R - K$ colors (red and blue galaxies), or on the basis of preliminary spectroscopic classification, we find that red and early type galaxies dominate the bright-end of the LF, *suggesting that early massive galaxies were already in place at $z \simeq 1$.*

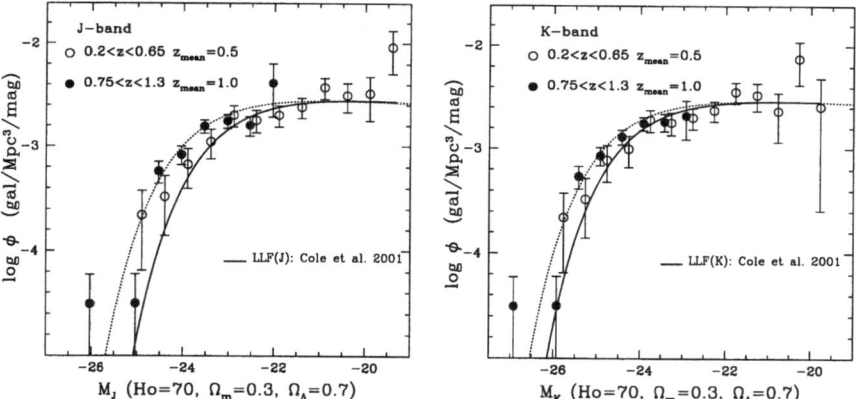

Fig. 1. Rest-frame J and K-band Luminosity Functions in two redshift bins: $z_{mean} = 0.5$ (empty circles) $z_{mean} = 1.0$ (filled circles). The solid curves are the LLF in the J and K-bands by Cole et al. 2001, while dotted curves are the same LF "evolved" to $z = 1$ assuming a mild luminosity evolution, $\Delta M_J = -0.6$ and $\Delta M_K = -0.4$ respectively

References

1. Cohen J.G., et al. 1999, ApJ, 512, 30
2. Cole S., et al. 2001, MNRAS, 326,255
3. Cowie L.L., et al. 1996, AJ, 112, 839

The Mass Function of Field Galaxies at $0.4 < z < 1.2$ as Derived from the MUNICS K-Selected Sample

Niv Drory[1], Ralf Bender[1], Jan Snigula[1], Georg Feulner[1], Ulrich Hopp[1], Claudia Maraston[1], Gary J. Hill[2], and Claudia Mendes de Oliveira[3]

[1] Universitäts-Sternwarte München, Scheinerstr. 1,D-81679 München, Germany
[2] University of Texas at Austin, Austin, Texas 78712, USA
[3] Instituto Astronômico e Geofísico, Av Miguel Stéfano 4200, 04301-904, São Paulo, Brazil

Abstract. We derive the number density evolution of massive field galaxies in the redshift range $0.4 < z < 1.2$ using the K-band selected field galaxy sample from the Munich Near-IR Cluster Survey (MUNICS). We rely on spectroscopically calibrated photometric redshifts to determine distances and absolute magnitudes in the rest-frame K-band. To assign mass-to-light ratios, we use two different approaches. First, we use an approach which maximizes the stellar mass for any K-band luminosity at any redshift. We take the mass-to-light ratio, $\mathcal{M}/L_K$, of a Simple Stellar Population (SSP) which is as old as the universe at the galaxy's redshift as a likely upper limit. Second, we assign each galaxy a mass-to-light ratio by fitting the galaxy's colours against a grid of composite stellar population models and taking their $\mathcal{M}/L_K$. We compute the number density of galaxies more massive than $2 \times 10^{10} h^{-2} \mathcal{M}_\odot$, $5 \times 10^{10} h^{-2} \mathcal{M}_\odot$, and $1 \times 10^{11} h^{-2} \mathcal{M}_\odot$, finding that the integrated stellar mass function is roughly constant for the lowest mass limit and that it decreases with redshift by a factor of ~ 3 and by a factor of ~ 6 for the two higher mass limits, respectively. This finding is in qualitative agreement with models of hierarchical galaxy formation, which predict that the number density of $\sim M^*$ objects is fairly constant while it decreases faster for more massive systems over the redshift range our data probe.

1 Introduction

The traditional observables used to characterise galaxies are unsuitable for studying the assembly history of galaxies, one of the most fundamental predictions of CDM models, since these observables may be transient. The best observable for this aim is, in principle, total mass, which is on the other hand very hard to measure. It has been argued that the best available surrogate accessible to direct observation is the near-IR K-band luminosity of a galaxy which reflects the mass of the underlying stellar population and is least sensitive to bursts of star formation and dust extinction [11,7,2]. The main uncertainty involved in the conversion of K-band light to mass is due to the age of the population, amounting to only a factor of two in mass uncertainty for populations older than ~ 3 Gyr.

The galaxy sample used here is a subsample of the MUNICS survey, selected for best photometric homogeneity, good seeing, and similar depth. Furthermore, in each of the remaining survey patches, areas close to the image borders in any passband, areas around bright stars, and regions suffering from blooming are excluded. The subsample covers 0.27 square degrees in V (23.5), R (23.5), I (22.5), J (21.5), and K (19.5); the magnitudes are in the Vega system and refer to 50% completeness for point sources.

The final catalog covers an area of 997.7 square arc minutes and contains 5132 galaxies. The fields included in this analysis are S2, S3f5-8, S5, S6, and S7f5-8. See Table 1 in [5] for nomenclature and further information on the survey fields.

The distances to the galaxies are derived using spectroscopically calibrated photometric redshifts. A comparison of spectroscopic and photometric redshifts is shown in [4].

We derive stellar masses by converting rest-frame K-band luminosities to mass using two different approaches to model the mass-to-light ratios of the galaxies. We discuss the resulting integrated stellar mass functions at different mass limits and their evolution with redshift.

We assume $\Omega_M = 0.3$, $\Omega_\Lambda = 0.7$ throughout this work. We write Hubble's Constant as $H_0 = 100\,h$ km s^{-1} Mpc^{-1}, using $h = 0.65$ unless the quantities in question can be written in a form explicitly depending on h.

2   The Maximum PLE Model

The integrated stellar mass function $n(\mathcal{M} > \mathcal{M}_{\text{lim}})$, the comoving number density of objects having stellar mass exceeding $\mathcal{M}_{\text{lim}}$, is computed using the V_{max} formalism as described in [4].

To compute the stellar mass of a galaxy, we first use an approach which maximises the stellar mass for any K-band luminosity at any redshift.

Noting that $\mathcal{M}/L_K$ is a monotonically rising function of age for Simple Stellar Populations (SSPs), we find that the likely upper limit for $\mathcal{M}/L_K$ is the mass-to-light ratio of a SSP which is as old as the universe at the galaxy's redshift. This is the most extreme case of passive luminosity evolution (PLE) one can adopt. It corresponds to a situation where all massive galaxies would be of either elliptical, S0, or Sa type.

We take the mass-to-light ratios from the SSP models published by [9], using a Salpeter IMF. Similar dependencies on age are obtained from the models of [12] and [2] although the absolute values of $\mathcal{M}/L_K$ vary somewhat, partly due to differences in the models themselves but mostly due to the way stellar remnants are treated by the different authors.

The resulting integrated mass functions for $\mathcal{M}_{\text{lim}} = 2 \times 10^{10} h^{-2} \mathcal{M}_\odot$, $\mathcal{M}_{\text{lim}} = 5 \times 10^{10} h^{-2} \mathcal{M}_\odot$, and $\mathcal{M}_{\text{lim}} = 1 \times 10^{11} h^{-2} \mathcal{M}_\odot$ are shown in Fig. 2 along with the integrated luminosity functions for comparison.

The mean values of $\mathcal{M}/L_K$ in the maximum PLE model in the four redshift bins are 0.99, 0.83, 0.73, and 0.65, as computed from the look-back time in our cosmology. With these mean values the mass limits correspond to absolute K-band magnitudes of -22.43, -22.63, -22.77, and -22.90, respectively, for $\mathcal{M}_{\text{lim}} = 2 \times 10^{10} h^{-2} \mathcal{M}_\odot$. For $\mathcal{M}_{\text{lim}} = 5 \times 10^{10} h^{-2} \mathcal{M}_\odot$ the numbers are -23.42,

-23.62, -23.76, and -23.89. Finally, for $\mathcal{M}_{\mathrm{lim}} = 1 \times 10^{11} h^{-2} \mathcal{M}_\odot$ we have -24.18, -24.38, -24.51, and -24.64 (magnitudes with respect to $h = 1$).

The upper and middle panels of Fig. 2 compare the evolution of the integrated luminosity to the integrated mass. It is evident that the number density of *luminous* K-band selected galaxies does not evolve significantly (given our uncertainties) to $z = 1.2$. However, because of the inevitable evolution of the mass-to-light ratio with z, the number density of *massive* systems does change. Transforming luminosities into masses with our maximum PLE scheme yields a roughly constant number density for our lowest mass limit, $2 \times 10^{10} h^{-2} \mathcal{M}_\odot$, and a decrease of the number density with redshift by a factor of ~ 3 for a mass limit of $5 \times 10^{10} h^{-2} \mathcal{M}_\odot$, and by a factor ~ 6 for objects more massive than $1 \times 10^{11} h^{-2} \mathcal{M}_\odot$. As the true $\mathcal{M}/L_K$ at high redshift will most likely be lower than in our maximum PLE model, the true number densities are likely to decrease more rapidly with redshift.

The steepening of the curves with increasing limiting mass in the maximum PLE models (despite them all having the same mass-to-light ratios at any given redshift) is due to the invariance of the LF with redshift and its steepness at the bright end. At increasing limiting mass, one is moving down the steepening bright end of the LF, so that the same change in the mass-to-light ratio yields a higher change in the number density.

To investigate the effect the uncertainties in the photometric redshifts have on the values of the integrated mass function, we have performed Monte-Carlo simulations. The errors of the mean values of the integrated mass function (size of open symbols in Fig. 2) are derived by repeating the mass function analysis using subsamples of the template SED library, as deficiencies in the templates are the main source of concern for the accuracy of the redshifts during the photometric redshift determination.

3   The Fitted Mass-to-Light-Ratio Model

To obtain a more realistic estimate of $\mathcal{M}/L_K$, we used our VRIJK color information and the photometric redshift to fit the age and SFR of each galaxy using a grid of composite stellar populations (CSP) with 9 exponential star formation timescales, τ, ranging from 0.1 to 10 Gyr with spectra extracted for 28 ages, t, between 0.04 Gyr and 15 Gyr for each τ. The input SSPs for constructing the composite stellar populations models are taken from [10], again using Salpeter IMF.

Fig. 1 shows the evolution of the K-band mass-to-light ratios as a function of age for each value of τ in the grid. Except for the two largest values of τ, the slope at ages $t > 2$ Gyr is remarkably independent of the actual star formation timescale. Moreover, the slope of the time evolution of $\mathcal{M}/L_K$ is the same even for the shortest value of τ, 0.1 Gyr, which essentially represents an SSP. Let aside normalisation effects, we therefore may expect a similar result for the mass function as obtained with the PLE model.

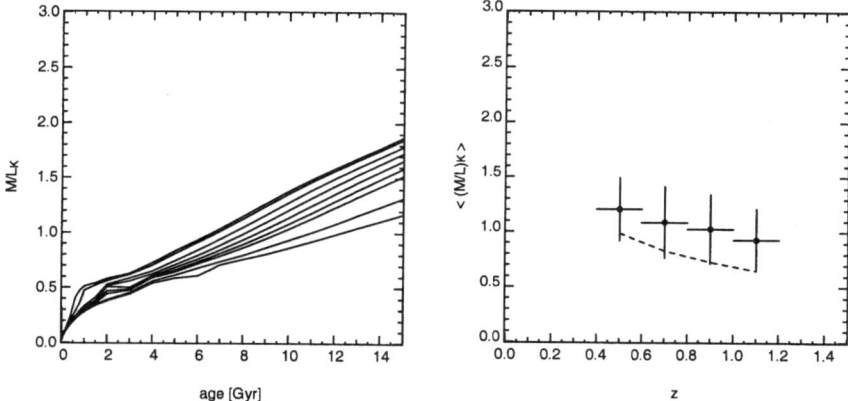

Fig. 1. Evolution of the K-band mass-to-light ratio, $\mathcal{M}/L_K$, with age for different star formation histories (left panel; see text). The average K-band mass-to-light ratio, $\mathcal{M}/L_K$, of the MUNICS sample as a function of redshift as determined from fitting CSP models to the V, R, I, J, K colour data base. The dashed line denotes the mass-to-light ration of the maximum PLE model (see text; right panel)

The average K-band mass-to-light ratio of the galaxy population determined by applying this fitting procedure is shown in Fig. 1. The figure also shows the PLE mass-to-light ratio as a function of z. Apart from the different normalisation, the evolution with redshift is very similar, a consequence of the insensitivity of $\mathcal{M}/L_K$ on the star formation history.

Finally, the lower panel of Fig.2 shows the integrated mass function for the same mass limits as those applied above, using the individually fitted $\mathcal{M}/L_K$ values.

The most striking feature of Fig. 2 is the similarity of the maximum PLE and the CSP-fitted curves. Note that there is a difference in the normalisation of the two, and due to the log scaling of the figure, the slope appears to be different in the plot.

If star formation played an important role at $z \sim 1$, the presence of young populations would have pushed $\mathcal{M}/L_K$ down, and therefore the CSP-fitted curves would be expected to be steeper than the maximum PLE curve, which assumes no star formation happens at all after $z = \infty$.

Nevertheless, the number density of massive systems seems to decline, with this decline being stronger for more massive systems, and therefore one is inclined to think that merging does play an important role. Indeed, [8] derive a number of 0.6 to 1.8 major mergers per L^* galaxy since $z \sim 1$ from HST-based pair counts of galaxies with known redshifts selected from the CFRS. We observe a decline in the number density by a factor of > 2 for somewhat more massive systems, and almost no significant density evolution at L^*.

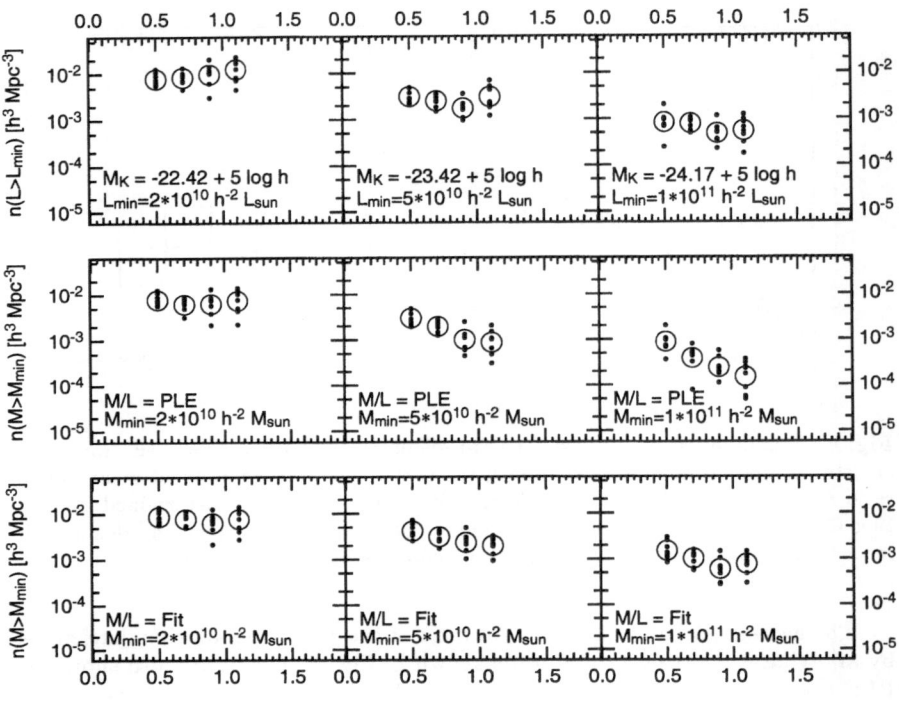

Fig. 2. Comoving number density of objects having rest-frame K-band luminosities exceeding $-22.42 + 5\log h$ ($2 \times 10^{10} h^{-2} L_\odot$), $-23.42 + 5\log h$ ($2 \times 10^{10} h^{-2} L_\odot$), and $-24.17 + 5\log h$ ($2 \times 10^{10} h^{-2} L_\odot$) (upper panels) and comoving number density of objects having stellar masses exceeding $\mathcal{M}_{\mathrm{lim}} = 2 \times 10^{10} h^{-2} \mathcal{M}_\odot$, $\mathcal{M}_{\mathrm{lim}} = 5 \times 10^{10} h^{-2} \mathcal{M}_\odot$, and $\mathcal{M}_{\mathrm{lim}} = 1 \times 10^{11} h^{-2} \mathcal{M}_\odot$ (integrated stellar mass functions; middle panels). Mass to light ratios are assigned to maximise the stellar mass at a given luminosity (see text), and thus are likely upper limits. The lower panel shows the integrated mass function for the same mass limits, this time individually determining $\mathcal{M}/L_K$ for each object by fitting against a grid of CSP models (see text). The solid points denote the values measured separately in each survey field, the open circles denote the mean values over the whole survey area. The size of the open circles is chosen to represent our estimate of the total uncertainty in the mean values

Therefore, we are inclined to think that if merging is the dominant factor in increasing the mass of these K-selected massive galaxies, most of the merging has to be dissipationless, involving rather low star formation activity.

The main uncertainty in these conclusions is still the field to field variation, in spite of the relatively large area surveyed, followed by the choice of SED templates used in the photometric redshift code (see above). The size of the open symbols in Fig. 2 represents our estimate of the total uncertainty of the mean values. If we assume a Gould IMF [6] instead of a Salpeter IMF, the evolving $\mathcal{M}/L_K$ curve becomes lower in its normalization as the mass-to-light ratio becomes smaller due to the reduced number of low-mass stars. The slope does not change significantly.

The observed density evolution as a function of mass is qualitatively consistent with the expectation from hierarchical galaxy formation models. Most rapid evolution is predicted for the number density of the most massive galaxies while the number density of L^*-galaxies should evolve much less. E.g. [1] predict that the number density of galaxies of a stellar mass of $10^{10}h^{-1}\mathcal{M}_\odot$ decreases by a factor of ~ 3.1 over redshift range $0.4 < z < 1.2$ (for the cosmological parameters as used here). Though this agreement is encouraging, both more elaborated models and improved sets of data are required. The latter can be obtained by larger and deeper samples, and more realistic estimates of $\mathcal{M}/L_K$ based on spectroscopic observations of the galaxies..

Acknowledgments

We would like to thank the Calar Alto staff for their long-standing support. This work was partly supported by the Deutsche Forschungsgemeinschaft, grant SFB 375 "Astroteilchenphysik" and the German Federal Ministry of Education and Research (BMBF), grant 05 AV9WM1/2.

References

1. Baugh, C. M., Cole, S., Frenk, C. S., & Lacey, C. G. 1998, ApJ, 498, 504
2. Brinchmann, J., & Ellis, R. S. 2000, ApJ, 536, L77
3. Bruzual, G. A., & Charlot, S. 1993, ApJ, 405, 538
4. Drory, N., Bender, R., Snigula, J., Feulner, G., Hopp, U., Maraston, C., Hill, G. J., & de Oliveira, C. M. 2001a, ApJ, 562, L111
5. Drory, N., Feulner, G., Bender, R., Botzler, C. S., Hopp, U., Maraston, C., Mendes de Oliveira, C., & Snigula, J. 2001b, MNRAS, 325, 550
6. Gould, A., Flynn, C., & Bahcall, J. N. 1998, ApJ, 503, 798
7. Kauffmann, G., & Charlot, S. 1998, MNRAS, 297, L23
8. Le Fèvre, O., et al. 2000, MNRAS, 311, 565
9. Maraston, C. 1998, MNRAS, 300, 872
10. Maraston, C. 2002, MNRAS, submitted
11. Rix, H., & Rieke, M. J. 1993, ApJ, 418, 123
12. Worthey, G. 1994, ApJS, 95, 107

Third Deepest Hubble Field
and Ground Based IR Follow-Up

James W. Colbert, Michael Rich, and Matthew A. Malkan

UCLA Dept. of Physics & Astronomy, University of California,
Los Angeles, CA 90095, USA

Abstract. Assembling archival data, we have produced an HST deep field which is the third deepest HST field imaged to date, centered on the z=2.39 radio galaxy 53w002 and an associated cluster of possible protogalactic clumps. We find that the summed images in F450W, F606W and F814W reach within 0.5-1 mag of the depth of the Hubble Deep Field North. To this we add fully reduced ground-based JHK imaging (to K=21.5), producing a full set of multi-wavelength photometry. Using this database, we have obtained photometric redshifts for all galaxies with both infrared and HST detections. We have discovered three Extremely Red Objects (EROs) with V-K > 6.

Some of the first ultra-deep imaging with the refurbished WFPC2 was undertaken in GO programs 5308, 5985, and 7459 (PIs=Windhorst and Keel). Their target was a radio galaxy at z=2.39, and the galaxies around it. Subsequent deep Lyman α imaging in the same field revealed 18 associated galaxies at nearly the same redshift, which may be protogalactic clumps that could eventually merge (Pascarelle et al. 1996, Nature, 383, 45). In all, the 53w002 field has over 60 hours of integration with WFPC2 and NICMOS, with 20 nights of JHK infrared images from the 2.4m MDM/TIFKAM telescope covering the entire field.

The HST data reaches within 0.5-1 mag of the HDF-N, and within 0.3-0.8 mag of the HDF-S. This can be seen in the F450W, F606W, and F814W (Vega magnitude system) number counts from our reductions which reach to 1-σ. Our IR data (obtained at the MDM observatory) reaches a 4-σ detection limit of J=23.3 and K=21.5, within 0.6 mag of the accompanying ground-based IR data for HDF-N (Connolly et al. 1997, ApJ, 486, L11).

Better statistics for the Bright Ages: Most of what we think we know about high-redshift galaxies rests on the limited statistics of the Hubble Deep Fields, mostly HDF-N. The HDF-N has 150 galaxies with photometric redshifts in the Bright Ages (1.8 < z < 2.6). However, only 8 of these are spectroscopically confirmed after extensive efforts,and no Bright Ages galaxy candidates in HDF-S have confirming spectra yet. Because LBG surveys have found large field-to-field variations–up to a factor of four in surface density–due to large-scale structures, it is dangerous to draw too many conclusions from a single 5 arcmin^2 image of the Universe.

Photometric redshifts from infrared photometry: We estimate the photometric redshifts of candidate galaxies by fitting the measured galaxy spectral energy distributions (SEDs) to redshifted galaxy templates (e.g. Koo 1985, AJ, 90, 418; Fernandez-Soto et al. 1999, ApJ, 513, 34). Adding the IR fluxes to

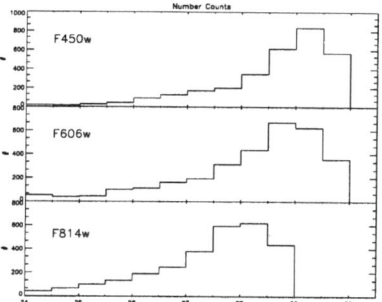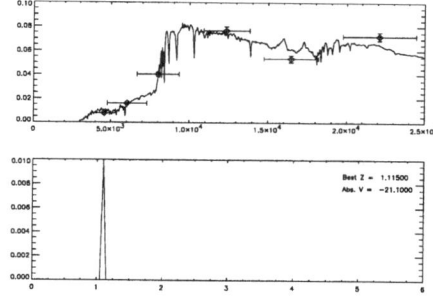

Fig. 1. a: (left) Histogram of object counts from the reduced WFPC2 B, V, and I images, as a function of Vega magnitudes, down to 1-σ. b: (right) BVIJHK photometry we measured for the galaxy shown. The best-fitting spectral template, at z=1.115 is over-plotted in the top panel. It includes a 0.5 Gyr burst of star formation reddened by A_V = 1.2 mag. The lower panel shows the probability distribution of acceptable redshifts fitted by the photometric redshift engine

the optical data gives us much improved spectral energy distributions (SEDs). A key spectral feature for these fits is the 3600/4000Å Balmer break, which moves out of the optical and into the IR at redshifts above $z \approx 1.5$.

To demonstrate the power of this method, we present a six band (B through K) picture of an extreme red object (ERO) taken from the data and its photometric redshift fit in Figures 1b and 2. The fit was produced using the photometric redshift code *hyperz* (Bolzonella et al. 2000, A&A, 363, 476). The infrared bands were crucial not only in identifying this as an interesting object, but allowing a fit to data beyond the 4000Å break. The fit estimates its redshift, 1.12, and also that it is a 0.5 Gyr old burst of star formation with AV = 1.2 mag of dust.

EROs: We have found three EROs in our WFPC2 field. Our photometric redshift fits place them all around z=1.2. Two of the three have very compact morphology through all bands. The third appears patchy and disk-like at shorter wavelengths, but with a bulge that becomes progressively more dominant with longer wavelength. The density of EROs down to K=21 in the 53w002 field is 0.5 per arcmin².

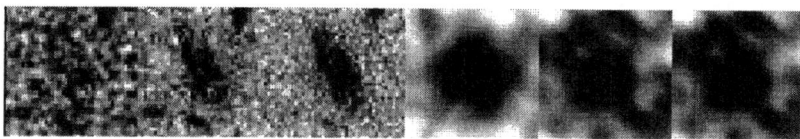

Fig. 2. Mosaic of reduced multicolor images of a $4'' \times 4''$ box around a galaxy in the 53W002 field with photometric redshift z=1.12. Wavelength increases from B at the left to K on the right; the object can qualify as an "Extremely Red Object" (ERO), with V-K>6. The smooth round bulge becomes progressively more dominant at long wavelengths, while at short wavelengths the galaxy appears patchy with irregular arms

The Mass of Radio Galaxies from Low to High Redshift

Matt J. Jarvis[1], Steve Rawlings[2], Steve Eales[3], Katherine M. Blundell[2], and Chris J. Willott[2]

[1] Sterrewacht Leiden, Postbus 9513, 2300 RA Leiden, The Netherlands
[2] Astrophysics, Department of Physics, Keble Road, Oxford, OX1 3RH, UK
[3] Department of Physics and Astronomy, University of Wales College of Cardiff, P.O.Box 913, Cardiff, CF2 3YB, UK

Abstract. Using a new radio sample, 6C* designed to find radio galaxies at $z > 4$ along with the complete 3CRR and 6CE sample we extend the radio galaxy $K - z$ relation to $z \sim 4.5$. The 6C* $K - z$ data significantly improve delineation of the $K - z$ relation for radio galaxies at high redshift ($z > 2$). In a spatially flat universe with a cosmological constant ($\Omega_M = 0.3$ and $\Omega_\Lambda = 0.7$), the most luminous radio sources appear to be associated with galaxies with a luminosity distribution with a high mean ($\approx 5L^*$), and a low dispersion ($\sigma \sim 0.5$ mag) which formed their stars at epochs corresponding to $z \gtrsim 2.5$)

1 The Advantages of Using Radio Galaxies

Radio galaxies provide the most direct method of investigating the host galaxies of quasars if orientation based unified schemes are correct. The nuclear light which dominates the optical/near-infrared emission in quasars is obscured by the dusty torus in radio galaxies, therefore difficult psf modelling and subtraction are not required to determine the properties of the underlying host galaxy. Unfortunately compiling samples of radio loud AGN is a long process, because of the radio selection there is no intrinsic optical magnitude limitation, making follow-up observations extremely time consuming, especially when dealing with the faintest of these objects. However, low-frequency selected radio samples do now exist with the completion of 3CRR (Laing, Riley & Longair 1983) along with 6CE (Eales et al. 1997; Rawlings et al. 2001) and the filtered 6C* sample (Blundell et al. 1998; Jarvis et al. 2001a; 2001b). We can now use these radio samples to investigate the underlying stellar populations through the radio galaxy $K - z$ Hubble diagram.

2 Previous Radio Samples and the $K - z$ Hubble Diagram

There has been much interest in the $K - z$ relation for radio galaxies in the past decade. Dunlop & Peacock (1993) using radio galaxies from the 3CRR sample along with fainter radio sources from the Parkes selected regions demonstrated that there exists a correlation between radio luminosity and the K−band emission. Whether this is due to a radio luminosity dependent contribution from a non-stellar source or because the galaxies hosting the most powerful radio sources are indeed more massive galaxies has yet to be resolved. Eales et al. (1997) confirmed this result and also found that the dispersion in the K−band magnitude from the fitted straight line increases with redshift. This result, along with the departure to brighter magnitudes of the sources at high redshift led Eales et al. to conclude that we are beginning to probe the epoch of formation of these massive galaxies. Using the highest redshift radio galaxies from ultra-steep samples of radio sources van Breugel et al. (1998) found that the near infrared colours of radio galaxies at $z > 3$ are very blue, consistent with young stellar populations. They also suggest that the size of the radio structure is comparable with the size of the near infrared region, and the alignment of this region with the radio structure is also more pronounced at $z > 3$. Lacy et al. (2000) using the 7C-III sample found evidence that the hosts of radio galaxies become more luminous with redshift and are consistent with a passively evolving population which formed at high redshift ($z > 3$). Thus, all of this work points to a radio galaxy population which formed at high redshift and has undergone simple passive evolution since. However, all of these studies were made with only a few high-redshift ($z > 2$) sources. With the 6C* sample we are now able to probe this high-redshift regime with increased numbers from samples with well-defined selection criteria.

3   The 6C* Filtered Sample

The 6C* sample is a low-frequency radio sample ($0.96\,\mathrm{Jy} \leq S_{151} \leq 2.00\,\mathrm{Jy}$) which was originally designed to find radio sources at $z > 4$ using filtering criteria based on the radio properties of steep spectral index and small angular size. The discovery of 6C*0140+326 at $z = 4.41$ (Rawlings et al. 1996) and 6C*0032+412 at $z = 3.66$ (Jarvis et al. 2001a) from a sample of just 29 objects showed that this filtering was indeed effective in finding high-redshift objects. Indeed, the median redshift of the 6C* sample is $z \sim 1.9$ whereas for complete samples at similar flux-density levels the median redshift is $z \sim 1.1$. We can now use this sample to push the radio galaxy $K - z$ diagram to high redshift ($z > 2$) where it has not yet been probed with any significant number of sources. Fig. 1 shows the radio luminosity-redshift plane for the three samples used in this analysis.

4   Emission-Line Contamination

The most-luminous sources at high redshift may be contaminated by the bright optical emission lines redshifted into the infrared. This is particularly true for sources in radio flux-density limited samples. The high-redshift sources in these samples are inevitably some of the most luminous, and we also know there is a strong correlation between low-frequency radio luminosity and emission-line strength (e.g. Rawlings & Saunders 1991; Willott et al. 1999; Jarvis et al. 2001a)

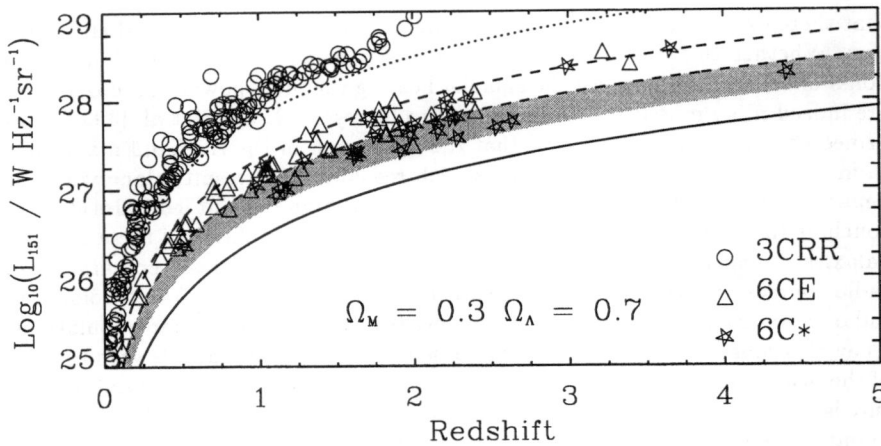

Fig. 1. Rest-frame 151 MHz luminosity (L_{151}) versus redshift z plane for the 3CRR (circles), 6CE (triangles) and 6C* (stars) samples. The rest-frame 151 MHz luminosity L_{151} has been calculated according to a polynomial fit to the radio spectrum (relevant radio data from Blundell et al. 1998). The curved lines show the lower flux-density limit for the 3CRR sample (dotted line; Laing et al. 1983) and the 7CRS (solid line; Blundell et al. in prep; Willott et al. in prep). The dashed lines correspond to the limits for the 6CE sample (Rawlings et al. 2001) and the shaded region shows the 6C* flux-density limits (all assuming a low-frequency radio spectral index of 0.5). Note that the area between the 3CRR sources and 6CE sources contains no sources, this is the area which corresponds to the absence of a flux-density limited sample between the 6CE ($S_{151} \leq 3.93\,\mathrm{Jy}$) and 3CRR ($S_{178} \geq 10.9\,\mathrm{Jy}$) samples. The reason why some of the sources lie very close to or below the flux-density limit of the samples represented by the curved lines is because the spectral indices lie very close to or below the assumed spectral index of the curves of $\alpha = 0.5$

which will increase the contribution to the measured K−band magnitudes from the emission lines in the most radio luminous sources.

To subtract this contribution we use the correlation between [OII] emission-line luminosity $L_{\mathrm{[OII]}}$ and the low-frequency radio luminosity L_{151} from Willott (2000), where $L_{\mathrm{[OII]}} \propto L_{151}^{1.00\pm0.04}$. Then by using the emission-line flux ratios for radio galaxies (e.g. McCarthy 1993) we are able to determine the contribution to the K−band magnitude from all of the other emission lines. This is illustrated in Fig. 2 where the emission-line contamination to the K−band flux is shown for various radio flux-density limits and a range of redshifts.

5   The $K - z$ Relation

In Fig. 3 we plot the $K - z$ diagram for all of the sources in our dataset. We also show four synthetic galaxy evolution models from the 'Galaxy Isochrone Synthesis Spectral Evolution Library' (GISSEL) of Bruzual & Charlot (1993)

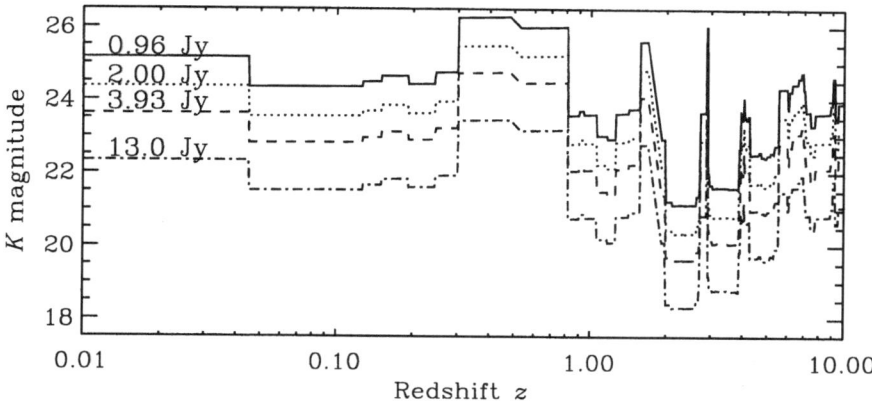

Fig. 2. Emission line contribution to the K−band magnitudes for various radio flux-densities assuming the power-law relation of $L_{[OII]} \propto L_{151}^{1.00}$

and a curve representing a galaxy which undergoes no-evolution. The GISSEL files that we have used are ones in which there is an instantaneous burst of star formation and one in which the burst of star formation lasts 1 Gyr, with a Salpeter IMF with a lower mass cut-off of 0.1 $M_\odot$ and an upper mass cut-off of 125 $M_\odot$. We use two different assumptions about the star formation history, one in which the burst of star formation begins at $z = 5$ and one in which the burst occurs at $z = 10$.

The no-evolution curve was constructed by taking the spectral energy distribution template from the GISSEL library that was found to fit the observed spectral energy distribution of a radio galaxy at $z = 0$, and which also reproduced the near-infrared colours. All of the curves are normalised to pass through the low-redshift ($z < 0.3$) points.

With our data on the 6C* sample in addition to the 6CE and 7C-III samples we find that in a low-density universe ($\Omega_M = 0.3$ and $\Omega_\Lambda = 0.7$) the data are predominantly brighter than the no-evolution curve and are consistent with a passively evolving stellar population with a high-formation redshift. If this passively evolving scenario is correct then hierarchical growth at $z < 2.5$ is not a required ingredient. However, this brightening may not just be due to passive evolution of the stellar population. Non-stellar contributions from the central AGN may also contribute a higher fraction of light at these redshifts. All of the studies to measure the non-stellar contribution to the K−band flux conducted to date (e.g. Leyshon & Eales 1998; Simpson, Rawlings & Lacy 1999), have concentrated on the most radio luminous 3CR sources at $z \simeq 1$, and may have little bearing on the properties of the high-redshift sources considered here. If it turns out that the high-redshift 6C sources have $\gtrsim$ non-stellar contamination to those of the $z \sim 1$ 3C sources (which have the same radio luminosity) then hierarchical build up may be necessary. Note that K−band observations of the

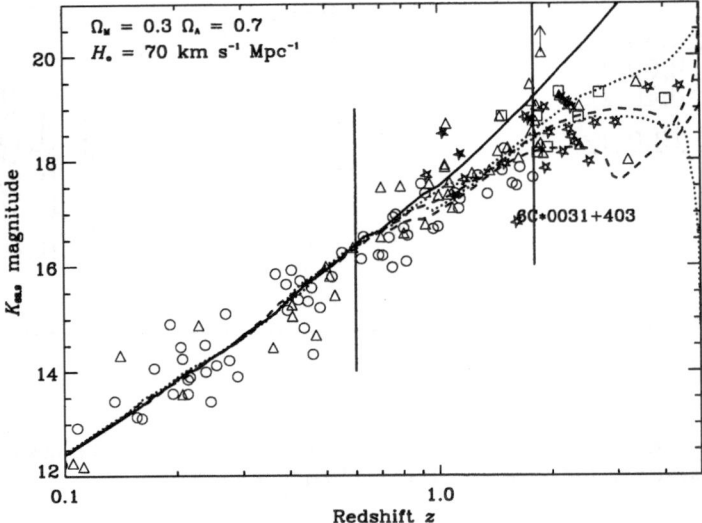

Fig. 3. The $K - z$ Hubble diagram for radio galaxies from the 3CRR (circles), 6CE (triangles), 6C* (stars) and 7C-III (squares) samples. $K_{63.9}$ denotes the K−band magnitude within a comoving metric aperture of 63.9 kpc (c.f. Eales et al. 1997; Jarvis et al. 2001b). The two vertical lines show the redshift above which the alignment effect begins to be seen ($z > 0.6$) and the higher redshift at which we chose to split the data ($z = 1.8$). The solid curved line is the predicted curve for galaxies which do not evolve (as described in the text). The dashed lines are models for a star-burst lasting 1 Gyr starting at $z = 5$ (lower) and $z = 10$ (upper). The two dotted curves represent the models of an instantaneous (0.1 Gyr) burst beginning at $z = 5$ (lower) and $z = 10$ (upper)

high-redshift 6C sources will be at shorter wavelengths than those of the 3C sources.

However, separate arguments lead us to conclude that the dominant factor is passive evolution of a stellar population which formed at $z \gtrsim 2.5$. First, recent sub-mm observations with SCUBA have shown that the dust masses in radio galaxies are larger at $z \simeq 3$ than in galaxies with similar radio luminosities at lower redshift (Archibald et al. 2001). This implies that the majority of star-formation activity in these galaxies is occurring at high redshift. Second, the discovery of six extremely red objects at $1 < z < 2$ in the 7C Redshift Survey (Willott, Rawlings & Blundell 2001) with inferred ages of a few Gyrs, implies that these objects formed the bulk of their stellar population at $z \simeq 5$. Third, detailed modelling of the optical spectrum of the weak radio source LBDS 53W091 at $z = 1.552$ has shown that this object is most plausibly an old elliptical, with an inferred age of $\gtrsim 3.5$ Gyr (Dunlop et al. 1996). The further discovery of LBDS 53W069 at $z = 1.43$, with an inferred age of ~ 4 Gyr (Dunlop 1999) suggests that there exists a population of evolved, radio weak ellipticals which formed at $z \gtrsim 5$. Therefore, the new data on the 6C* sample presented in here is consistent with the results from various other observational studies of radio galaxies in which these radio-luminous systems formed most of their stars at epochs corresponding to very high redshifts ($z \gtrsim 2.5$), and have undergone simple passive stellar evolution since. Willott et al. (2001) have pointed out that such galaxies probably undergo at least two active phases: one, at epochs corresponding to $z \gtrsim 5$, when the black hole and stellar spheroid formed, and another, at e.g. $z \sim 2$, when powerful jet activity is triggered, or perhaps re-triggered, by an event such as an interaction or a merger. The small scatter in the $K - z$ relation (Jarvis et al. 2001b) and sub-mm results (Archibald et al. 2001) suggest that the second active phase has little influence on the stellar mass of the final elliptical galaxy

We have shown that the powerful radio galaxies in our samples are consistent with having passively evolving stellar populations. If we now compare the masses of these powerful radio galaxies to the derived value of $M_K^\star$ for nearby elliptical galaxies [$M_K^\star = -24.3$ for $H_0 = 70\,\mathrm{km\,s^{-1}\,Mpc^{-1}}$ (Kochanek et al. 2000)], we find, if passive evolution is accounted for, that the powerful radio galaxies considered in this here are consistent with being $\approx 5L^\star$ throughout the redshift range $0 < z \lesssim 2.5$.

References

1. Archibald E.N., et al., 2001, MNRAS, 323, 417
2. Blundell K.M., et al., 1998, MNRAS, 295, 265
3. Bruzual G., Charlot S., 1993, ApJ, 405, 438
4. Dunlop J.S., 1999, in 'The most distant radio galaxies' KNAW Colloquium, Amsterdam, October 1997, eds Röttgering et al., Kluwer
5. Dunlop J.S., Peacock J.A., 1993, MNRAS, 263, 936
6. Dunlop J.S., Peacock J.A., Spinrad H., Dey A., Jimenez R., Stern D., Windhorst R.A., 1996, Nature, 381, 581
7. Eales S.A., Rawlings S., Law-Green D., Cotter G., Lacy M., 1997, MNRAS, 291, 593
8. Jarvis M.J., et al., 2001a, MNRAS, 326, 1563
9. Jarvis M.J., et al., 2001b, MNRAS, 326, 1585
10. Kochanek C.S., et al., 2000, ApJ, 543, 131
11. Lacy M., Bunker A.J., Ridgway S.E., 2000, AJ, 120, 68
12. Laing R.A., Riley J.M., Longair M.S., 1983, MNRAS, 204, 151
13. Leyshon G., Eales S.A., 1998, MNRAS, 295, 10
14. McCarthy P.J., 1993, ARAA, 31, 639
15. Rawlings S., Eales S.A., Lacy M., 2001, MNRAS, 322, 523
16. Rawlings S., et al., 1996, Nature, 383, 502
17. Rawlings S. & Saunders R., 1991, Nature, 349, 138
18. Simpson C., Rawlings S., Lacy M., 1999, MNRAS, 306, 828
19. van Breugel W.J.M., et al., 1998, ApJ, 502, 614
20. Willott C.J., 2000, to appear in Proc. "AGN in their Cosmic Environment", Eds. B. Rocca-Volmerange & H. Sol, EDPS Conf. Series (astro-ph/0007467)
21. Willott C.J., Rawlings S., Blundell K.M., 2001, MNRAS, 324, 1
22. Willott C.J., Rawlings S., Blundell K.M., Lacy M., 1999, MNRAS, 309, 1017

The Fundamental Plane of Radio Galaxies

Daniela Bettoni[1], Renato Falomo[1], Giovanni Fasano[1], Federica Govoni[2], Marilena Salvo[1], and Riccardo Scarpa[3]

[1] Osservatorio Astronomico di Padova
[2] Dipartimento di Astronomia, Universitá di Bologna and IRA, Bologna,
[3] ESO

1 Introduction

The global properties of early–type galaxies are fairly well described through a linear relation (the Fundamental Plane, FP [2],[3]) in the space defined by three observables: the effective radius r_e, the average surface brightness μ_e and the central velocity dispersion σ_c. The FP has been shown to be close to the plane defining the virial equilibrium condition, over the assumption of a rigorous homology among galaxies [4]. The systematic deviation of the FP from the one defined by the virial equilibrium, has been interpreted as due to changes of the Mass to Light ratio (M/L) (due to different stellar content or to differences in dark matter distribution) or to the breakdown of the homology assumption. In this context it is therefore important to determine whether the FP has universal validity among ellipticals. In particular it is not yet clear if active and non-active galaxies do follow the same FP, as it would be expected in view of the recent results on Black Holes (BH) demography in galaxies [5],[8] that suggest all, inactive and active, luminous ellipticals contain a massive BH.

Here we investigate the FP of low redshift radio galaxies (LzRG). Comparison of the optical properties [7] of radio galaxies with those of normal (non radio) ellipticals have shown that the probability to exhibit radio emission increases with the optical luminosity of the galaxies and also that the radio morphology (typically classes FR I and FR II) may depend on the absolute magnitude of the host galaxy. This could be related to the different BH mass (proportional to the galaxy luminosity) and to the jet power [6].

2 Results

To investigate the FP of LzRG we constructed a data set of 73 objects at $z < 0.2$, (22 new observations and 51 collected from the literature). New observations were obtained at ESO (1.5 Danish+DFOSC), while most of literature values were derived from the Hypercat database [9]. All values of μ_e (in the R filter), r_e, and σ_c have been processed in order to make them homogeneous. In addition we have estimated the ratio between the mass of the galaxy and that of the central BH using the relationships: $M_{Host}=5\sigma^2 r_e/G$ [1], and $M_{BH} = 1.48 \times 10^8 (\sigma/200)^{4.65}$ [8] respectively. The main conclusions of this study are:

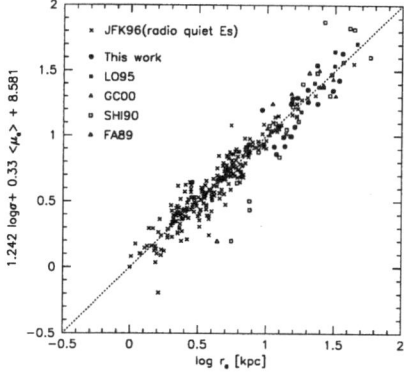 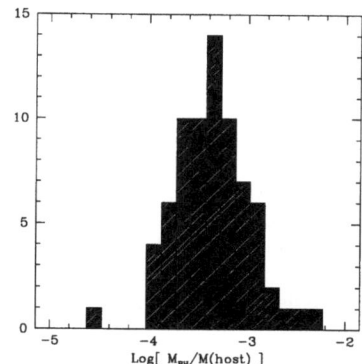

Fig. 1. *Left panel*: The Fundamental Plane of LzRG and normal ellipticals (log $r_e^{kpc} =$ 1.27log σ +0.326< μ_e >$_R$ - 8.56). *Right panel*: The distribution of Log(M_{BH}/M_{Host}) for Radio Galaxies in our sample

- The fundamental plane of radio galaxies is consistent with the one defined by normal galaxies (see Fig 1, left panel). This indicates that radio and non radio ellipticals have indistinguishable global properties, irrespective of their nuclear activity. The nuclear activity could be therefore triggered and fueled without a significant alteration of the status of equilibrium of galaxies.
- We confirm the dependence of M/L on σ, and thus on the total mass of the galaxy. This relation is partly due to a progressive reddening of the stellar population going from small to big galaxies, but most of dependence remains unexplained. It looks like that ellipticals pass from being baryon dominated to dark matter dominated with increasing luminosity, with a "fine tuning" required by the small and basically constant dispersion around the FP.
- We found that, the ratio between the mass of the BH (M_{BH}) and the mass of the host galaxy (M_{Host}) in our sample of LzRG ranges between 10^{-4} to 10^{-3} (average < M_{BH}/M_{Host} >= 6×10^{-4}) (see Fig 1, right panel).

References

1. R. Bender, D. Burstein, S.M. Faber: *ApJ*, **399** pp. 462–480 (1992)
2. S. Djorgovski, M. Davis: *ApJ*, **313** pp. 59–68 (1987)
3. A. Dressler et al.: *ApJ*, **313** pp. 42-56 (1987)
4. S.M. Faber, et al.: *ApJSS*, **69** pp. 763–808 (1989)
5. K. Gebhardt et al. *ApJL*, **539** pp. L13–L16 (2000)
6. G. Ghisellini, A. Celotti, *AA*, **379** pp. L1–L4 (2001)
7. F. Govoni, R. Falomo, R. Fasano, R. Scarpa, *AA*, **353** pp. 507–527 (2000)
8. D. Merrit, L. Ferrarese: astro-ph0101344 (2001)
9. Ph. Prugniel, G. Maubon, in *Dynamics of Galaxies: from the Early Universe to the Present*, ASP Conf. Ser. **197**, pp. 403–405 (2000)

The Hubble $K-z$ Diagram
of Radio and Near-IR Selected Galaxies

Carlos De Breuck[1], Wil van Breugel[2], Adam Stanford[2,3], Huub Röttgering[4], George Miley[4], and Daniel Stern[5]

[1] Institut d'Astrophysique de Paris, 98bis Boulevard Arago, 75014 Paris, France
[2] IGPP/LLNL, L-413, 7000 East Ave, Livermore, CA 94550, USA
[3] Physics Department, University of California, Davis, CA 95616, USA
[4] Sterrewacht Leiden, Postbus 9513, 2300 RA Leiden, The Netherlands
[5] Jet Propulsion Laboratory, Caltech, Mail stop 169-327, Pasadena, CA 91109, USA

Abstract. We give a short overview of the Hubble $K-z$ diagram of powerful radio galaxies, and compare it with the same diagram for $K-$band selected galaxies.

Ever since it was first constructed by Lilly & Longair (1984), the Hubble $K-z$ diagram of radio galaxies has remained one of the main points of interest in the study of the evolution of powerful radio galaxies. One of the main drivers of the $K-z$ diagram was its use as a tool to determine Cosmological parameters. Although the opposite effects of galaxy evolution have made the $K-z$ diagram almost useless for these kinds of studies, it can still provide some independent constraints on these parameters (e. g. Inskip et al. 2002).

One of the most remarkable characteristics of the $K-z$ diagram is that displays a tight correlation out to redshifts where the *observed* $K-$band samples the *rest-frame* $U-$band ($z = 5.19$; van Breugel et al. 1999). Using data from the 3CR and 6C surveys, Eales et al. (1997) reported an increase in the scatter at higher redshifts ($z > 2$). However, this is likely due to limited S/N with 4m-class telescopes, as van Breugel et al. (1998) detect a remarkably tight correlation out to $z = 4.4$ using near-IR imaging data from the Keck telescope.

Part of the scatter could be caused by the contribution of emission lines such as Hα, [OIII] $\lambda5007$Å, or [OII] $\lambda3727$Å to the observed $K-$band. However, the increasing number of radio galaxies with $K-$band spectroscopy, and a calculation using the radio-power versus emission-line luminosity (Jarvis et al. 2001) shows that this contribution is $<50\%$, even for these bright lines.

The only parameter known to affect the $K-z$ diagram is the radio power. Eales et al. (1997) found that the host galaxies of 6C radio galaxies are on average ~ 0.6 magnitudes fainter than those of the 3CR, which is most likely due to the difference of about an order of magnitude in radio power.

To examine this further, we have compared this radio-selected $K-z$ diagram with $K-$band selected 'field' galaxies. Such a comparison is now possible also at high redshifts with the redshift determinations in the Hubble deep fields. We have used the KPNO/IRIM $K-$band images of the HDF-N (Dickinson et al. 2002), which are approximately total magnitudes. For the lower redshift galaxies, we have used data from the Hawaii sample (Cowie et al. 1994, Songaila et al. 1994).

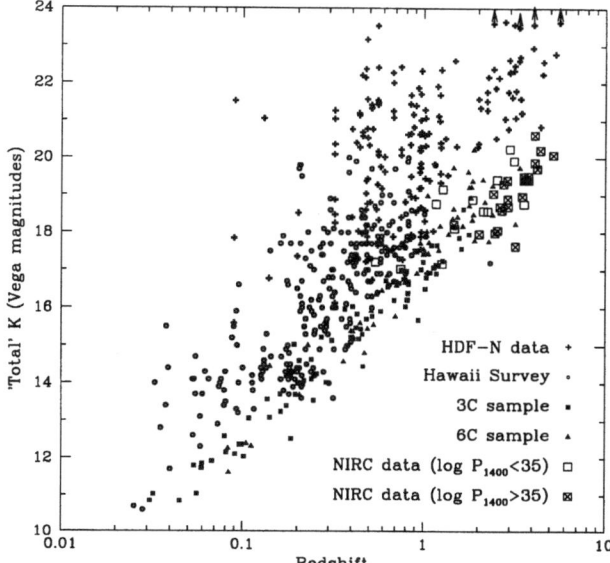

Fig. 1. Composite Hubble $K-z$ diagram. The open squares and crosses represent near-IR selected galaxies from the Hawaii survey and the HDF, respectively

The resulting composite $K-z$ diagram (Fig. 1) clearly shows that the radio-selected galaxies trace the bright envelope of the near-IR selected galaxies out to $z \sim 1$, while at higher redshifts, the radio galaxies are more than 2 magnitudes brighter than the field galaxies.

Fig. 1 shows that powerful radio sources trace the most luminous (and most massive; see e. g. Rocca-Volmerange, these proceedings) galaxies at each given redshift. Moreover, the tight correlation between redshift and $K-$band magnitude can be used as a redshift estimator. We have successfully used this technique to find the most distant radio galaxies (De Breuck et al. 2002).

References

1. Cowie, L., et al. 1994a, ApJ, 432, L83
2. De Breuck, et al. 2002, AJ, in press, astro-ph/0109540
3. Dickinson, et al. 2002, in preparation
4. Eales, S., et al. 1997, MNRAS, 291, 593
5. Inskip, K, Best, P., Longair, M, & MacKay, D. 2002, MNRAS, 329, 277
6. Jarvis, M., et al. 2001, MNRAS, 326, 1585
7. Lilly, S., & Longair, M. 1984, MNRAS, 211, 833
8. Songaila, A., Cowie, L., Hu, E., & Gardner, J. 1994, ApJSup, 94, 461
9. van Breugel et al. 1998, ApJ, 502, 614
10. van Breugel et al. 1999, ApJL, 518, 61

A Radio-Quiet Radio Galaxy at High Redshift

Daniel Stern[1] and Edward C. Moran[2]

[1] Jet Propulsion Laboratory/Caltech, Pasadena, CA 91109, USA
[2] Dept. of Astronomy, Wesleyan University, Middleton, CT 06459, USA

Abstract. We report on observations of CXO52, a Type 2 quasar at redshift $z =$ 3.288 identified as a hard X-ray source in a deep observation with the *Chandra X-ray Observatory*. Optical and near-infrared Keck spectroscopy of CXO52 show high-ionization, relatively-narrow emission lines with velocity widths ~ 1000 km s^{-1} and flux ratios similar to a Seyfert 2 galaxy or radio galaxy. The latter are the only class of high-redshift, Type 2 luminous AGN which have been extensively studied to date. Unlike radio galaxies, however, CXO52 is radio quiet, remaining undetected in deep radio observations, $f_{4.8\text{GHz}} < 40\mu$Jy. This discovery marks amongst the first high-redshift, radio-quiet Type 2 quasars identified to date. Such systems are expected from unification models of active galaxies and long-thought necessary to explain the X-ray background.

Surveys of bright galaxies have characterized the local demographics of AGN: the local ratio of obscured (Type 2) to unobscured (Type 1) Seyferts is approximately 4:1. Higher-luminosity Type I quasars are well-characterized, with the luminosity function of these broad-lined sources measured out to $z \sim 5$ and individual examples known out to redshift $z = 6.28$ (Fan et al. 2001). Our knowledge of Type 2, or narrow-lined quasars, however, is observationally sparse with many fewer examples known. Type 2 quasars are not identified at high redshift from shallow large-area sky surveys. Hard ($2-10$ kev) X-ray surveys have identified several examples (e.g., Dawson et al. 2001, Norman et al. 2002). Importantly, but surprisingly neglected in much of the X-ray Type 2 quasar literature, radio surveys have been detecting the radio-loud end of the Type 2 quasar population for several decades: these sources are called radio galaxies and they have been identified beyond redshift $z = 5$.

CXO52 was identified in our 185 ks *Chandra* observation of the Lynx field (Stern et al. 2002a). The Lynx field, which contains three known X-ray emitting clusters out to redshift $z = 1.27$, is one of four fields which comprise the $BRIzJK_s$ SPICES survey. CXO52 was identified independently both as an optical, color-selected, high-redshift source and from optical follow-up of X-ray sources in the Lynx field. CXO52 is amongst the hardest X-ray sources detected in the Lynx field. Keck spectra of CXO52 are presented in Fig. 1.

With high-ionization emission lines with widths $\gtrsim 1000$ km s^{-1} and a spatially-resolved morphology, CXO52 has very similar properties to high-redshift radio galaxies (HzRGs; e.g., McCarthy 1993). As opposed to quasars which have broad permitted lines and optical morphologies dominated by an unresolved nucleus, HzRGs have narrow (FWHM < 2000 km s^{-1}) permitted lines and host galaxies

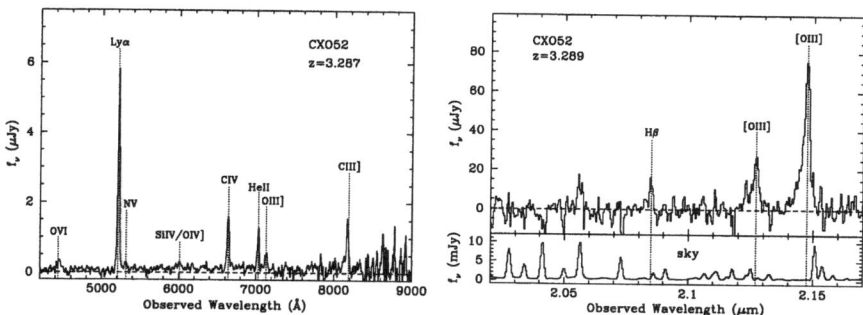

Fig. 1. Optical and near-infrared spectra of CXO52 obtained with the Keck telescopes

which are spatially extended. Equivalent widths of forbidden lines are larger in HzRGs than in quasars.

HeIIλ1640 is strongly detected in CXO52 with CIVλ1549/HeIIλ1640 ≈ 2. This is typical of HzRGs and atypical of Type I quasars: composite HzRGs spectra show CIVλ1549/HeIIλ1640 ≃ 1.5 (McCarthy 1993), while quasar spectra show CIVλ1549/HeIIλ1640 ≈ 10 − 50 (Vanden Berk et al. 2001). Assuming an [OII]λ3727/[OIII]λ5007 ratio similar to HzRGs, the [OII]λ3727 emission line luminosity of CXO52 is ≈ 5 × 10⁴² ergs s⁻¹, typical of bright 3CR HzRGs (McCarthy 1993). We reiterate that CXO52 remains *undetected* in our radio images which, at 4.8 GHz, probe *four* orders of magnitude deeper than classical HzRG surveys.

Besides (1) X-ray selection, (2) radio selection, and (3) color-selection, high-redshift, Type 2 quasars are also identifiable from (4) narrow-band imaging surveys and (5) mid-IR surveys. This population of misdirected or obscured quasars, long thought necessary to explain the X-ray background and demanded by unified models of AGN, are finally being identified in both their radio-quiet and radio-loud flavors. They are likely more populous than the well-studied unobscured, Type I quasar population.

These results are more extensively discussed in Stern et al. (2002b).

References

1. S. Dawson, D. Stern, et al.: AJ **122**, 598 (2001)
2. X. Fan, et al.: AJ **121**, 54 (2001)
3. P.J. McCarthy: ARAA **31**, 639 (1993)
4. C. Norman, et al.: ApJ, submitted (astro-ph/0103198; 2002)
5. D. Stern, P. Tozzi, et al.: AJ, submitted (2002a)
6. D. Stern, E.C. Moran, et al.: ApJ **568**, in press (astro-ph/0111513; March 2002b)
7. D.E. VandenBerk, et al.: AJ **122**, 549 (2001)

The Universal Equilibrium of CDM Halos: Making Tracks on the Cosmic Virial Plane

Ilian T. Iliev[1] and Paul R. Shapiro[2]

[1] Osservatorio Astrofisico di Arcetri, Largo Enrico Fermi 5, 50125 Firenze, Italy
[2] Department of Astronomy, University of Texas, Austin, 78712, USA

Abstract. Dark-matter halos are the scaffolding around which galaxies and clusters are built. They form when the gravitational instability of primordial density fluctuations causes regions which are denser than average to slow their cosmic expansion, recollapse, and virialize. Objects as different in size and mass as dwarf spheroidal galaxies and galaxy clusters are predicted by the CDM model to have halos with a universal, self-similar equilibrium structure whose parameters are determined by the halo's total mass and collapse redshift. These latter two are statistically correlated, however, since halos of the same mass form on average at the same epoch, with small-mass objects forming first and then merging hierarchically. The structural properties of dark-matter dominated halos of different masses, therefore, should reflect this statistical correlation, an imprint of the statistical properties of the primordial density fluctuations which formed them. Current data reveal these correlations, providing a fundamental test of the CDM model which probes the shape of the power spectrum of primordial density fluctuations and the cosmological background parameters.

1 The Truncated Isothermal Sphere (TIS) Model

We have developed an analytical model for the postcollapse equilibrium structure of virialized objects which condense out of a cosmological background universe, either matter-dominated or flat with a cosmological constant [23,8]. The model is based upon the assumption that cosmological halos form from the collapse and virialization of "top-hat" density perturbations and are spherical, isotropic, and isothermal. This leads to a unique, nonsingular TIS, a particular solution of the Lane-Emden equation (suitably modified when $\Lambda \neq 0$). The size r_t and velocity dispersion σ_V are unique functions of the mass and redshift of formation of the object for a given background universe. Our TIS density profile flattens to a constant central value, ρ_0, which is roughly proportional to the critical density of the universe at the epoch of collapse, with a small core radius $r_0 \approx r_t/30$ (where $\sigma_V^2 = 4\pi G \rho_0 r_0^2$ and $r_0 \equiv r_{King}/3$, for the "King radius" r_{King}, defined by [1], p. 228). The density profiles for gas and dark matter are assumed to be the same (no bias), with gas temperature $T = \mu m_p \sigma_V^2 / k_B$.

These TIS results differ from those of the more familiar approximations in which the virialized sphere resulting from a top-hat perturbation is assumed to be either the standard uniform sphere (SUS) or else a singular isothermal sphere (SIS). We summarize their comparison in Fig. 1 and Table 1, where η is the final radius of the virialized sphere in units of the top-hat radius r_m at maximum expansion (i.e. $\eta_{SUS} = 0.5$), $\rho_t \equiv \rho(r_t)$, $\langle \rho \rangle$ is the average density

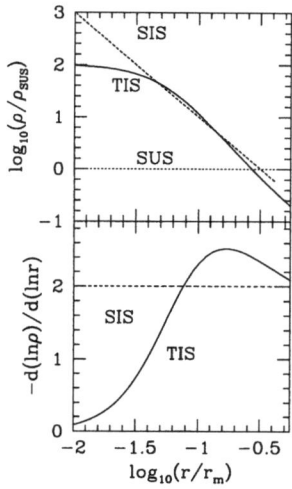

Table 1

	SUS	SIS	TIS[a]
η/η_{SUS}	1	0.833	1.11;1.07
T/T_{SUS}	1	3	2.16;2.19
ρ_0/ρ_t	1	∞	514;530
$\langle\rho\rangle/\rho_t$	1	3	3.73;3.68
r_t/r_0	– NA –	∞	29.4;30.04
$\Delta_c/\Delta_{c,SUS}$	1	1.728	0.735;0.774
$K/|W|$	0.5	0.75	0.683;0.690

[a] The two values refer to flat universe with $\Lambda = 0$ (left value) and $\Omega_0 = 0.3$, $\lambda_0 = 0.7$ (right value).

Fig. 1. (top) Density profile of TIS in a matter-dominated universe. Radius r is in units of r_m - the top-hat radius at maximum expansion. Density ρ is in terms of the density ρ_{SUS} of the SUS approximation for the virialized, post-collapse top-hat. (bottom) Logarithmic slope of density profile

of the virialized spheres, $\Delta_c = \langle\rho\rangle/\rho_{\text{crit}}(t_{\text{coll}})$, and $K/|W|$ is the ratio of total kinetic (i.e. thermal) to gravitational potential energy of the spheres.

2   TIS Model vs. Numerical CDM Simulations

The TIS model reproduces many of the average properties of the halos in numerical CDM simulations quite well, suggesting that it is a useful approximation for the halos which result from more realistic initial conditions:

(1) The TIS mass profile agrees well with the fit to N-body simulations by [18] ("NFW") (i.e. fractional deviation of $\sim 20\%$ or less) at all radii outside of a few TIS core radii (i.e. outside King radius or so), for NFW concentration parameters $4 \leq c_{\text{NFW}} \leq 7$ (Fig. 2). The flat density core of the TIS halo differs from the singular cusp of the NFW profile at small radii, but this involves only a small fraction of the halo mass, thus not affecting their good agreement outside the core. As a result, the TIS central density ρ_0 can be used to characterize the core density of cosmological halos, even if the latter have singular profiles like that of NFW, as long as we interpret ρ_0, in that case, as an average over the innermost region.

(2) The TIS halo model predicts the internal structure of X-ray clusters found by gas-dynamical/N-body simulations of cluster formation in the CDM model. Our TIS model predictions, for example, agree astonishingly well with the mass-temperature and radius-temperature virial relations and integrated mass profiles derived empirically from the simulations of cluster formation by [4,16]

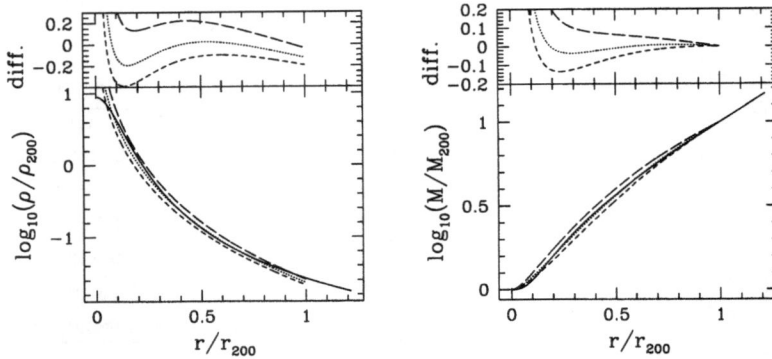

Fig. 2. Profiles of density (left) and integrated mass (right), for TIS (solid) and NFW with $c_{NFW} = 4$ (short-dashed), 5 (dotted) and 7 (long-dashed) with same (r_{200}, M_{200})

(EMN). Apparently, these simulation results are not sensitive to the discrepancy between our prediction of a small, finite density core and the N-body predictions of a density cusp for clusters in CDM. Let X be the average overdensity inside radius r (in units of the cosmic mean density) $X \equiv \langle \rho(r) \rangle / \rho_b$. The radius-temperature virial relation is defined as $r_X \equiv r_{10}(X)(T/10\,\text{keV})^{1/2}$ Mpc, and the mass-temperature virial relation by $M_X \equiv M_{10}(X)(T/10\,\text{keV})^{1/2} h^{-1} 10^{15} M_\odot$. A comparison between our predictions of $r_{10}(X)$ and the results of EMN is given in Fig. 3. EMN obtain $M_{10}(500) = 1.11 \pm 0.16$ and $M_{10}(200) = 1.45$, while our TIS solution yields $M_{10}(500) = 1.11$ and $M_{10}(200) = 1.55$.

(3) The TIS halo model also successfully reproduces the mass - velocity dispersion relation for clusters in CDM N-body simulations and its dependence on redshift for different background cosmologies. N-body simulation of the Hubble volume $[(1000\,\text{Mpc})^3]$ by the Virgo Consortium (reported by Evrard at the U. of Victoria meeting, August 2000) yields the following empirical relation:

$$\sigma_V = (1080 \pm 65) \left[h(z) M_{200}/10^{15} M_\odot \right]^{0.33} \text{km/s}, \tag{1}$$

where M_{200} is the mass within a sphere with average density 200 times the cosmic mean density, and $h(z) = h_0 \sqrt{\Omega_0 (1+z)^3 + \lambda_0}$ is the redshift-dependent Hubble constant. Our TIS model predicts:

$$\sigma_V = 1103 \left[h(z) M_{200}/10^{15} M_\odot \right]^{1/3} \text{km/s}. \tag{2}$$

(4) The TIS model successfully predicts the average virial ratio, $K/|W|$, of halos in CDM simulations. An equivalent TIS quantity, $GM_{200}/(r_{200}\sigma_V^2) = 2.176$, is plotted for dwarf galaxy minihalos at $z = 9$ in Fig. 3(b), from [10], showing good agreement between TIS and N-body halo results. A similar plot, but of $K/|W|$ for such halos, was shown by [10] based upon N-body simulations, in which the average $K/|W|$ is close to 0.7, as predicted by the TIS model (Table 1). Those authors were apparently unaware of this TIS prediction since they

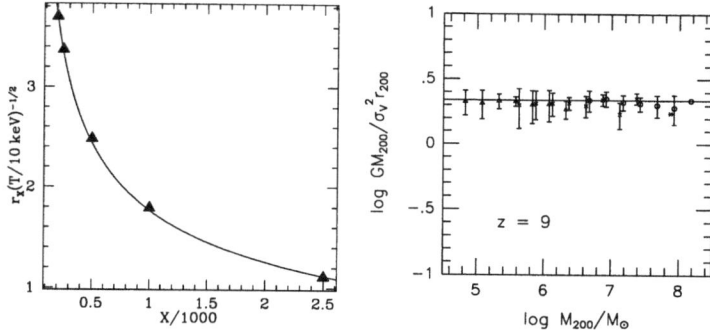

Fig. 3. (a) (left) (triangles) Cluster radius-temperature virial relation for CDM simulation results (at $z = 0$) as fit by EMN and as predicted by TIS (solid curve). (b) $GM_{200}/(\sigma_V^2 r_{200})$ vs. mass for halos from N-body simulations [with 1σ error bars]. Horizontal line is analytical prediction of TIS model

compared their results with the SUS value of $K/|W|$, 0.5, and interpreted the discrepancy incorrectly as an indication that their halos were not in equilibrium.

3   TIS Model vs. Observed Halos

(1) The TIS profile matches the observed mass profiles of dark-matter-dominated dwarf galaxies [7]. The observed rotation curves of dwarf galaxies are well fit by a density profile with a finite density core given by

$$\rho(r) = \frac{\rho_{0,B}}{(r/r_c + 1)(r^2/r_c^2 + 1)} \tag{3}$$

[2]. The TIS model gives a nearly perfect fit to this profile (Fig. 4(a)), with best fit parameters $\rho_{0,B}/\rho_{0,TIS} = 1.216$, $r_c/r_{0,TIS} = 3.134$. This best-fit TIS profile correctly predicts $v_{\max}$, the maximum rotation velocity, and the radius, $r_{\max}$, at which it occurs in the Burkert profile: $r_{\max,B}/r_{\max,TIS} = 1.13$, and $v_{\max,B}/v_{\max,TIS} = 1.01$.

(2) The TIS halo model can explain the mass profile with a flat density core measured by [24] for cluster CL 0024+1654 at $z = 0.39$, using the strong gravitational lensing of background galaxies by the cluster to infer the cluster mass distribution [21]. The TIS model not only provides a good fit to the projected surface mass density distribution of this cluster within the arcs (Fig. 4b), but also predicts the overall mass, and a cluster velocity dispersion in close agreement with the value $\sigma_v = 1150$ km/s measured by [3].

4   Making Tracks on the Cosmic Virial Plane

The TIS model yields $(\rho_0, \sigma_V, r_t, r_0)$ uniquely as functions of (M, z_{coll}). This defines a "cosmic virial plane" in (ρ_0, r_0, σ_V)-space and determines halo size,

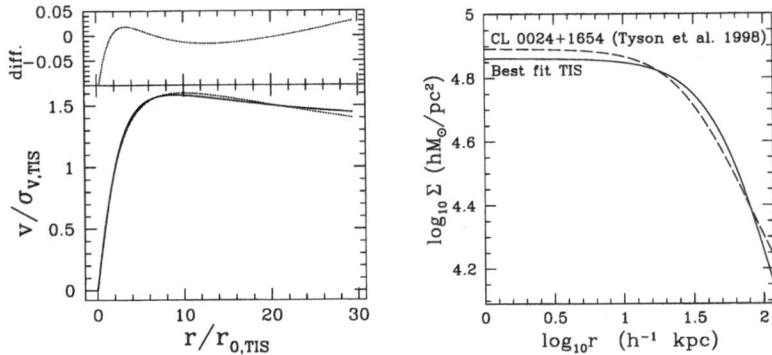

Fig. 4. (a) (left) Rotation Curve Fit. Solid line = Best fit TIS; Dashed line = Burkert profile (b) (right) Projected surface density of cluster CL 0024+1654 inferred from lensing measurements, together with the best-fit TIS model

mass, and collapse redshift for each point on the plane. In hierarchical clustering models like CDM, M is statistically correlated with z_{coll}. This determines the distribution of points on the cosmic virial plane. We can combine the TIS model with the Press-Schechter (PS) approximation for $z_{coll}(M)$ – typical collapse epoch for halo of mass M – to predict correlations of observed halo properties.

(1) The combined (TIS+PS) model explains the observed $v_{max} - r_{max}$ correlation of dwarf spiral and LSB galaxies, with preference for the currently favoured ΛCDM model [7] (Fig. 5(a)).

(2) The TIS+PS model also predicts the correlations of central mass and phase-space densities, ρ_0 and $Q \equiv \rho_0/\sigma_V^3$, of dark matter halos with their velocity dispersions σ_V, with data for low-redshift dwarf spheroidals to X-ray clusters again most consistent with ΛCDM [22] (Fig. 5 (b,c)). There have been recent claims that ρ_0 =const for all cosmological halos, independent of their mass, as expected for certain types of SIDM [5,11]. This claim, however does not seem to be supported by most current data (Fig. 5 (c)).

This work was supported by European Community RTN contract HPRN-CT2000-00126 RG29185, grants NASA NAG5-10825 and TARP 3658-0624-1999.

References

1. Binney, J., & Tremaine, S.: *Galactic Dynamics* (Princeton University Press, Princeton 1987)
2. Burkert, A.: ApJ, **447**, L25 (1995)
3. Dressler, A., & Smail, I., Poggianti, B.M., Butcher, H., Couch, W.J., Ellis, R.S., Oemler, A., Jr.: APJS, **122**, 51 (1999)
4. Evrard, A.E., Metzler, C.A., & Navarro, J.F.: ApJ, **469**, 494 (1996)
5. Firmani, C., D'Onghia, E., Chincarini, G., Hernandes, X., & Avila-Reese, V.: MNRAS, **321**, 713 (2000)

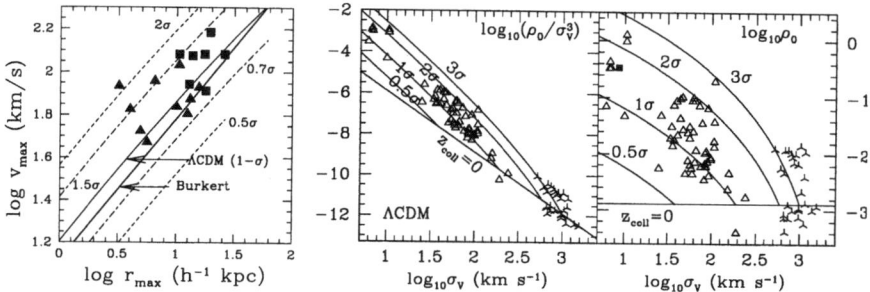

Fig. 5. Correlations predicted by (TIS+PS) model for ΛCDM [COBE normalized, $\Omega_0 = 1 - \lambda_0 = 0.3, h = 0.65$; no tilt), for halos formed from $\nu - \sigma$ fluctuations [i.e. $\nu \equiv \delta_{\rm crit}(z)/\sigma(M, z)$, $\sigma(M, z)$ = standard deviation of linear density fluctuations filtered on mass scale M; typical $M(z_{\rm coll}) = M(\nu = 1)$], as labelled with ν-values. All curves are (TIS+PS) results, except curve in (a) labelled "Burkert" is a fit to data [2]. (a) (left) $v_{\rm max}$–$r_{\rm max}$ correlation. Observed dwarf galaxies (triangles) and LSB galaxies (squares) from [14]; (b) (middle) Q–σ_V correlation. Line representing halos of different mass which collapse at the same redshift is shown for the case $z_{\rm coll} = 0$. Data points for galaxies and clusters are from the following: (1) 49 late-type spirals of type Sc-Im and 7 dSph galaxies from [12,13] (open triangles); (2) Local Group dSph Leo I from [15] (filled square); (3) 28 nearby clusters, σ_V from [6] and [9], and ρ_0 from [17] (crosses). (c) (right) Same as (b), except for ρ_0 vs. σ_V

6. Girardi, M., Giuricin, G., Mardirossian, F., Mezzetti, M., Boschin, W.: ApJ, **505**, 74 (1998)
7. Iliev, I.T. & Shapiro, P.R.: ApJ, **546**, L5 (2001a)
8. Iliev, I.T., & Shapiro, P.R.: MNRAS, **325**, 468 (2001b) (Paper II)
9. Jones, C. & Forman, W.: ApJ, **511**, 65 (1999)
10. Jang-Condell, H., & Hernquist: ApJ, 548, 68 (2001)
11. Kaplinghat, M., Knox, L., & Turner, M.S.: Phys.Rev.Lett., **85**, 3335 (2000)
12. Kormendy, J. & Freeman, K.C.: 'Scaling laws for dark matter halos in late-type and dwarf spheroidal galaxies'. In *Ringberg Proceedings 1996 Workshop*, eds. R. Bender, T. Buchert, P. Schneider, & F. von Feilitzsch (MPI, Munich 1996), pp.13-15
13. Kormendy, J. & Freeman K.C. 2001, in preparation
14. Kravtsov, A.V., Klypin, A.A., Bullock, J.S., & Primack, J.R.: ApJ, **502**, 48 (1998)
15. Mateo, M., Olszewski, E.W., Vogt, S.S., & Keane, M.J.: ApJ, **116**, 2315 (1998)
16. Mathiesen, B.F., & Evrard A.E.: ApJ, **546**, 100 (2001)
17. Mohr, J.J., Mathiesen, B.F., & Evrard A.E.: ApJ, **517**, 627 (1999)
18. Navarro, J., Frenk, C. S., White, S. D. M.: ApJ, **462**, 563 (1996)
19. Press, W.H., & Schechter, P.: ApJ, **187**, 425 (1974)
20. Shapiro P.R.: 'Cosmological Reionization'. In: Proceedings of the 20th Texas Symposium on Relativistic Astrophysics and Cosmology, eds. H. Martel and J. C. Wheeler, (AIP Conference Series, v.586, 2001), pp. 219-232
21. Shapiro, P.R., & Iliev, I.T.: ApJ, **542**, L1 (2001)
22. Shapiro, P.R., & Iliev, I.T.: ApJL, accepted (2002) (astro-ph/0107442)
23. Shapiro, P.R., Iliev, I.T., & Raga, A.C.: MNRAS, **307**, 203 (1999) (Paper I)
24. Tyson J.A., Kochanski, G.P., and Dell'Antonio I.P.: ApJ, **498**, L107 (1998)

Resolving the Spin Crisis: Mergers and Feedback

Avishai Dekel and Ariyeh H. Maller

Racah Institute of Physics, The Hebrew University, Jerusalem 91904, Israel

Abstract. We model in simple terms the angular momentum (J) problem of galaxy formation in CDM, and identify the key elements of a scenario that can solve it. The buildup of J is modeled via dynamical friction and tidal stripping in mergers. This reveals how over-cooling in incoming halos leads to transfer of J from baryons to dark matter (DM), in conflict with observations. By incorporating a simple recipe of supernova feedback, we match the observed J distribution in disks. Gas removal from small incoming halos, which make the low-J component of the product, eliminates the low-J baryons. Partial heating and puffing-up of the gas in larger incoming halos, combined with tidal stripping, reduces the J loss of baryons. This implies a higher baryonic spin for lower mass halos. The observed low baryonic fraction in dwarf galaxies is used to calibrate the characteristic velocity associated with supernova feedback, yielding $V_{fb} \sim 100\,\mathrm{km\,s^{-1}}$, within the range of theoretical expectations. The model then reproduces the observed distribution of spin parameter among dwarf and bright galaxies, as well as the J distribution inside these galaxies. This suggests that the model captures the main features of a full scenario for resolving the spin crisis.

1 Introduction

The 'standard' model of cosmology, CDM, which assumes hierarchical buildup of structure, is facing difficulties in explaining observed properties of galaxies, such as the number density of dwarfs and the inner density profile of halos. Standing out is the angular-momentum problem, that is the apparent failure of the theory, via simulations, to reproduce the large sizes of disk galaxies and their structure. We make progress by first reproducing the problem via a simple model in which the important physical elements are spelled out, and then incorporating in this model the key process which may cure the problem – feedback.

The sizes of disks are commonly linked to the spins of their parent halos as measured in N-body simulations [10]. The assumptions that the baryons and DM share the same distribution of specific angular momentum j and that the baryons conserve their j while contracting to a disk lead to disk sizes comparable to those observed. However, simulations that incorporate gas find that most of the baryonic j is transfered to the DM, resulting in disk sizes smaller by an order of magnitude [e.g. 14,15], and thus leading to a *spin catastrophe*.

In addition, there is a *mismatch of j profiles*. The j distribution (or profile) within simulated halos has been found to scatter about a universal shape, with an excess of low-j (and high-j) material compared to the exponential disks observed [1, BD]. This mismatch has been demonstrated in an observed sample of

14 dwarf galaxies [18, BBS], which serves as the target for our modeling effort. BBS used for each halo the measured rotation curve and an assumed NFW profile to determine the halo virial quantities, with an average $\langle V_{\mathrm{vir}} \rangle \simeq 60 \, \mathrm{km\,s^{-1}}$. They then determined the baryonic spin parameter, averaging to $\langle \lambda'_{\mathrm{b}} \rangle \sim 0.07$, significantly larger than the $\langle \lambda'_{\mathrm{dm}} \rangle \sim 0.035$ of simulated halos, and then demonstrated the j-profile mismatch case by case. BBS also estimated the ratio of disk to DM mass to be $\langle f_{\mathrm{d}} \rangle \sim 0.04$, about a factor of 3 smaller than the universal fraction, indicating significant gas loss.

The spin catastrophe is commonly being associated with "over-cooling", that the gas rapidly cools and becomes tightly bound in small halos. When such a halo spirals into a bigger halo, the baryonic component survives intact all the way to the center and thus transfers all its orbital j to the DM. It has therefore been speculated that energy feedback from supernova may remedy the problem by balancing the early cooling [e.g. 9]. A key idea is that the spin segregation between baryons and DM can go either way. While gas cooling tends to lower the baryonic spin, heating due to feedback would reduce this effect, and gas removal from small halos would even lead to higher baryonic spin. However, a realistic implementation of feedback has proved challenging [e.g. 17]. The feedback process has not yet been studied or implemented in satisfactory detail. We do not know yet whether they can indeed solve the CDM problems, and how. This motivates our attempt to first understand how the feedback scenario may work using a very simple semi-analytic model. Knowing that in a hierarchical scenario the halo formation can be largely interpreted as a sequence of mergers, our model is based on a simple algorithm for the buildup of halo spin by adding up the orbital angular momenta of merging satellites [13, MDS; 19]. It matches well the spin distribution among halos in N-body simulations as well as the j profile within halos. This makes it a useful tool for understanding the over-cooling origin of the spin problem and for the attempt to cure it via feedback effects. Our work is described in more detail in [12, MD].

2   Buildup of Halo Spin by Mergers

We characterize the angular momentum J of a galaxy by the modified spin parameter bull:01 $\lambda' = (J/M)/(\sqrt{2} V_{\mathrm{vir}} R_{\mathrm{vir}})$. This quantity, which equals the standard λ for an isothermal sphere and for an NFW profile with concentration $c \sim 5$, is straightforward to compute separately for the baryons and for the DM, λ'_{b} and λ'_{dm}. The distribution of $\ln \lambda'$ in the simulations is normal, with an average corresponding to $\lambda'_0 \simeq 0.035$ (compared to $\lambda_0 \simeq 0.042$) and a standard deviation $\sigma_{\lambda'} \simeq 0.5$. The "orbital-merger" model of MDS reproduces this spin distribution. To materialize this model we generate many random realizations of merger histories based on the Extended Press Schechter formalism [11] with slight adjustments, and for each merger tree we create random realizations of the orbital $\mathbf{J}$ added in each merger. The encounter parameters are taken to mimic typical mergers, with the directions of the orbits drawn at random (or fine-tuned for a slight correlation between successive mergers as seen in simulations [5,16].

The resultant distribution of halo spins matches the log-normal distribution obtained in the simulations.

The cumulative mass distribution of j within simulated halos is fit by the universal function $M(<j) = M_v \mu j/(j_0 + j)$, with $\mu > 1$ and $j \leq j_{max} = j_0/(\mu - 1)$ (BD). This is a simple power law, $M(<j) \propto j$, for at least half the mass, with a possible bend characterized by μ. The other parameter, j_0 or j_{max}, can be replaced by λ'. The distribution of μ is Gaussian in $\ln(\mu - 1)$, with a mean -0.6. The model also recovers these simulated j profiles. We create an $M(<j)$ profile for each of the EPS model realizations by dividing the mass growth of the halo into bins and assigning to each the corresponding orbital J. A sample of profiles produced by this procedure and the distribution of μ values are shown in Figs. 1 and 2 of MD, demonstrating the match with the simulation results of BD. The model also reproduces the insensitivity to halo mass and redshift.

The successes of the simple model in recovering both the distribution of spins and the j profiles makes it a useful tool for studying the j buildup. A new feature revealed by the model, which provides an interesting clue, is that the final halo spin is predominantly determined by the last major merger, while the many smaller satellites come in at different directions and therefore tend to sum up to a low j. If small satellites would lose gas before they merge into the halo, then much of the galactic gas would originate in big satellites, the final gas would lack the low-j component, and the baryonic spin would end up higher than the DM.

3   Reproducing the Baryonic Spin Loss

We can understand the j loss of baryons via a simple model including gas cooling, dynamical friction and tidal stripping for how the orbital j is converted into halo spin. First, the dynamical friction exerted by the halo on the satellite brings the satellite towards the halo center and thus transfers j from the orbit to the halo. Second, once satellite particles are tidally stripped they retain their j at the stripping point and add it directly to the halo.

The mass loss at halo radius r can be estimated by evaluating the tidal radius ℓ_t of the satellite at r via the resonance condition, $m(\ell_t)/\ell_t^3 = M(r)/r^3$, where $m(\ell)$ and $M(r)$ are the mass profiles of the satellite and halo. If these two are self-similar, then this implies $\ell_t/\ell_{vir} = r/R_{vir}$, and the bound mass of the satellite is $m[\ell_t(r)] \propto M(r)$. A more accurate recipe for tidal stripping, tested with merger simulations, reveals that this is a good approximation in general [6], so we adopt it in our model.

When exploring the effect of cooling, one assumes that initially the baryons follow the DM. As the gas cools, it contracts to a more compact configuration of radius $R_b < R_{dm}$. This spatial segregation in the satellite implies that the j-rich mass stripped at the early stages of the merger in the outer halo is dominated by DM, while the more compact baryons penetrate into the inner halo and lose more of their j via dynamical friction. The result is a net spin transfer from the baryons to the DM. Using the stripping recipe $m(r) \propto M(r)$, we obtain for the final baryonic spin $J_b/J_{dm} = (R_b/R_{dm})$. In the case of maximum cooling, the baryons dominate the halo center, $R_b = f_b R_{dm}$, where $f_b \simeq 0.13$ is the universal baryon fraction. Fig. 1 shows the resultant baryonic spin distribution according to this model; there is a shift down to $\lambda'_0 = 0.005$, reproducing the spin crisis. The role of feedback would be to delay the cooling, increase R_b, and thus reduce the baryonic spin loss.

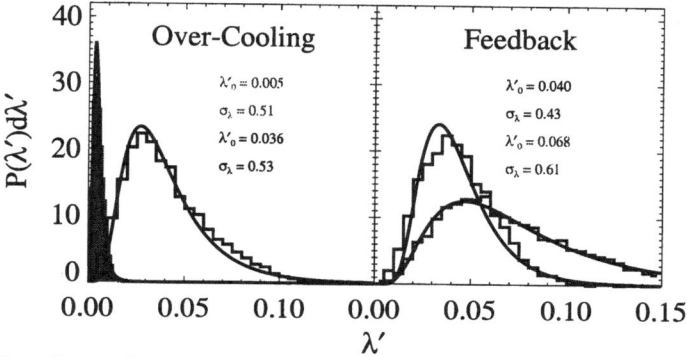

Fig. 1. The effects of over-cooling and feedback on the spin distribution of baryons compared to the DM, for dwarf and bright galaxies. Log-normal fits are shown, with the mean and scatter quoted. **Left:** λ'_b is shifted down by an order of magnitude compared to λ'_{dm}, reproducing the spin catastrophe. **Right:** λ'_b in bright galaxies is boosted up by heating and partial blowout in incoming halos and roughly matches λ'_{dm}, while in dwarf galaxies λ'_b is boosted up further by the blowout in small satellites

4   Feedback

Our approach here is to avoid the details of star formation and feedback and rather appeal to a very simple prescription for the effect of feedback as a function of the satellite's virial velocity, V_{sat}. Following the general idea of Dekel & Silk [8] we assume that the feedback by supernova-driven winds pumps energy into the gas and heats it uniformly to a temperature corresponding to a characteristic velocity V_{fb}, on the order of $100\,\mathrm{km\,s^{-1}}$. We therefore assume that the spatial extent of the baryons is determined by the ratio $\eta \equiv V_{fb}/V_{sat}$. The limit $\eta \ll 1$, of massive, deep potential wells, corresponds to maximum cooling, $R_b \ll R_{dm}$. In smaller halos, Still with $\eta \simeq 1$, we expect the heating to balance the cooling and yield $R_b \simeq R_{dm}$. Our model is therefore an interpolation between these limits, $R_b = \eta^{\gamma_1} R_{dm}$, with γ_1 an arbitrary exponent, which we set for now to be unity.

If V_{fb} is larger than V_{sat}, the feedback can cause gas blowout. We assume that partial blowout starts occurring once V_{fb} exceeds the escape velocity of the satellite, $\sim \sqrt{2}V_{sat}$, while total blowout is expected for $\eta \gg 1$. We therefore parameterize the amount of gas that remains in the halo by another interpolation, $f_d = (\eta/\sqrt{2})^{\gamma_2}$, with γ_2 an arbitrary exponent tentatively set to unity. We report here the results for the simplest choice $\gamma_1 = \gamma_2 = 1$, and explore the robustness of the results to different choices of γ_1 and γ_2 in MD.

In Fig. 1 we demonstrate the effects of this feedback scheme, with $V_{fb} = 95\,\mathrm{km\,s^{-1}}$, on the distribution of λ'_b. We do it for two kinds of final halos, with $V_{vir} = 60$ and $220\,\mathrm{km\,s^{-1}}$, representing *dwarf* and *bright* galaxies. The baryons in the bright galaxies end up with spins comparable to their DM halos, with $\lambda'_0 = 0.042$, while in dwarfs they have significantly higher spins, with $\lambda'_0 = 0.067$. We learn that λ'_b in dwarfs, which are build up by small satellites, is dominated by the blowout from these satellites, and it ends up with $\lambda'_b > \lambda'_{dm}$. For bigger galaxies, which are largely made of bigger satellites, the dominant effect is the heating, with some contribution from blowout, together leading to a λ'_b distribution similar to λ'_{dm}, in general agreement with observations.

In MD we explore a range of values for the exponents γ_1 and γ_2. For each choice we determine V_{fb} such that for the dwarfs $\langle f_d \rangle = 0.04$ as in BBS. We find that our results for dwarfs remain practically unchanged, while the results for bright galaxies have a weak dependence on the value of γ_1 in the range $(0.5, 3)$.

5   Model versus Observations

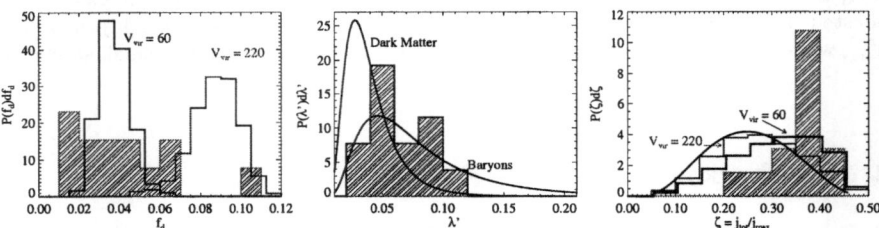

Fig. 2. Model realizations (with $V_{fb} = 95\,\mathrm{km\,s^{-1}}$) versus BBS observations of dwarf galaxies (shaded). Distribution of baryon fraction f_d, spin parameter λ', and j-profile shape parameter ζ. **Left:** The model prediction for $V_{vir} = 60\,\mathrm{km\,s^{-1}}$ dwarfs, with significant blowout, is in agreement with the BBS data, while the bright galaxies retain most of their baryons. **Middle:** The predicted λ'_b distribution is in agreement with the dwarf data. Shown for comparison is the simulation result for dark halos, which is similar to that of bright galaxies. **Right:** The predicted distribution of ζ for the baryons in dwarfs (heavy histogram) is shifted upwards compared to the DM (smooth curve) and the bright galaxies (light histogram), like the data, but its width is overestimated

The distribution of f_d for the dwarfs observed by BBS is displayed in Fig. 2, showing values significantly lower than the universal value of $f_b \simeq 0.13$ and thus consistent with baryonic blowout. Shown for comparison are the model predictions for dwarf and bright galaxies. We enforced a match of the means for the dwarfs at $\langle f_d \rangle = 0.04$ by choosing $V_{fb} = 95\,\mathrm{km\,s^{-1}}$, but the scatters are also in agreement. For bright galaxies, f_d is typically lower than the universal value by less than 50%, reflecting the limited fraction of small progenitors who lost their gas.

Next, we compare predicted and observed spin distributions for dwarfs. We convert each value of λ as quoted by BBS to λ', and show their distribution in Fig. 2. the observed spins are significantly higher than the λ' values of halos in cosmological simulations, with an average of $\lambda_0' \simeq 0.07$ compared to 0.035. Then shown is our model prediction with $V_{fb} = 95\,\mathrm{km\,s^{-1}}$ for the baryonic spin distribution in dwarfs. The effect of blowout brings the baryonic spins to a good agreement with the observed dwarfs.

Fig. 3 shows the average j profiles and the scatter about them for the observed BBS dwarfs and for the corresponding model realizations compared to the typical j profile in halos by BD. We construct the baryonic j profile in each of our model realizations following the same method used to produce DM j profiles but now including feedback effects. The BBS dwarfs show low baryonic fractions (indicated by the integral under the histogram) and significant deficits of j at the two ends of the distribution compared to the halos. The profile for model dwarf galaxies is similar to the observations except for the very lowest j bin which is a 2σ overestimate. The high-j tail tends to be reduced in the baryons because it is often the result of a small satellite that comes in with its orbital J aligned with the halo spin, and now has lost its gas.

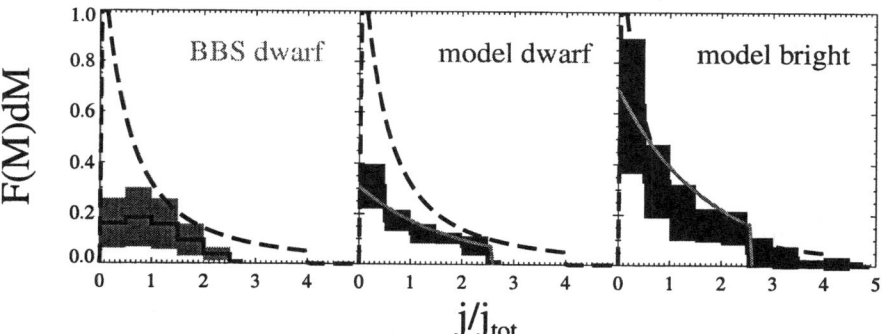

Fig. 3. Average j profiles (histogram) and the 1σ scatter (shaded) for the observed dwarfs in comparison with the model dwarf and bright galaxies ($V_{fb} = 95\,\mathrm{km\,s^{-1}}$). The integral under the histogram is f_d. Shown in comparison is the typical profile for DM halos by BD

Fig. 3 also shows the average model prediction for bright galaxies, $V_{vir} = 220\,\mathrm{km\,s^{-1}}$. They retain most of their baryons, so their profiles are less affected by blowout. The average model profile is in better agreement with an exponential disk, towards solving the j-profile discrepancy pointed our by BD.

A quantity used by BBS to characterize the shape of the j profile, as an alternative to the BD parameter μ, is $\zeta \equiv j_{tot}/j_{max}$. In Fig. 2 we also plot the distribution of this quantity in the BBS dataset in comparison with our model predictions for the baryons in dwarf and bright galaxies. The predicted ζ distribution for dwarfs is shifted upwards compared to the halos and the bright galaxies, in qualitative agreement with the BBS data, but the width is overestimated.

6   Conclusion

We devised a simple model to address the j problems of galaxy formation within CDM. By adding up the orbital $\mathbf{J}$ in random realizations of merger histories, the model successfully reproduces the simulated distribution of spins among halos (MDS) and the distribution of j within halos (MD). A simple analysis of how the merger orbital j turns into a spin profile provides a clue for how feedback effects in the satellite can resolve the spin problems. The idea is that the effective size of the gas component within the incoming halo determines its tidal stripping position in the big halo and thus its final remaining baryonic spin after the merger. The finding that the low-j material originates in many minor mergers, that tend to cancel each other's $\mathbf{J}$, provides the clue for a possible solution to the j-profile mismatch problem. The blowout of gas from small incoming halos, which is more pronounced in satellites of dwarfs, would eliminate the low-j baryons in the merger product and increase the spin parameter, as observed.

The feedback effects, including heating and blowout, are modeled as a function of halo virial velocity, with one free parameter – the characteristic velocity V_{fb} corresponding to the feedback energy from supernovae. To match the low baryonic fraction observed in dwarfs it has to be $V_{\text{fb}} \sim 100\,\text{km}\,\text{s}^{-1}$, consistent with the theoretical predictions [8]. This leads to an agreement between the model predictions and the observed disks, for the distribution of baryonic spin among galaxies and the baryonic j distribution within galaxies, both dwarfs and bright galaxies.

We attempt to resolve the problems within the successful cosmological framework of CDM, by appealing to inevitable feedback effects. Another approach is to appeal to the Warm Dark Matter (WDM) scenario, despite the fact that it requires fine-tuning of the particle mass to $\simeq 1\,keV$. The main feature of WDM is the suppression of small halos and the corresponding mergers. While an N-body simulation of WDM [2] indicates the same j properties of halos (the same properties can also be obtained as a general result of tidal-torque theory, see MDS), one expects the cooling to be less efficient in the absence of small halos, and thus the baryonic spin to be higher. However, the j profile is still expected to be a problem, and the weaker feedback effects in the absence of small halos may not be enough for resolving it. These issues are yet to be studied in hydro simulations of WDM.

Feedback effects may also provide the cure to the missing dwarf problem in CDM, where the predicted number of dwarf halos is much larger than the observed number of dwarf galaxies [3]. While the number of dwarfs is automatically suppressed in WDM, it seems that the inclusion of the minimum inevitable feedback effects would reduce the predicted number of dwarfs to significantly below the observed number and thus be an overkill (J. Bullock, private comm.). Finally, we find [7] that the key elements of our toy model – the tidal effects in mergers and the feedback in small halos – are also very relevant in understanding and resolving the third problem of CDM, where the halos in simulations typically show steep cusps in their inner profiles [14], while observations indicate flat cores at least in some galaxies [4]. An analysis of tidal effects explains the inevitable formation of an asymptotic cusp as long as satellites continue penetrating into the halo center. Feedback effects may puff up small satellites, make them disrupt in the outer halo and thus allow a stable core.

The success of our toy model in matching several independent observations indicates that it indeed captures the relevant elements of the complex processes involved, and in particular that feedback effects may indeed provide the cure to all three problems of galaxy formation in CDM. The next natural step should be to incorporate a more sophisticated feedback recipe into the model using semi-analytic models and then full-scale cosmological simulations.

This research has been supported by the Israel Science Foundation grant 546/98, by the US-Israel Binational Science Foundation grant 98-00217, and by the German-Israeli Science Foundation grant I-629-62.14/1999.

References

1. Bullock J. S., Dekel A., Kolatt T. S., Kravtsov A. V., Klypin A. A., Porciani C. & Primack J. R. 2001, ApJ, 555, 240 (BD)
2. Bullock J. S., Kravtsov A. V. & Colin P. 2001, ApJL, in press
3. Bullock J. S., Kravtsov A. V. & Weinberg D. H. 2000, ApJ, 539, 517
4. de Blok, W.J.G., McGaugh, S.S., Bosma, A. & Rubin, V.C. 2001, ApJL, 552, L23
5. Dekel, A., Bullock, J.S., Porciani, C., Kravtsov, A.V., Kolatt, T.S., Klypin, A.A., & Primack, J.R. 2001, in Galaxy Disks and Disk Galaxies, eds. J.G. Funes S.J. & E.M. Corsini (ASP Conference Series) (astro-ph/0011002)
6. Dekel, A. & Devor, J. 2002, in preparation
7. Dekel, A., Arad, I., et al. 2002, in preparation
8. Dekel A. & Silk J., 1986 ApJ, 303, 39
9. Efstatiou, G. 2000, MNRAS, 317 697
10. Fall S. M. & Efstathiou G. 1980, MNRAS, 193, 189
11. Lacey C. & Cole S. 1993, MNRAS, 262, 627
12. Maller A. H. & Dekel A. 2002, MNRAS, submitted (MD)
13. Maller A. H., Dekel A. & Somerville R. S., 2001, MNRAS, in press (astro-ph/0105168) (MDS)
14. Navarro J. F., Frenk C. S. & White S. D. M. 1995, MNRAS, 275, 56
15. Navarro J. F. & Steinmetz M., 2000, ApJ, 538, 477
16. Porciani, C., Dekel, A., & Hoffman, Y. 2001, MNRAS, in press (astro-ph/0105165)
17. Thacker R. J. & Couchman H. M. P. 2001, ApJL, 555, L17
18. van den Bosch F. C., Burkert A. & Swaters R. A. 2001, MNRAS, 326, 1205 (BBS)
19. Vitvitska M., Klypin A., Kravtsov A. V., Bullock J. S., Primack J. R. & Wechsler R. H. 2001, ApJ, in press

Correlating Galaxy Properties

Andreas Faltenbacher and Stefan Gottlöber

Astrophysikalisches Institut Potsdam, An der Sternwarte 16, 14482 Potsdam, Germany

Abstract. We study the correlation between galaxy clustering and their intrinsic properties. We find in high density regions an enhancement of the spin parameter as well as an over-abundance of old objects.

On the basis of numerical simulations we use here a new statistical technique with which clustering of galaxies can be described in dependence of their internal properties. The basic idea of this technique is to study the joint probability of finding objects with a certain property (mark) in a given distance. The *mark correlation functions* have been introduced into astrophysics quite recently [1]. For scalar marks the mean mark $k_m(r) \equiv \langle m_1 + m_2 \rangle_{\mathrm{P}}(r)/2\overline{m}$ quantifies the clustering of marks at a scale r. If $k_m(r) = 1$ mark segregation is absent, $k_m(r) > 1$ means preferred clustering of marks.

For discrete marks conditional cross–correlation functions can be constructed in a similar way. Supposing the marks of our objects belong to classes labeled with $i, j..$, they are given by $C_{ij}(r) \equiv \langle \delta_{1i}\delta_{2j} + (1 - \delta_{ij})\delta_{2i}\delta_{1j} \rangle_{\mathrm{P}}(r)$, with the Kronecker $\delta_{1i} = 1$ for $m_1 = i$ and zero otherwise. Mark segregation is indicated by $C_{ij} \neq 2\varrho_i \varrho_j / \varrho^2$ for $i \neq j$ and $C_{ii} \neq \varrho_i^2/\varrho^2$, where ϱ_i denotes the number density of points with label i. The C_{ij} are cross–correlation functions under the condition that two points are separated by a distance of r.

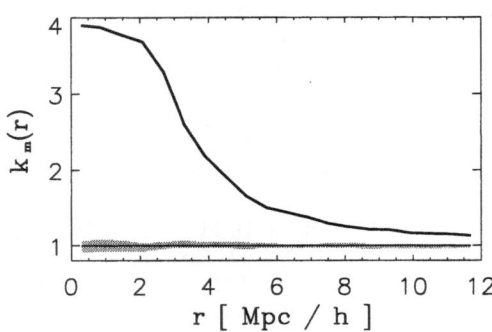

Fig. 1. Marked correlation function of halos with $v_{circ} > 100$ km/s. The spin parameter λ used as scalar mark. The shaded area is obtained by randomizing the mark among the halos

From a high-resolution N-body simulation (ΛCDM model; $\Omega_0 = 1 - \Omega_\Lambda = 0.3$) we extract dark matter halos hosting galaxies. First we use the spin parameter λ of the halos as a mark. In Fig. 1 $k_m > 1$ indicates mark segregation for halo pairs with a separation r, specifically their mean spin parameter is larger than the overall spin parameter average. This result reveals, that objects in high density regions tend to have higher spin parameters.

According to the age of the objects we can split the halo sample into two subsamples. In Fig. 2 we compare the distribution of halos formed after redshift $z < 5$ (sample: l) and halos formed earlier (sample: e). The increasing $C_{e,e}$ on scales below $3h^{-1}$Mpc indicates that on this distances pairs of older halos are relative over–abundant (lower panel), at the expense of younger halos (upper panel). This indicates that regions of high halo density (for example the central regions of clusters) should preferentially host old halos [2].

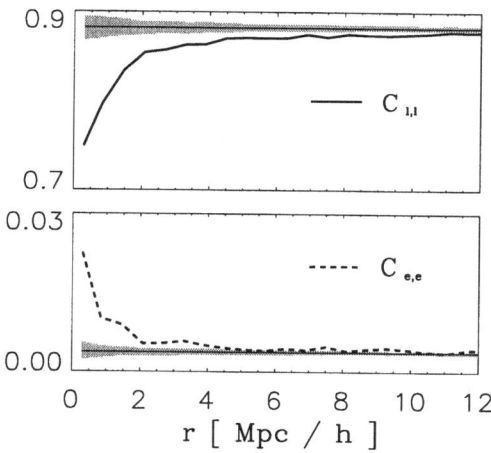

Fig. 2. Conditional cross–correlations of halos with $v_{circ} > 100km/s$. $C_{e,e}$ denotes pairs of old halos $z_{form} < 5$ $C_{l,l}$ such pairs with $z_{form} \geq 5$. The shaded area is obtained by randomizing the mark among the halos

We have demonstrated that marked correlation functions provide a very useful and simple tool for the statistical analysis of cosmological data. Since the marks can be chosen almost arbitrarily, this opens a huge scope of applications in observational astrophysics.

References

1. C. Beisbart, M. Kerscher: ApJ **546**, 6 (2000)
2. S. Gottlöber, M. Kerscher, A. Klypin, A. Kravtsov, V. Müller, A. Faltenbacher: *in preparation*

Induced Activity in Galaxies of Mixed Pairs (E+S): Surface Photometry

Alfredo Franco, Deborah Dultzin-Hacyan, and Héctor M. Hernández-Toledo

Instituto de Astronomía, UNAM, México

1 The Sample

The "Catalogue of Isolated Pairs of Galaxies of the North Hemisphere" of Karachentsev (1972) (CPG), was selected under a strict isolation criterion that excluded, as far as possible, the optical pairs. Measures of redshift were obtained for all the pairs. The sample is full of morphological signatures of interactions, such as tails and bridges that give evidence of their physical nature.

The catalogue consist of 600 members, of which 120 were considered as of mixed morphology.

E+S pairs represent an assembly of objects in which we see the effects of the interaction of a rich gas object: a spiral galaxy (S) in the presence of a relatively clean disturber, in this case an elliptical galaxy (E), simplifying the interpretation of the interaction phenomena.

2 Importance

If environmental factors play an important role in the formation of E+S pairs, then the significative number of mixed pairs existing suggest a change in the theories of formation and evolution of galaxies. A problem of these theories is that in de case of paired galaxies predict concordant morphology; in other words, initial conditions of gas content and angular moment must be similar, suggesting that E+S pairs are a non primordial product of encounter between galaxies.

On the contrary, the high percentage of mixed isolated pairs, as well as their enhanced level of optical and far-infrared emission (compared against a sample of isolate galaxies) (Xu & Sulentic 1991; Hernández-Toledo et al. 1999, 2001), suggests that most of them must be physical pairs. In support of this, numerical simulations indicate that the fraction of E+S pairs created by random encounters is of only 0.05% in regions with low density where these pairs have been observed (Chatterjee 1987).

3 Observations

We have carried out optical BVRI Johnson-Cousins CCD observations for 40 KPG E+S pairs, and 20 KIG isolated galaxies taken from the "Catalogue of Isolated Galaxies" from Karachentseva (1973)(CIG). The later set constitutes a control sample for our photometric study.

Images were obtained at San Pedro Mártir Observatory (SPMO) in two different telescopes with a 1024 x 1024 SI003 CCD and a 2x2 binning.
Each image is:
4.3$'$ × 4.3$'$ ($\sim$ 0.5 $''$/pixel) in 1.5 m. telescope, and
6.7$'$ × 6.7$'$ ($\sim$ 1.0 $''$/pixel) in 84 cm. telescope.

4 Surface Photometry

Surface brightness is a measure of intensity, or flux per solid angle (square arcsec), from a galaxy. With sufficient resolution, the variation of surface brightness as a function of position in a galaxy provides important information about the distribution of stellar populations and star formation.

The morphologic evaluation can be done in three parts:

1. - Visual identification of false pairs E+S.

2. - Evaluation of the geometric parameters for each component.

3. - Evaluation of all characteristics, mainly those related to structures like spiral arms, shells, rings, bars and large regions of stellar formation.

Important parameters like the size, the distribution of mass, the star formation rate, as well as the magnitude of nuclear activity can be determined with the help of our observations.

5 Preliminary Results

a) Most E+S pairs show signs of disturbance like bridges, tails, geometric and morphologic distortion, and in some cases nuclear activity, which are evidence of gravitational interactions.

b) Deep images show evidence of small faint companions.

c) In E+S pairs, "grand desing" spirals are common and luminous, whereas low-luminosity S members exhibit flocculent spiral structure.

d) Spirals in mixed pairs have bluer colors than normal spirals.

e) Boxy ellipticals tend to be more luminous than disky ones, in E+S pairs. Possibly boxy ellipticals are formed by mergers of (mostly) ancestral objects. On the other hand, disky ellipticals may have been formed from a single protogalaxy, or from the merger of mainly gaseous ancestral objects.

References

1. Chatterjee T., 1987, Ap&SS 135, 131.
2. Hernández-Toledo, H. M., Dultzin-Hacyan, D., Gonzàles, J. J., & Sulentic, J., 1999, AJ, 118:108-125.
3. Hernández-Toledo, H. M., Dultzin-Hacyan, D., & Sulentic, J., 2001, AJ, 121.
4. Karachentsev, I. D., 1972, Comm. Spec. Ap. Obs. 7, 1 (CGP).
5. Karachentseva, V., 1973, Comm. Spec. Ap. Obs. 8, 1 (CIG).
6. Kormeny, J., and S. Djorgovsli, 1989, ARA&A, 27:235.
7. Rampazzo, R., Sulentic, J. W., 1992 A&A 259, 43-60.
8. Xu C., Sulentic J.W., 1991, Apj 242, 469.

Detection of CDM Substructure

C.S. Kochanek[1] and N. Dalal[2]

[1] Harvard-Smithsonian Center for Astrophysics, Cambridge, MA 02420
[2] Dept. of Physics, UCSD, La Jolla, CA 92093

Abstract. The properties of multiple image gravitational lenses require a fractional surface mass density in satellites of $f_{sat} = 0.02$ ($0.006 \lesssim f_{sat} \lesssim 0.07$ at 90% confidence) that is consistent with the expectations for CDM. The characteristic satellite mass scale, 10^6-$10^9 M_\odot$, is also consistent with the expectations for CDM. The agreement between the observed and expected density of CDM substructure shows that most low mass galactic satellites fail to form stars, and this absence of star formation explains the discrepancy between the number of observed Galactic satellites and CDM predictions rather than any modification to the CDM theory such as self-interacting dark matter or a warm dark matter component.

1 Introduction

The existence of a "CDM crisis" (e.g. Moore 2001) rests on three pillars. First, the central rotation curves of some dwarf and low surface brightness galaxies may be inconsistent with the expectation of a central dark matter density cusp, as is discussed by Swaters and McGaugh in these proceedings. Second, the observed and predicted magnitudes and distributions of the baryonic angular momentum may be inconsistent, as discussed by Dekel and Burkert in these proceedings. Third, the Galaxy has far fewer satellites than expected for a CDM halo of its mass (e.g. Moore et al. 1999, Klypin et al. 1999). It is this third pillar of the crisis which we will undermine with our present analysis.

The satellite crisis essentially boils down to the observation that cluster and galaxy mass halos show similar amounts of substructure in CDM simulations, while observations appear to show that the Galaxy and the Coma cluster have very different amounts of substructure. The simplest solution is to suppress star formation in low mass halos relative to high mass halos (e.g. Bullock et al. 2000). A similar effect is required to explain the difference between the steep slope of the halo mass function ($dn/dM \sim M^{-2}$) and the shallow slope of the luminosity function ($dn/dL \sim L^{-1} \sim M^{-1}$) independent of the satellite problem (e.g. Scoccimarro et al. 2001, Kochanek 2001, Chiu et al. 2001). More exotic solutions are to destroy the satellites using self-interacting dark matter (e.g. Spergel & Steinhardt 2000) or to avoid creating them by adding warm dark matter (e.g. Bode et al. 2001, Colin et al. 2000). We can distinguish between these possibilities only if we have a probe which is sensitive to the presence of mass in the absence of light.

Moore et al. (1999) also realized that the only such probes we possess are gravitational lenses. Indeed, Mao & Schneider (1998) had already pointed out

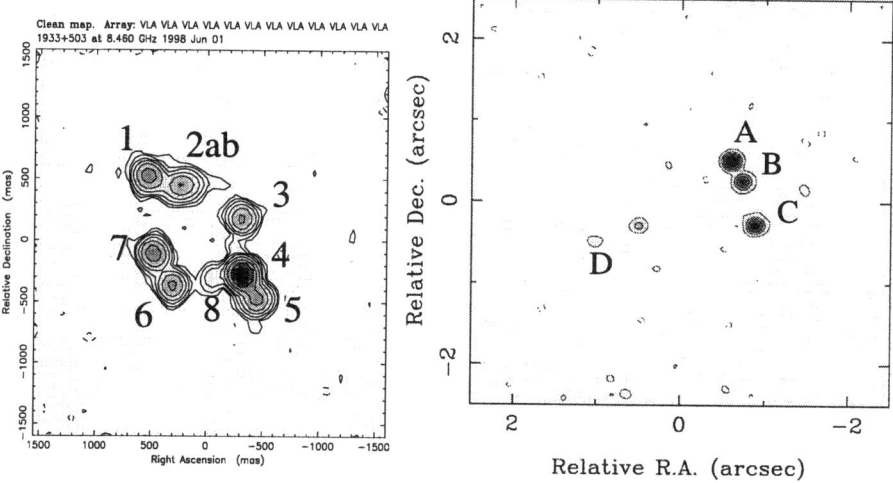

Fig. 1. Examples of lenses with anomalous flux ratios: the 4/136 images of B1933+503 (left, Sykes et al. 1998) and the A/B images of B2045+265 (right, Fassnacht et al. 1999)

that the anomalous flux ratios of close image pairs, where the fluxes expected for a mass distribution consisting only of the primary lens galaxy would be nearly equal, could be explained by the presence of small, satellite galaxies. More recently, Metcalf & Madau (2001) showed that these effects would be easily detected given the expected amount of substructure in CDM models, and Chiba (2001) showed that CDM substructure could explain the flux ratios in PG1115+080 and B1442+231. In Fig. 1 we show two less well-known examples of gravitational lenses with anomalous flux ratios. The problem, however, is to convert these arguments about plausibility into quantitative estimates of the surface density and properties of satellites.

In Dalal & Kochanek (2001) we developed a formalism for making these estimates and applied it to a sample of 7 four-image lenses. We do not repeat the mathematical development here, providing only a qualitative outline in §2. We present our results and and discuss the future of the method in §3.

2   An Outline of the Method

The fundamental problem in estimating the properties of any substructure in a gravitational lens is degeneracy between the effects of substructure and the primary lens. For example, equal radial deflection perturbations to the images are degenerate with a change in the mass of the macro model. Thus we must limit our analysis to systems with more constraints than reasonable models of the primary lens have parameters. This rules out two-image lenses, so we will analyze only lenses consisting of four images which have (in general) three degrees of freedom left after fitting a macro model consisting of a singular isothermal ellipsoid (SIE) in an external shear field.

For each of the 7 lenses in our sample we start from the best fit model for the lens and linearly expand the lens equations around each image including both changes in the lens parameters and the local perturbations due to substructure. For any model of the substructure we can then adjust the parameters of the macro model to compensate for its effects and compute a new goodness of fit χ^2. This provides us with an estimate of the probability $P(D_j|\delta_{ij})$ that substructure realization i for lens j provides a fit to the data D_j. Of course, with so few degrees of freedom left after fitting the macro model, many substructure realizations produce significant improvements in the fit statistic – we have too few constraints to uniquely determine the substructure near each lensed image.

We are interested in the statistical properties of the substructure, not the particular realizations, and we can estimate the statistical properties using a Bayesian analysis of the data. For satellites drawn from a population described by parameters $\boldsymbol{p}$, the probability of substructure realization δ_{ij} is $P(\delta_{ij}|\boldsymbol{p})$. Given a prior $P(\boldsymbol{p})$ for the parameters, the probability of $\boldsymbol{p}$ given the data after marginalizing over the actual realizations is

$$P(\boldsymbol{p}|D) \propto P(\boldsymbol{p})\Pi_j \sum_i P(D_j|\delta_{ij})P(\delta_{ij}|\boldsymbol{p}) \tag{1}$$

where the normalization is set by the constraint that $\int d\boldsymbol{p}P(\boldsymbol{p}|D) \equiv 1$ and we multiply the contributions from each lens j and sum (marginalize) over the substructure realizations i. Qualitatively, as we vary the parameters, the probability of finding substructure realizations which improve the fit from $\chi^2 \gg N_{dof}$ to $\chi^2 \sim N_{dof}$ varies, thereby allowing us to estimate the parameters.

We modeled the substructure as pseudo-Jaffe models (Munoz et al. 2001) with surface densities, in units of the critical surface density Σ_c,

$$\kappa = (b/2)\left[r^{-1} - (r^2 + a^2)^{-1/2}\right]. \tag{2}$$

The full model has three parameters: a mass scale b, a tidal truncation radius a, and the satellite surface mass density Σ. For isothermal models the tidal truncation radius is $a = (bb_0)^{1/2}$ where we fix $b_0 = 1\rlap{.}''0$ for the Einstein radius of the primary lens. The satellite mass is $M = \pi ab\Sigma_c$, and the fraction of the mass in substructure is $f_{sat} \simeq 2\Sigma/\Sigma_c$. Near the Einstein ring, where we see the images, most of the mass is dark matter. We model the substructure using random realizations of the satellite distributions. The perturbations are dominated by the variance in the shear and convergence (rather than the astrometry), with

$$\langle \kappa^2 \rangle^{1/2} \simeq \langle \gamma^2 \rangle^{1/2} \simeq 0.13 \left(\frac{10\Sigma}{\Sigma_c}\right)^{1/2} \left(\frac{10^3 b}{b_0}\right)^{1/4} \left(\frac{\ln \Lambda}{10}\right)^{1/2}. \tag{3}$$

There is a "Coulomb" logarithm $\ln \Lambda = \ln(a/s)$ where $s \ll a$ is an effective core radius to the lens.

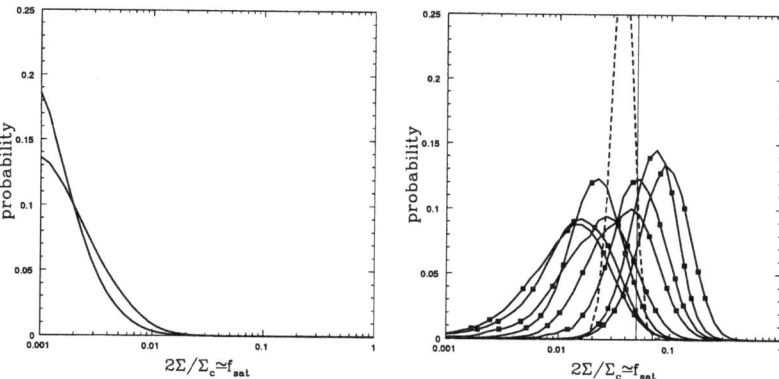

Fig. 2. (left) Null tests. The curves show the probability of fractional surface density $f_{sat} = 2\Sigma/\Sigma_c$ for two Monte Carlo realizations of 7 lenses with measurement errors but no substructure

Fig. 3. (right) Monte Carlo realizations with $f_{sat} = 0.05$ and $b = 0\!''\!001$ modeled with $b = 0\!''\!001$. The solid curves show the probability for each of the 8 realizations, and the dashed curve shows the joint probability for all 8 realizations (mimicking a sample of 56 lenses). The points correspond to the median of the distribution and the range encompassing 68% (1σ), 90% and 95% (2σ) of the probability. The vertical line marks the true surface density

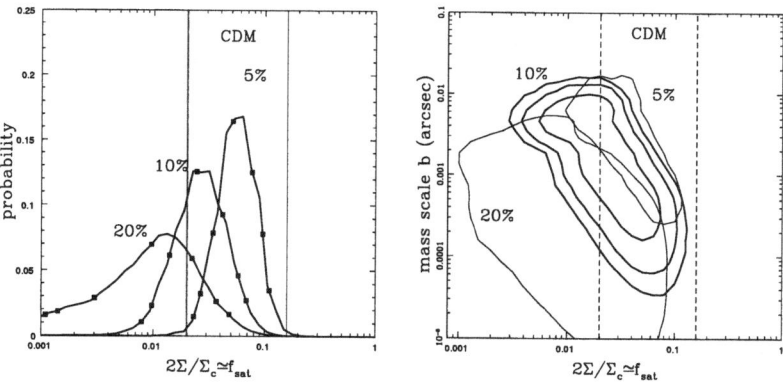

Fig. 4. (left) Estimates of f_{sat} in the real data for a fixed $b = 0\!''\!001$ mass scale assuming 5%, 10% or 20% errors in the image fluxes. The points on the heavy curves correspond to the median of the distribution and the range encompassing 68% (1σ), 90% and 95% (2σ) of the probability

Fig. 5. (right) Simultaneous estimates of f_{sat} and b for the real data. For the 10% flux error case (heavy lines) we show the isoprobability contours encompassing 68% (1σ), 90% and 95% (2σ) of the probability around the peak. For the 5% and 20% flux error cases, the light solid curve encompasses 90% of the probability

Figs. 2 and 3 show the results of two Monte Carlo tests of the algorithm. In the first test we take the macro models for our 7 lenses, add measurement errors and analyze the data for substructure. We find only upper bounds on the substructure density with $f_{sat} \lesssim 0.004$ and no preferred mass scale. In the second test we added perturbers with surface density $f_{sat} = 0.05$, mass scale $b = 0\rlap{.}''001$, and tidal radius $a = 0\rlap{.}''032$. Multiple trials with samples of 7 lenses recover the surface density and mass scale with reasonable accuracy given the sample size. In a third test, which we do not show, we find that we can determine both b and f_{sat} simultaneously.

3   The Surface Density of Satellites

We then fit our sample of 7 lenses either with $b = 0\rlap{.}''001$ (Fig. 4) or varying both b and f_{sat} (Fig. 5). Qualitatively, the probability curves look very similar to our Monte Carlo simulations. The dominant uncertainty is the degree to which the flux measurements in the real lenses are affected by systematic errors. To explore this, we assumed a standard flux uncertainty of 10%, but show results for both 5% and 20% flux errors. The errors are certainly smaller than 20%, probably smaller than 10% and unlikely to be smaller than 5%. In the models where we estimate both f_{sat} and b we find median estimates of $f_{sat} = 0.020$ and $b = 0\rlap{.}''0013$ for our standard model with 90% confidence regions of $0.0058 < f_{sat} < 0.068$ and $0\rlap{.}''0001 < b < 0\rlap{.}''007$. If we assume 5% errors, the surface density must be higher, $0.013 < f_{sat} < 0.078$, while if we assume 20% errors it must be lower, $0.0016 < f_{sat} < 0.051$. There is a relatively strong degeneracy between the values of b and f_{sat}, with higher mass scales allowing lower surface densities. The slope closely matches $b \propto f_{sat}^{-2}$, which corresponds to keeping the rms shear perturbation constant (see eqn. 3).

These estimates are consistent with the expectations for CDM models, where $0.02 \lesssim f_{sat} \lesssim 0.15$ (Moore et al. 1999, Klypin et al. 1999), and much larger than the expectations for normal satellite populations, $10^{-4} \lesssim f_{sat} \lesssim 10^{-3}$ (Mao & Schneider 1998, Chiba 2001). For a $dn/dM \propto M^{-2}$ substructure mass function over the range $M_{low} < M < M_{high}$, our mass scale $M = \pi ab\Sigma_c$ provides a crude estimate of the upper mass $M_{high} \sim 10M \sim 10^6 M_\odot$ to $10^9 M_\odot$ which is also consistent with the CDM scenario. Thus, our results are most naturally explained as a detection of the satellite galaxies expected in the CDM model.

While one could debate whether we detect "satellites" or "CDM satellites," alternative explanations do not appear to be viable. There are systematic uncertainties in the data, but the anomalous flux ratios which produce the result are generic and seen in repeated observations spread over long time scales (years) and a range of wavelengths. They can be misinterpreted but not eliminated. Coherent structures in the primary lens galaxy (e.g. spiral arms) are not seen in deep HST images of the lenses, and would need to be far larger fractional mass perturbations to produce the same effects because, unlike satellites, they cannot perturb individual images. Stellar microlensing, while it is the same physical phenomenon, is ruled out as a full explanation. Radio sources are generally too large to suffer from microlensing and the flux ratio anomalies are too long lived. While microlensing has been seen in one radio lens (see Koopmans & de Bruyn 2000), it only affects a superluminal subcomponent, it has a small rms amplitude, and it has a very short fluctuation time scale. Moreover, our estimate of the mass scale is consistent with that of satellites rather than stars.

There is considerable work to be done in the future. First, there is no reason that careful observation and monitoring of the lenses cannot measure the image fluxes to 1% accuracy including the effects of variability and any microlensing. Not only would this allow us to estimate the properties of the CDM substructure more accurately, but it would also supply a large sample of accurate time delay measurements for determining H_0 without the problematic systematic errors of the local distance scale (e.g. Schechter 1999). Second, deep high resolution imaging with both HST and the VLBA is needed. Images of the radio source's host galaxy can constrain the macro model without being affected by substructure because of the larger angular size of the host. Deep VLBA images to map extended structures can be used to measure the substructure mass scale accurately. Extended radio structure also can be used to estimate the surface density over larger areas of the lens than point sources, and can be used to study the internal structure of the satellites. Finally, more lenses will reduce the statistical uncertainties. This includes finding more lenses and more detailed studies of existing lenses to find enough constraints on the macro model to allow a substructure analysis.

Acknowledgments: CSK was supported by the Smithsonian Institution and NASA grants NAG5-8831 and NAG5-9265. ND was supported by the Smithsonian Institution Short Term Visitor Program, DOE grant DOE-FG03-97-ER 40546 and the ARCS Foundation.

References

1. Bode, P., Ostriker, J. P. and Turok, N. 2001, ApJ, 556, 93
2. Bullock, J. S., Kravtsov, A. V. and Weinberg, D. H. 2000, ApJ, 539, 517.
3. Chiba, M., 2001, astro-ph/0109499
4. Chiu, W.A., Gnedin, N.Y., & Ostriker, J.P., 2001, astro-ph/0103359
5. Colin, P., Avila-Reese, V., & Valenzuela, O., 2000, ApJ, 542, 622
6. Dalal, N., & Kochanek, C.S., 2001, ApJ submitted
7. Fassnacht, C.D., et al. 1996, ApJL, 460, 103
8. Fassnacht, C.D., Blandford, R.D., Cohen, J.G., et al., 1999, AJ, 117, 658
9. Klypin, A., Kravtsov, A.V., Valenzuela, O., & Prada, F., 1999, ApJ, 522, 82.
10. Kochanek, C.S., 2001, The Dark Universe, M. Livio, ed., also astro-ph/0108160
11. Koopmans, L.V.E., & de Bruyn, A.G., 2000, A&A, 358, 793
12. Mao, S., & Schneider, P., 1998, MNRAS, 295, 587
13. Metcalf, R.B., & Madau, P., 2001, astro-ph/0108224
14. Moore, B., et al., 1999, ApJ, 524, L19
15. Munoz, J.A., Kochanek, C.S., & Keeton, C.R., 2001, ApJ, 558, 657
16. Schechter, P., 1999, IAU 201, A.N. Lasenby & A. Wilkinson, eds., astro-ph/0009048
17. Scoccimarro R., Sheth R.K., Hui L., Jain, B., 2001, ApJ, 546, 20
18. Spergel, D. N. and Steinhardt, P. J. 2000, PRL, 84, 3760
19. Sykes, C.M., Browne, I.W.A., Jackson, N.J., et al., 1998, MNRAS, 301, 310

Galaxy-Galaxy Lensing in the FORS Deep Field

Stella Seitz[1], Thomas Erben[2], Ralf Bender[1], and the FDF-Team[3]

[1] Munich University Observatory (USM), Scheinerstr. 1, 81679 München
[2] Institut f. Astrophysik und Extraterrestrische Forschung, Bonn
[3] Landessternwarte Heidelberg, Universitätssternwarte Göttingen,
Universitätssternwarte München

Abstract. We demonstrate the presence of a galaxy-galaxy lensing signal of $z = 0.5 - 1.0$ foreground galaxies by analyzing the distortion of $z = 2.5 - 4.5$ background galaxies in the FORS Deep Field. The signal decreases when fainter foreground galaxies are considered in agreement with expectations. Also, the lensing signal of the faintest foreground sample agrees with that found in the HDF by [5] for $z < 0.85$ foreground galaxies with $< z >= 0.6$.

1 Introduction and Earlier Results

Galaxy-galaxy lensing is a powerful tool to determine the halo-parameters (mass, size and flattening) of field and cluster galaxies [3]. However, with unknown individual galaxy distances, a large number of foreground-background pairs (and thus a large area) is necessary to constrain the halo parameters (eg., compare [2] to [5]). For the investigation of medium redshift galaxy-halos – where the data have to be deep to obtain reliable shapes and measure the distortions for the background galaxies, and therefore fields are still small – the photometric redshift technique is most appropriate for selecting foreground objects (irrespective of their SED and morphology) and background objects. Hudson et al. ([8]) have demonstrated this for the HDFN-data, they have constrained the size and rotational-velocity of L_*-galaxies at redshift $z \approx 0.6$ and compared that to the Tully-Fisher relation at the same redshift. The fairly large error bars are caused by the small field size of the HDF-N, which yields only about 700 galaxies with shape and phot-z measurements, (208 of them are foreground with $z < 0.85$ and $< z > \approx 0.6$.)

2 Galaxy-Galaxy Lensing in the FORS Deep Field

The FORS-team (*FDF is a common project of the Landessternwarte Heidelberg and the University Observatories in Göttingen and Munich*) has carried out deep imaging in the U-, B-, g-, R- and I-broad-band and z-narrow-band filters using FORS1&2 at the VLT and in the J- and K'-filters using SOFI at NTT. The Data are shallower than the HDF's but exceed their area by a factor of 8, which is essential for the measurement of the galaxy-galaxy lensing signal on larger scales where the halo size of field galaxies can possibly be measured. For details concerning filters, field position, integration times, FWHM and depth of the

FDF data see [6]. The data-reduction, including correction of the field distortion of FORS, are described in more detail in [7] Object detection was performed with SExtractor, and about 7000 objects were detected in the I-band image with 6.8h total integration time and a coadded seeing of about $< .55''$. The colors of these objects were used for photometric redshifts estimates similar as described in [1]. The photometric redshifts proved to be accurate to an rms of $[z_{spec} - z_{phot}]/(1 + z_{spec}) < 0.05$ as a comparison to the roughly 300 spectra obtained in this field up to now shows.

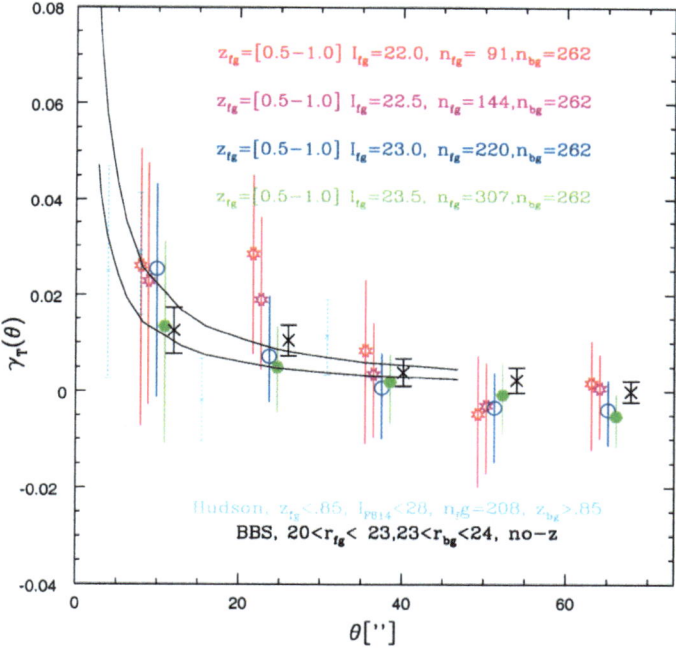

Fig. 1. Tangential shear of background galaxies with $z = 2.5 - 4.5$ relative to foreground galaxies at $z = 0.5 - 1.0$ as a function of foreground-background separation. The faint-magnitude cut-off of the foreground galaxies is increased from $I = 22$ to $I = 23.5$. (these data points are shifted by 1 arcsecond horizontally for clarity). We have obtained our errorbars from bootstrapping of the background galaxies. The number of background and foreground objects in indicated by n_{fg} and n_{bg}, respectively. For comparison we also add the tangential shear of '$r = 23 - 24$ background' galaxies relative to $r < 23$, more local foreground galaxies obtained by BBS (black data points). The results from Hudson et al in the HDFN are shown as cyan data points

Shape analysis, PSF-smearing and anisotropy correction of the background objects were carried out on the I-band image with the seeing of about $< 0.55''$

using the IMCAT software (and modifications) analogously to [4]. Then the photometric redshift catalog (containing roughly 7000 objects) and the shape catalog were merged. Stars were excluded conservatively using half-light radius, stellar classification and comparison to stellar SED-templates.

For this proceedings we discuss only the tangential shear signal in the FDF caused by galaxies with $z = 0.5 - 1.0$ and different apparent brightness limits in the I-band. As background galaxies we define all galaxies with $z = 2.5 - 4.5$ with sufficient S/N to allow shape determination. The number of galaxies in every foreground sample and in the background sample can be seen in the Figure. The tangential shear of the background galaxies follows the expected functional form (roughly $\propto 1/\theta$). The signal is largest for foreground galaxies with $I < 22$ and decreases down to $I < 23.5$ when less luminous and thus less massive galaxies enter the foreground sample.

We have included the data points of Hudson et al.[8] in the Figure (cyan color). Their foreground sample is similar to ours with $I < 23.5$.

The tangential shear of isothermal spheres at $z \approx 0.5$ with rotational velocities of 210km/s (which fits the Hudson data best) and one with 380 km/s are shown as solid lines to guide the eye. We conclude that the rotational velocity $v \propto \sqrt{\gamma_T}$ of the halos of the brightest subsample is about 1.8 higher than that of the faintest sample. This is in excellent agreement with the upper limit of 2 obtained from the magnitude difference and the Tully-Fisher/Faber-Jackson relation.

A maximum-likelihood analysis of the galaxy-galaxy-lensing signal in the FDF that of will be presented in the near future.

References

1. Bender, R. 2001: The FORS Deep Field: Photometric Data and Photometric Redshifts, ESO Workshop on "Deep Fields"
2. Brainerd et al. 1996, APJ, 466, 623 (BBS)
3. Brainerd, T.G. 2001, ASP Conference Proceedings, Astronomical Society of the Pacific, ISBN: 1-58381-074-9, 2001, p.379
4. Erben, T. et al. 2001, A&A, 366, 717
5. Fischer, P. et al. 2000, AJ, 120, 1198
6. Heidt, J., I. Appenzeller, Bender, R., Böhm, A., Drory, N., Fricke, K.J., Gabash, A., Hopp, U., Jäger, K., Kümmel, M., Mehlert, D., Möllenhoff, C., Moorwoord, A., Nicklas, H., Noll, S., Saglia, R., Seigert, W., Seitz, S., Stahl, O., Sutorius, E., Szeifert, T., Wagner, S.J., Ziegler, B., 2001, "The FORS Deep Field", AG Tagung, Bremen, RvMA, 14, 209
7. Heidt, J. et al 2002, in preparation
8. Hudson, M. et al. 1998, APJ, 503, 531
9. Smith, D.R. et al. 2001, APJ, 551, 643
10. Wilson, G. et al. 2001, APJ, 555, 572

Lens Galaxies vs. CDM

Charles R. Keeton[1,2,3]

[1] Steward Observatory, University of Arizona, 933 N. Cherry Ave., Tucson, AZ 85721
[2] Astronomy and Astrophysics Department, University of Chicago,
   5640 S. Ellis Ave., Chicago, IL 60637
[3] Hubble Fellow

Abstract. By directly probing mass distributions, gravitational lensing offers several new tests of the CDM paradigm. Lens statistics place upper limits on the dark matter content of elliptical galaxies. Galaxies built from CDM mass distributions are too concentrated to satisfy these limits, so lensing extends the "concentration problem" in CDM to elliptical galaxies. The central densities of the model galaxies are too low on ~ 10 pc scales to agree with the lack of central images in observed lenses. The flux ratios of four-image lenses imply a substantial population of dark matter clumps with a typical mass $\sim 10^6$ $M_\odot$. Thus, lensing implies the need for a mechanism that reduces dark matter densities on kiloparsec scales without erasing structure on smaller scales.

1   Introduction

The popular Cold Dark Matter (CDM) paradigm is facing several challenges on small scales (e.g., [26]). The dynamics of spiral galaxies, especially rotation curves and fast-rotating bars, suggest that in observed galaxies dark matter halos are much less concentrated than predicted by CDM (e.g., [11], [13]), although this conclusion is still controversial (e.g., [32]). The number of satellite dwarf galaxies in the Local Group is much smaller than the number of subhalos in CDM simulations [20], [25], although the discrepancy may be explained by the astrophysics of star formation rather than by the physics of the dark matter particle [6]. These tests of CDM are limited, however, by uncertainties in interpreting luminous tracers of the potential. Gravitational lensing offers a different test that probes mass distributions directly. Strong lensing by galaxies robustly determines the total mass in the inner 5–10 kpc of lens galaxies, which are predominantly elliptical galaxies. It also offers the possibility to detect small-scale mass concentrations in galaxy halos [8], [10], [19], [22], [24]. Lensing thus offers new tests of CDM that avoid dynamical uncertainties and extend the tests from spiral galaxies to ellipticals.

2   Star+Halo Models

I construct new models for lens statistics that include both stellar and dark matter components (see [18] for details). In principle, I take a CDM dark matter halo, add baryons, let the baryons condense into a galaxy, and use the adiabatic contraction formalism [3] to compute how the dark matter distribution is modified by the baryons.[2] In practice, I fix the stellar galaxies and use the models to place dark matter halos around them. The stellar components are treated as Hernquist models for elliptical galaxies, normalized by observed galaxy luminosity functions [21], Fundamental Plane relations [29], and Bruzual & Charlot [1] model mass-to-light ratios (which are reliable for the old stellar components of elliptical galaxies).

Two free parameters apply to the dark matter halos. First, halos with the Navarro, Frenk & White [27] dark matter profile are described by a concentration parameter. A halo's concentration is determined by its mass and redshift, but with a scatter of 0.18 dex [7]. I include the scatter and take the median concentration to be a free parameter. Second, to relate the total, virial mass of the dark matter halo (M_d) to the mass of the stellar component (M_s), I define the "cooled mass fraction" $f_{cool} = M_s/(M_d + M_s)$. I take the cooled mass fraction to be the second free parameter in the models, assuming only that it is smaller than the global baryon fraction, $f_{cool} \leq \Omega_b/\Omega_M$.

3   Lens Statistics and Galaxy Masses

Lens statistics can be used to test the CDM models, because changes to galaxy dark matter halos affect the number of lenses and the distribution of lens image separations. Figure 1a demonstrates the test by comparing the model predictions with the data from the Cosmic Lens All-Sky Survey (CLASS; e.g., [16]), which is the largest homogeneous survey for lenses. Increasing the concentration of dark matter halos raises the amount of dark matter in the inner parts of galaxies, leading the models to predict more and larger lenses. Because the stellar components of the galaxies are fixed, *decreasing* the cooled mass fraction *increases* the amount of dark matter, again leading to more and larger lenses.

Using statistical tests to compare the models to the data leads to confidence intervals in the (C, f_{cool}) plane, as shown in Fig. 1b. Lensing requires the models to have low concentrations or high cooled mass fractions. Adding the constraint on f_{cool} from the baryon content of the universe leaves only a small region of parameter space where the models are acceptable. Fiducial CDM models predict a median concentration $C \simeq 7.7$ for galaxies (indicated in Fig. 1b). This value is allowed by lens statistics only if galaxies are nearly 100% efficient at cooling their baryons ($f_{cool} \simeq \Omega_b/\Omega_M$), which is implausible (e.g., [2]). The constraints in Fig. 1b are conservative, because most of the systematic effects in the lensing analysis strengthen the lensing constraints (see [18]). Changing the cosmology (increasing Ω_M) has little effect on the lensing analysis but reduces the upper limit $f_{cool} \leq \Omega_b/\Omega_M$.

Translating the constraints into enclosed mass leads to the conclusion that dark matter can account for no more than 33% of the mass within $1 R_e$ and 40%

---

[2] Gottbrath [15] shows that adiabatic contraction agrees remarkably well with detailed gasdynamical simulations even in the merger scenarios thought to produce elliptical galaxies.

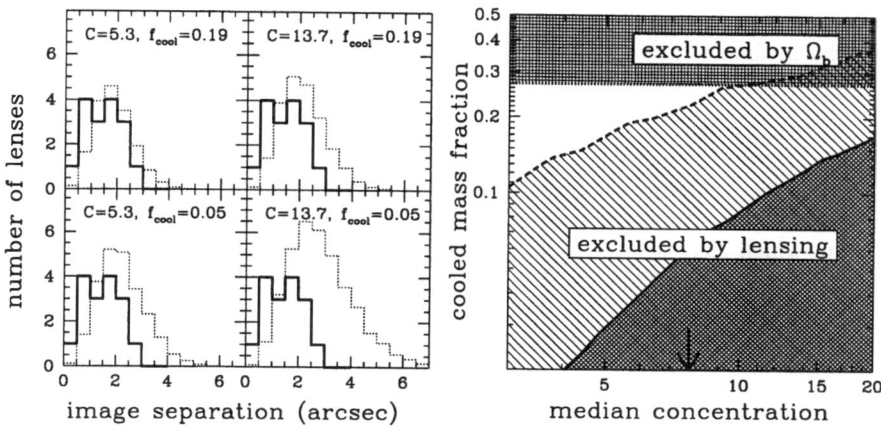

Fig. 1. *(a, left)* Image separation histograms for the CLASS data (solid lines) and for sample models (dotted lines). The model concentration C and cooled mass fraction f_{cool} are indicated in each panel. *(b, right)* Confidence intervals in the (C, f_{cool}) plane. The hatched region is excluded at 95% confidence by the distribution of lens image separations, and the cross-hatched region is further excluded by the number of lenses. The shaded region at the top is excluded at 95% confidence by measurements of Ω_b (e.g., [12], [31]). All results are shown for a cosmology with $\Omega_M = 0.2$ and $\Omega_\Lambda = 0.8$. See [18] for more discussion

of the mass within $2\,R_e$ (95% confidence limits on average mass fractions). Note that these limits are for the mass in *spheres*, whereas lensing limits on the mass in *cylinders* indicate that dark matter halos are still important in ellipticals. The lensing limits are consistent with the mass estimates from dynamical analyses of nearby elliptical galaxies [14]. By contrast, the CDM models predict dark matter mass fractions of $\sim 28\%$ inside $1\,R_e$ if baryon cooling is 100% efficient, and even higher fractions for more reasonable cooling efficiencies.

4   Odd Images and Galaxy Centers

Nearly all observed lenses have an even number of images (usually two or four). Lens theory, by contrast, predicts that each lens should have an additional "odd" image located near the center of the lens galaxy, although it is demagnified by high central density of the lens galaxy. At optical wavelengths an odd image would be swamped by light from the lens galaxy, but in a radio lens an odd image should be detectable. The lack of odd images in radio lenses thus places strong lower limits on the central densities of lens galaxies [28].

The CDM model galaxies predict that $\gtrsim 30\%$ of (radio) lenses should have detectable odd images, implying that the model densities are much too low on ~ 10 pc scales (see [18] for details). Steep central cusps ($\rho \propto r^{-\alpha}$ with $\alpha \simeq 2$) and/or central black holes can help suppress odd images, but for realistic parameter ranges neither offers an attractive solution. The lack of odd images in observed lenses thus remains a puzzle whose resolution will reveal interesting new constraints on the very inner parts of distant galaxies.

5   Lensing and CDM Substructure

One claimed problem with CDM is that the number of subhalos in CDM model galaxies is much larger than the number of satellite dwarf galaxies in the Local Group, which suggests that CDM overpredicts the amount of substructure in galaxy-mass halos [20], [25]. Two solutions have been proposed. On the one hand, changing the nature of the dark matter could reduce the power on small scales and eliminate the substructure [4], [9]. On the other hand, astrophysical processes such as photoionization could inhibit star formation in low mass systems, meaning that the CDM subhalos exist but are dark [6]. Dwarf galaxy surveys cannot distinguish between these scenarios. Tidal streams offer an alternate test, because they can be disrupted by encounters with subhalos [17], [23], but the observational evidence is not yet available.

Lensing offers a better test by being directly sensitive to mass in subhalos. Mass clumps in the lens galaxy introduce small-scale variations in the lensing potential that alter the flux ratios of the lensed images [8], [22], [24]. Dalal & Kochanek [10] show that the incidence of "anomalous" flux ratios[3] in 4-image lenses requires that $\sim 2\%$ of the mass be in small clumps on the scale $\sim 10^4$–10^8 $M_\odot$, which is in good agreement with the amount of substructure predicted by CDM. In other words, lensing strongly supports the scenario in which many subhalos exist but lack stars, and opposes changes to the nature of the dark matter that eliminates substructure.

To complement statistical analyses like [10], I have studied a single 4-image lens in detail using data at a variety of wavelengths to obtain constraints on individual mass clumps [19]. In B1422+231, the optical A/C flux ratio is largely consistent with smooth lens models while the radio A/C flux ratio is not (Fig. 2). Simultaneously explaining the optical and radio flux ratios and the shape of the radio image requires a mass clump in front of image A. A highly concentrated, point mass clump must have a mass $\sim 10^4$–10^5 $M_\odot$, while a more extended isothermal sphere must have a mass $\sim 10^6$–10^7 $M_\odot$. This is the first measurement of a particular clump lying in a distant galaxy ($z_l = 0.34$) and detected by its mass. Interestingly, there also appears to be a clump passing in front of image B, but this clump is probably just a star in the lens galaxy. In the future, detailed analyses of individual clumps as in B1422+231 will be combined with statistical analyses like [10] to constrain not only the substructure mass fraction but also the masses, densities, and sizes of dark subhalos, and the substructure mass function.

---

[3] Flux ratios that cannot be explained by smooth lens models.

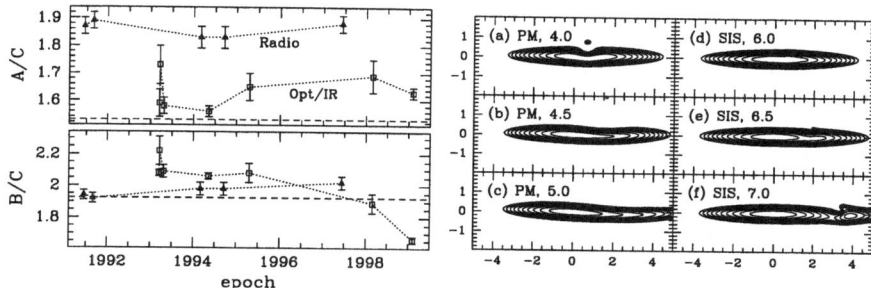

Fig. 2. *(Left)* Flux ratios for B1422+231 as a function of time. The dashed lines indicate the flux ratios predicted by a smooth lens model. See [19] for details (including references for the data). *(Right)* Maps of the radio image predicted by various sublensing models for image A, assuming infinite resolution. Results are shown for models with the clump treated as a point mass (PM) or a singular isothermal sphere (SIS); each panel gives the clump mass as $\log M$ (in $h^{-1} M_\odot$). The axes are labeled in mas; the contours are spaced by 0.2 dex

6   Conclusions

Lens statistics imply that the dark matter densities in the inner parts of elliptical galaxies are lower than predicted by CDM, in agreement with the conclusion from dynamical analyses of spiral galaxies. The CDM paradigm must therefore be modified to reduce dark matter densities on kiloparsec scales. Various mechanisms have been proposed ranging from astrophysics (disk bars that erase dark matter cusps [33]) to cosmology (a tilted power spectrum [1]) to particle physics (dark matter that is not collisionless and cold [4], [9], [30]).

Lensing also implies that lens galaxies have high densities on small scales ($\lesssim 10$ pc). The central densities of galaxies must be much higher than predicted in CDM model galaxies to explain the absence of central or "odd" images in observed lenses. The flux ratios in four-images lenses imply that a substantial fraction of the dark matter ($\sim 2\%$) lies in small-scale clumps rather than a smooth halo component [10], and B1422+231 suggests that a typical clump mass is $\sim 10^6 M_\odot$ [19]. Thus, while lensing supports other evidence that a mechanism is needed to reduce dark matter densities on kiloparsec scales, it also suggests that the mechanism must *not* remove structure on small scales – which argues against changing the nature of the dark matter particle.

Acknowledgments

Support for this work was provided by Steward Observatory at the University of Arizona, and by NASA through Hubble Fellowship grant HST-HF-01141.01-A from the Space Telescope Science Institute, which is operated by the Association of Universities for Research in Astronomy, Inc., under NASA contract NAS5-26555.

References

1. S. M. K. Alam, J. S. Bullock, D. H. Weinberg: preprint astro-ph/0109392 (2001)
2. M. L. Balogh et al.: MNRAS 326, 1228 (2001)
3. G. Blumenthal, S. Faber, R. Flores, J. Primack: ApJ 301, 27 (1986)
4. P. Bode, J. P. Ostriker, N. Turok: ApJ 556, 93 (2001)
5. G. Bruzual, S. Charlot: ApJ 405, 538 (1993)
6. J. .S. Bullock, A. V. Kravtsov, D. H. Weinberg: ApJ 539, 517 (2000)
7. J. .S. Bullock et al.: MNRAS 321, 559 (2001)
8. M. Chiba: preprint astro-ph/0109499 (2001)
9. P. Colin, V. Avila-Reese, O. Valenzuela: ApJ 542, 622 (2000)
10. N. Dalal, C. S. Kochanek: preprint astro-ph/0111456 (2001)
11. V. P. Debattista, J. A. Sellwood: ApJL 493, L5 (1998)
12. P. de Bernardis et al.: preprint astro-ph/0105296 (2001)
13. W. J. G. de Blok et al.: ApJ 552, L23 (2001)
14. O. Gerhard, A. Kronawitter, R. P. Saglia, R. Bender: AJ 121, 1936 (2001)
15. C. Gottbrath: MS thesis, University of Arizona (2000)
16. P. Helbig: preprint astro-ph/0008197 (2000)
17. R. A. Ibata, G. F. Lewis, M. J. Irwin: preprint astro-ph/0110690 (2001)
18. C. R. Keeton: ApJ 561, 46 (2001)
19. C. R. Keeton: preprint astro-ph/0111595 (2001)
20. A. Klypin, A. V. Kravtsov, O. Valenzuela, F. Prada: ApJ 522, 82 (1999)
21. H. Lin et al.: ApJ 518, 533 (1999)
22. S. Mao, P. Schneider: MNRAS 295, 587 (1998)
23. L. Mayer et al.: preprint astro-ph/0110386 (2001)
24. R. B. Metcalf & P. Madau: preprint astro-ph/0108224 (2001)
25. B. Moore et al.: ApJL 524, L19 (1999)
26. B. Moore: preprint astro-ph/0103100 (2001)
27. J. F. Navarro, C. S. Frenk, S. D. M. White: ApJ 462, 563 (1996)
28. D. Rusin, C.-P. Ma: ApJL 549, L33 (2001)
29. D. Schade, L. F. Barrientos, O. López-Cruz: ApJL 477, L17 (1997)
30. D. N. Spergel, P. J. Steinhardt: PRL 84, 3760 (2000)
31. D. Tytler, J. M. O'Meara, N. Suzuki, D. Lubin: Phys.Rep. 333, 409 (2000)
32. F. C. van den Bosch, R. A. Swaters: MNRAS 325, 1017 (2001)
33. M. D. Weinberg, N. Katz: preprint astro-ph/0110632 (2001)

Baryonic Dark Matter and Microlensing

Philippe Jetzer

Institute of Theoretical Physics, University of Zürich, Winterthurerstrasse 190, CH-8057 Zürich, Switzerland,   E-mail: jetzer@physik.unizh.ch

Abstract. The nature of the dark matter in the halo of our Galaxy is still largely unknown. The microlensing events found so far towards the Large Magellanic Cloud (LMC) suggest that at most about 20% of the halo dark matter is in the form of MACHOs (Massive Astrophysical Compact Halo Objects). The dark matter could also, at least partially, consist of cold molecular clouds (mainly H_2).

1   Microlensing Towards the LMC

The microlensing events found so far towards the LMC [1,2] lead to an optical depth of $\tau = 1.2^{+0.4}_{-0.3} \times 10^{-7}$, which implies that at most $\sim 20\%$ of the halo dark matter is in the form of MACHOs. The issue of the origin of MACHOs detected since 1993 remains still controversial. Although the events detected towards the Small Magellanic Cloud seem to be a self-lensing phenomenon [3,4], a similar interpretation of all the events discovered towards the LMC looks unlikely [1,5]. Indeed, the optical depth for LMC self-lensing is at most $(3 - 4) \times 10^{-8}$, so that not all MACHO events can be due to self-lensing [5]. Yet – even if most of the MACHOs are dark matter candidates lying in the galactic halo – their physical nature is unclear, since their average mass strongly depends on the still uncertain galactic model, ranging from $\sim 0.1\ M_\odot$ for a maximal disk up to $\sim 0.5\ M_\odot$ for a standard isothermal sphere. At first glance white dwarfs look as the best explanation, but the resulting excessive metallicity of the halo makes this option untenable, unless their contribution to halo dark matter is not substantial [6]. So, some variations on the theme of brown dwarfs have been explored [7]. Clearly, more microlensing observations are needed in order to be able to clarify the location of the MACHOs.

2 H_2 as Halo Dark Matter

The dark matter could also, at least partially, be made of cold molecular clouds (mainly H_2). Indeed, a few years ago we proposed a scenario [8–10], which predicts that dark clusters made of brown dwarfs and cold H_2 clouds should lurk in the galactic halo at galactocentric distances larger than $10 - 20$ kpc. Accordingly, the inner halo is populated by globular clusters, whereas the outer halo is dominated by dark clusters. In spite of the fact that the dark clusters resemble in many respects globular clusters, an important difference exists. Since practically no nuclear reactions occur in the brown dwarfs, strong stellar winds are presently lacking. Therefore, the leftover gas - which is ordinarily expected to exceed 60% of the original amount - is not expelled from the dark clusters but remains confined inside them. Thus, also cold gas clouds are clumped into the dark clusters. Although these clouds are primarily made of H_2, they should be surrounded by an atomic layer and a photo-ionized "skin".

An important prediction of our model is that high-energy cosmic-ray (CR) protons in the galactic halo scattering on the clouds should give rise to a detectable diffuse γ-ray flux from the halo of our galaxy. Indeed, an analysis of EGRET data has led to the discovery of a statistically significant diffuse γ-ray emission from the galactic halo. Dixon et al. [11] analysed the EGRET data concerning the diffuse γ-ray flux with a wavelet-based technique, using the expected (galactic plus isotropic) emission as a null hypothesis, and found a statistically significant diffuse emission from an extended halo surrounding the Milky Way. This emission traces a somewhat flattened halo and its intensity at high-galactic latitude is: $\Phi_\gamma(E_\gamma > 1\text{GeV}) \simeq 10^{-7} - 10^{-6} \gamma \text{ cm}^{-2} \text{ s}^{-1} \text{ sr}^{-1}$. This value is consistent with the one predicted by our model for the halo γ-ray flux at high-galactic latitude. Moreover, the comparison of our computed overall shape of the contour lines with the corresponding ones of Figure 3 in Dixon et al. [11] suggests that models with flatness parameter $q \sim 0.8$ are in better agreement with the data, thereby implying that most likely the halo dark matter in form of H_2 clouds is not spherically distributed.

Nevertheless, given the large uncertainties both in the data and in the model parameters one might also explain the observations with a nonstandard Inverse Compton (IC) mechanism, whereby γ-ray photons are produced by IC scattering of high-energy CR electrons off galactic background photons. Our calculation [12,13], however, points out that the corresponding IC contour lines decrease much more rapidly than the observed ones for the halo γ-ray emission (see Figure 3 in Dixon et al. [11]).

As M31 resembles our galaxy, the discovery of Dixon et al. [11] naturally leads to the expectation that the halo of M31 should give rise to a γ-ray emission as well [13]. Clearly, a good angular resolution of about one degree or less is necessary in order to distinguish between the halo and disk emission from M31. So, the next generation of γ-ray satellites like GLAST will hopefully be able to test the predictions of our model.

References

1. C. Alcock et al., *Astrophys. J.* **542**, 281 (2000)
2. A. Milsztajn and A. Lasserre, *Nucl. Phys.* **B** (Proc. Suppl.) **91**, 413 (2001)
3. P. Salati et al., *Astron. Astrophys.* **350**, L57 (1999)
4. G. Gyuk, N. Dalal and K. Griest, *Astrophys. J.* **535**, 90 (2000)
5. Ph. Jetzer, L. Mancini and G. Scarpetta, submitted (2001)
6. B. Gibson and J. Mould, *Astrophys. J.* **482**, 98 (1997)
7. F. De Paolis, G. Ingrosso, Ph. Jetzer and M. Roncadelli, *Mon. Not. R. Astron. Soc.* **294**, 283 (1998)

8. F. De Paolis, G. Ingrosso, Ph. Jetzer and M. Roncadelli, *Phys. Rev. Lett.* **74**, 14 (1995)
9. F. De Paolis, G. Ingrosso, Ph. Jetzer and M. Roncadelli, *Astron. Astrophys.* **295**, 567 (1995)
10. F. De Paolis, G. Ingrosso, Ph. Jetzer and M. Roncadelli, *Astrophys. J.* **500**, 59 (1998)
11. D.D. Dixon et al., *New Astronomy* **3**, 539 (1998)
12. F. De Paolis, G. Ingrosso, Ph. Jetzer and M. Roncadelli, *Astrophys. J.* **510**, L103 (1999)
13. F. De Paolis, G. Ingrosso, Ph. Jetzer and M. Roncadelli, *New Journal of Physics* **2**, 12.1-12.18 (2000)

Detecting Disc-Like Structures
in Early-Type Galaxies with Lensing

Ole Möller[1], Paul Hewett[2], and Andrew Blain[3]

[1] Kapteyn Institute, P.O. Box 800, 9700 AV Groningen, Netherlands
[2] Institute of Astronomy, Cambridge, UK
[3] California Institute of Technology, Pasadena, USA

The most striking difference between late- and early-type galaxies is the presence of pronounced discs in late-type galaxies. This difference is predicted by standard scenarios of galaxy formation. However, the possibility exists that small disc-like structures may also form in early-type galaxies; recent simulations by Naab & Burkert (2001) show that discs of sizes between 1 and 2 times the effective radius and masses up to 20% of that of the bulge may be present in ellipticals as a result of merging processes. Recent optical studies of the member galaxies of CL 1358+62 by Kelson et al. (2000) have shown that the two dimensional light profiles of many early-type galaxies are not fit well by a pure de Vaucouleurs law, and that the fit is improved considerably if an exponential, disc-like structure is included. The strong lensing properties of late-type galaxies are very sensitive to the properties of the disc component (e.g. Maller, Flores & Primack 1997, Möller & Blain 1998, Blain, Möller & Maller 1999). Using a single example, we show here how lensing could be used to detect disc-like structures in early-type galaxies and how their properties can be determined from lens modelling. A much more thorough investigation is presented in Möller, Hewett & Blain (in preparation).

We modelled elliptical lens systems as de Vaucouleur profiles, with scale length r_b and a total mass M_b. We add an exponential, thin disc of scale length r_d and total mass M_d. The disc is inclined to the line of sight by an angle θ_d and the bulge has ellipticity e. Assuming that both the disc and bulge axes are aligned along the x-axis, the total surface mass density can be written as

$$\Sigma = \Sigma_b \exp\left\{-7.67\left[\xi/r_b\right]^{0.25} - 1\right\} + \Sigma_d \exp\left[-\xi/r_d\right],$$

where Σ_b and Σ_d are central surface brightnesses for bulge and disc, respectively. The coordinate $\xi(x,y) = \sqrt{x^2 + y^2/(1 - e^2)}$. We assume a constant mass to light ratio which is normalised in such a way that the image separations are $\approx 2''$.

In the Figure we show an example fit to the simulated images for one of the galaxies in Kelson's catalogue (gal. id 242, Table 2 in Kelson et al. 2000). The lensing galaxy at redshift $z_l = 0.3$ has a total bulge mass of $M_b = 8.1 \times 10^{11} M_\odot$, a disc mass of $M_d = 4.2 \times 10^{10} M_\odot$ and scale lengths $r_b = 11.4$ kpc and $r_d = 0.97$ kpc for the bulge and disc, respectively. The disc is inclined at $\theta_d = 75$ deg and the ellipticity of the bulge $e = 0.3$. Both disc and bulge axes are aligned at 45 deg to the x-axis. In this model, the disc:bulge mass fraction within an Einstein radius is less than 5%.

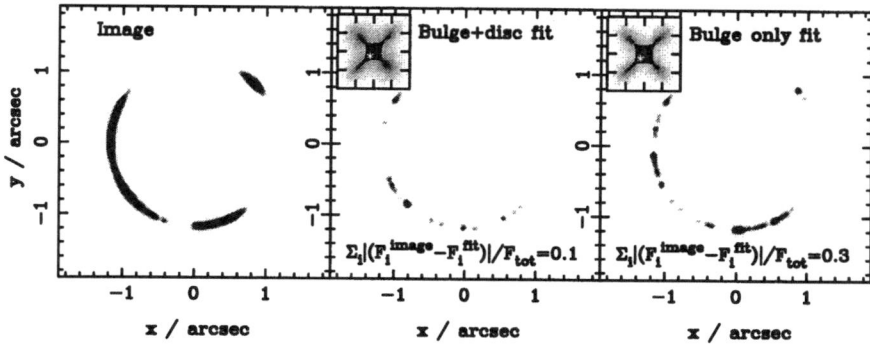

The left panel in the figure shows the simulated, 'true', image of a lensed, small ($\approx$ 0.01″ FWHM) gaussian source at redshift $z_s = 1$. The difference between the true image and the image predicted from the best fit model is shown for a bulge+disc model (middle panel) and a pure bulge model (right panel). The insets show the magnification maps on the source plane with darker shades corresponding to higher magnifications. The fits are obtained by fitting the positions and magnifications of the 4 magnified images. The assumed accuracy in image positions is ±0.05″, and that of the magnifications is ±0.1. To simulate effects of seeing we convolved both actual and reconstructed images with a PSF of 0.1″ FWHM. Comparison of the middle and right panel show quite clearly that the fit of a bulge+disc model to the actual image is much better than a pure bulge fit. For the bulge+disc model 10% of the total flux is reconstructed incorrectly, whereas for a bulge only model this fraction rises to 30%. In the case of the bulge+disc model the reconstructed total mass is within 10% of the true value and all the other parameters are within 20% of the true values. The correspondence in the case of the bulge only fit is much worse, scale length and ellipticity are off by more than 50% and the reconstructed total mass of the best fit model is half the true value. We also find that statistical properties, like the distribution of magnification ratios, are affected by the presence of discs in ellipticals. It also seems likely that the presence of discs in ellipticals may be able to explain some of the unusual magnification ratios observed; this will be investigated in more detail in Möller, Hewett & Blain (in preparation).

We would like to thank Steve Gull for providing us with the *bayesys2* minimisation code. This work was supported by the TMR LensNet network.

References

1. A.W. Blain, O. Möller, A.H. Maller: MNRAS **303**, 423 (1999)
2. D. Kelson, G.D. Illingworth, P.G. van Dokkum, M. Franx: ApJ **531**, 137 (2000)
3. A.H. Maller, Flores, Primack: ApJ **486**, 681 (1997)
4. O. Möller, A.W. Blain: MNRAS **299**, 845 (1998)
5. T. Naab, A. Burkert: ApJ **555**, L91 (2001)

Kinematics of Distant Galaxies from Keck

David C. Koo

UCO/Lick Observatory and Department of Astronomy and Astrophysics,
University of California,
Santa Cruz, CA 95064, USA

Abstract. DEEP is a two-phase spectral survey of faint field galaxies with the Keck Telescopes. The goals include exploring galaxy formation and evolution, mapping distant large scale structures, and constraining cosmology. DEEP, since its inception in the early 1990's, has been distinguished by an emphasis on studying the kinematics and masses of distant galaxies. The major DEEP survey in the second phase (DEEP2) is scheduled to begin in 2002 and will mainly aim for a sample of 50,000 galaxies to $I \sim 23$. Until then, the first phase of DEEP science programs will have been concentrating on using existing Keck spectrographs to undertake spectral surveys of over 1000 galaxies that have also been observed with HST. I will highlight the study of rotation curves of distant spirals; the fundamental plane of faint, high-redshift E/S0s; the narrow velocity widths seen in luminous blue compact galaxies; and the diversity of kinematics seen in a small sample of high redshift ($z \sim 3$) galaxies. These DEEP pilot programs have clearly demonstrated the feasibility, importance, and potential of using kinematics to better understand distant galaxies.

1 What is DEEP?

The first decade of the 21st century promises many new surveys of distant galaxies, especially with the advent of a suite of new 8-10 m class, ground-based, optical telescopes. Besides adding critical redshifts to data from space and other wavebands, the higher S/N and spectral-resolutions affordable with 8-10m telescopes provide three new and quite powerful diagnostics for the analysis of distant galaxies: internal velocities (i.e., kinematics and hence dynamical masses when size is also measured); chemical abundances; and star-formation-rate and stellar-population-age estimates. Compared to the traditional parameters of counts, colors, luminosities, and clustering properties of distant galaxies, these new diagnostics yield independent probes of galaxy properties in the early universe and have solid links to theoretical simulations of galaxy formation. Moreover, since both galaxy evolution and their large scale patterns involve a complex interplay of diverse galaxy classes, environments, and physical mechanisms and because precision cosmology via the volume test requires averaging over the fluctuations due to large-scale clustering, very large samples are essential to extract reliable results.

To meet the challenge, DEEP[1] was initiated over 9 years ago as a spectral survey of 50,000 [2] faint field galaxies, using the Keck II 10-m Telescope with a new spectrograph DEIMOS [2] [3]. The use of DEIMOS provides a clean division of DEEP into two parts or phases. The first is a set of pilot-style surveys of relatively small samples (10's - 1000) of galaxies. These pilot surveys exploit the pre-DEIMOS spectrographs available on Keck and were designed to determine feasibility and to refine the scope of DEEP2. DEEP is distinguished by aiming to gather internal kinematic data in the form of rotation curves or linewidths, as well as spectral-line measurements sensitive to star formation rates, gas conditions, stellar-population ages, and metallicity.

2   DEEP Highlights on Kinematics

To maximize the scientific returns for the small samples from our phase-one, pilot surveys, we only observed fields where HST WFPC2 images already exist, including the HDF and flanking fields [6], [13], [19], [18]; the Groth Strip Survey (GSS) [10] [20] [28] [31]; and Selected Area 68. Such HST images provide not only morphology and photometry but also the structure, size, and inclination data needed to convert kinematic observations from Keck into direct measures of dynamical mass.

The DEEP data reach to $I \sim 24$ and confirms that DEEP2 is feasible and that kinematics and masses will be worth the extra effort [11]. I will highlight our studies of distant spirals via rotation curves; luminous red spheroids and blue compact galaxies via velocity dispersions; and a few high redshift ($z \sim 3$) "Lyman-drop" galaxies for which spatially-resolved kinematics is possible.

2.1   Rotation Curves of Distant Spirals

As seen in Fig.1, we have clearly demonstrated that emission-line rotation curves of likely spirals can be measured with Keck's low resolution spectrograph (LRIS: [17]) to redshifts near $z \sim 1$ for galaxies as faint as $I \sim 22$ with 1–2 hour exposures [18]. Our new sample of about 100 rotation curves [30] [31] support the original conclusion that the optical Tully-Fisher relation for spirals near redshifts $z \sim 1$ show only modest (< 0.6mag) changes relative to that seen locally[28], [18]. These results appear on the surface to disagree with the claims for more extensive evolution of 1.5 to 2.0 mag [21] [23] [15], but the differences may reflect the selection criteria adopted. While our high-quality Tully-Fisher sample included galaxies that were slightly elongated and resolved along the slit, i.e. larger disk systems, the other samples were generally limited to very

---

[1] DEEP: Deep Extragalactic Evolutionary Probe; more details on participants and programs of DEEP can be found at URL: http://www.ucolick.org/~deep

[2] our original goal of 10,000 has been revised upwards to improve significantly the reliability of cosmological tests and of large scale structure studies

[3] DEIMOS: DEep Imaging Multi-Object Spectrograph; more information is provided at URL: http://www.ucolick.org/~loen/Deimos/deimos.html

blue or strong emission line targets, some of which may be very compact. We are currently expanding the completeness of our kinematic sample by including emission-line velocity widths in our studies (see contribution by B. Weiner). More solid conclusions on the Tully-Fisher relation (velocity vs. luminosity), as well as that of other scaling relations when one adds size or surface brightness, are critical to test whether various theories of disk formation are correct[3].

2.2   Spheroid/Bulge Evolution

We have undertaken several approaches to explore the evolution of luminous early-type galaxies, two of which exploit kinematics. In the first study[8], the luminosity function of over 100 early-type galaxies (E/S0) was derived, where the early-type class was selected on the basis of B/T being larger than 0.4; low levels of asymmetries; and high levels of smoothness using the GIM2D software for structural parameter extractions [24]. Keck redshifts were complemented by photometric redshifts. The main result is that there is evidence for roughly 1 magnitude of luminosity brightening back to redshifts $z \sim 1$, but otherwise no evidence for any dramatic drop in number density [8]. This result supports hierarchical models of galaxy formation in an open or accelerating universe rather than a high-Ω one (e.g., SCDM), in which early-type galaxies are presumed to have formed via merging of spirals in significant numbers since $z \sim 1$. In the second work [7], a more detailed study was made of the blue E/S0s. Kinematic information, including some very high resolution velocity widths as measured with an Echelle system (ESI: [22]), were used and we found nearly all to be low-mass systems, rather than bona fide massive E/S0 galaxies undergoing intense star formation.

In the third study[4], we explored the fundamental plane (luminosity:size:velocity dispersions) of 35 galaxies using LRIS data and found evidence for nearly

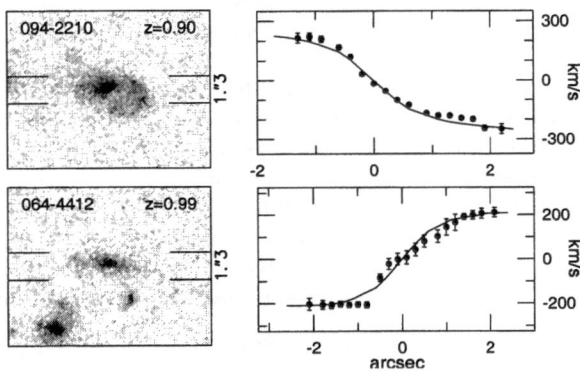

Fig. 1. Examples of the rotation curves measured for two high redshift galaxies [18], the upper with total $I \sim 21.4$ and the lower with $I \sim 22.4$. Besides the ID and redshift at the top, the images show the orientation and width of the slit

2 mag of luminosity brightening back to redshifts $z \sim 1$. This amount would nominally suggest recent formation, if one adopts only passive evolution after a single burst of star formation, but oddly, the colors of these galaxies were moderately red. This result was confirmed by a fourth[12], purely photometric study of the luminosities and colors of luminous bulge subcomponents rather than the entire galaxy. While the size-luminosity relation indicated nearly the same amount of luminosity evolution (~ 1.5mag), virtually all bulges were nevertheless found to be very red (restframe $U - B \sim 0.5$). A possible explanation for this paradox would be that a very metal-rich, old stellar population is later contaminated by small amounts of bluer young or metal-poor stars.

2.3   Luminous Blue Compact Galaxies

Though spatially-resolved rotation curves are preferred, most faint galaxies are too small to yield more than linewidths as kinematic data. Except for galaxies bright enough to yield *absorption* linewidths, emission lines are used (see contribution by B. Weiner). Assuming linewidths are reliable measures of the true gravitational potential (after an upward correction of 40% [21],[27] to match HI or Hα values of kinematics), and adding HST sizes, we are able to obtain masses.

The masses of blue compact galaxies have been found to be especially interesting. For some, we find that luminosity can be a poor gauge of their masses. Though many have the luminosities of massive galaxies(L^*), their velocity widths (σ) may be smaller than 30 km-s^{-1} as seen in the line profiles shown in Fig.2 [9], which were measured using the high resolution echelle spectrograph (HIRES: [32]). The resultant masses are very small and yield M/L ratios that span a wide

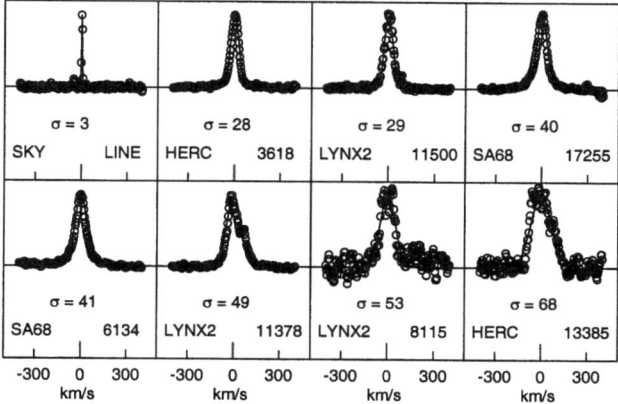

Fig. 2. Panel of HIRES [32] emission line profiles from a sample of luminous (L^*) blue compact galaxies at redshifts $z \sim 0.1$–0.7 [5]. The σ are the FWHM/2.35 velocity widths in km s^{-1}. The dynamic range of the M/L in rest-frame B for these compact galaxies spans a factor of 45. The HIRES instrumental profile is shown in the upper-left panel

range [5], [6], [19]. Our results suggest that some luminous blue compacts may be the progenitors of quiescent spheroidals today; that the down-sizing scenario [1] may apply to these galaxies [6]; and that such galaxies seen at $z < 1$ may be lower-redshift counterparts to the Lyman-drop galaxies seen at higher redshifts $z \sim 3$ [25]. A key point is that optical luminosities and mass are seen to be poorly correlated for this sample, i.e. stable and constant M/L may be a poor assumption at least for some classes of galaxies. This result clearly demonstrates the necessity, usefulness, and promise of kinematics as an important new dimension to discern the evolution of different galaxy populations.

2.4   High Redshift $z \sim 3$ Galaxies

A major advance with Keck has been the dramatic demonstration that galaxies chosen by multicolor photometry to be at very high redshift ($z \sim 3$) are confirmed to be so spectroscopically [25]. The DEEP team has extended the pioneering efforts in the Hubble Deep Field (HDF) [26] by pushing over one magnitude fainter; using redder "dropouts" to reach higher redshifts and higher levels of completeness; and adopting higher spectral resolutions to improve kinematic measurements [13]. Small motions observed from spatially-unresolved velocity widths [13] [17] and spatially-resolved kinematics indicate that some $z \sim 3$ galaxies appear to be small-mass systems that become dwarfs today or that later merge to form more massive galaxies [13], instead of being only the cores of massive galaxies in formation (which should yield high motions).

3   Summary

The main theme that arises from our DEEP pilot programs is that galaxy evolution is a complex problem. Galaxies are diverse in size, luminosity, structure, etc.; are composed of subcomponents which may experience different star formation and dynamical histories and evolution; and reside in a wide range of environments involving different physical mechanisms for their evolution. We have established that kinematics are both feasible with 8-10 m class telescopes and valuable for understanding distant galaxies. For example, we find relatively little evolution in the Tully-Fisher relation or disk surface brightness [23] to redshifts $z \sim 1$, as well as little evidence for evolution in the fundamental plane, volume density, or luminosity beyond that expected from passive evolution for early-type galaxies to $z \sim 1$. The colors of the spheroids and bulges are, however, redder than expected and thus a puzzle. On the other hand, luminous blue compact galaxies appear, whether at low redshifts $z < 1$ or at high redshifts $z \sim 3$, to have very low dynamical masses and are suggested to be possible progenitors of quiescent low-mass spheroidals today or the building blocks of larger galaxies rather than massive ellipticals undergoing formation via monolithic collapse.

The lessons from our DEEP phase-one pilot programs indicate great promise for our main survey DEEP2 of 50,000 galaxies. Such large numbers are vital for analysis after subdivision of the full sample by a wide range in luminosity, size,

M/L, structure, redshift, and environment. More relevant to this conference, we are optimistic that the kinematic data will yield new studies that rely on mass, including mass functions; M/L functions; Tully-Fisher, Fundamental Plane, and other scaling law evolution; mergers rates; dark matter distributions (halo vs disk; large scale structure vs mass); and precision cosmology (velocity function vs volume tests; estimates of the equation of state [8]).

Acknowledgements: DEEP was initiated by the Berkeley Center for Particle Astrophysics (CfPA), and has been supported by various other NSF, NASA, UC, and STScI grants. The senior members of DEEP have managed the project, but I would like to give special thanks to our talented pool of more junior astronomers over the years (see DEEP URL for names), without whom the results presented here would not have been possible.

References

1. Cowie, L. L., et al. 1996, *AJ*, **112**, 839
2. Davis, M., & Faber, S. M. 1998, *Wide Field Surveys in Cosmology*, eds. S. Colombi, Y. Mellier, & B. Raban (Edition Frontiers: Gif-sur-Yvette), 161-164
3. Faber, S. M., et al. 2001, *ASP Conf. Ser.*,**230**, 517-526
4. Gebhardt, K., et al. 2001, *ApJ*, submitted
5. Guzmán, R. et al. 1996, *ApJ*, **460**, L5
6. Guzmán, R. et al. 1997, *ApJ*, **489**, 559
7. Im, M., et al. 2001, *AJ*, **122**, 750
8. Im, M., et al. 2002, *ApJ*, in press
9. Koo, D. C. et al. 1995, *ApJ*, **440**, L49
10. Koo, D. C. et al. 1996, *ApJ*, **469**, 535
11. Koo, D. C. 2000, *Building Galaxies from the Primordial Universe to the Present*, eds. F. Hammer, et al. (Edition Frontieres: Gif-sur-Yvette) 279-287
12. Koo, D. C. 2002, *ApJ*, in preparation
13. Lowenthal, J. D. et al. 1997, *ApJ*, **481**, 673
14. Lowenthal, J. D. et al. 1998, *ASP Conf. Ser.*, **215**, 271
15. Malleń-Ornelas, G. et al. 1999, *ApJ*, **518**, L83
16. Newman, J. A. & Davis, M. 2000, *ApJ*, **534**, 11
17. Oke, J. B. et al. 1995, *PASP*, **107**, 375
18. Pettini, M. et al. 2001, *ApJ* **554**, 981
19. Phillips, A. C. et al. 1997, *ApJ*, **489**, 543
20. Phillips, A. C. et al. 2002, *ApJ*, in preparation
21. Rix, H.-W., et al. 1997, *MNRAS*, **285**, 779
22. Sheinis, A. I., et al. 2000, *Proc. SPIE*, **4008**, 522
23. Simard, L., et al. 1999, *ApJ*, **519**, 563
24. Simard, L., et al. 2002, *ApJ*, submitted
25. Steidel, C. C., et al. 1996, *AJ*, **112**, 352
26. Steidel, C. C., et al. 1996b, *ApJ*, **462**, L17
27. Telles, E., & Terlevich, R. 1993, *Ap&SS*, **205**, 49
28. Vogt, N. P. et al. 1996, *ApJ*, **465**, L15
29. Vogt, N. P. et al. 1997, *ApJ*, **479**, L121
30. Vogt, N. P. 1999, *ASP Conf. Ser.*,**193**, 145
31. Vogt, N. P. et al. 2002, *ApJ*, in preparation
32. Vogt, S., et al. 1994, *Proc. SPIE*, **2198**, 362

Galaxy Evolution from Emission Linewidths

Stéphane Courteau[1] and Young-Jong Sohn[2]

[1] Univ. of British Columbia, Dept. of Physics & Astronomy, BC V6T 1Z1 Canada
[2] Yonsei University, Center for Space Astrophysics, Seoul 120-749, Korea

Abstract. The major thrust of the Tully-Fisher (TF) surveys of distant galaxies is the measurement of *linewidths* rather than mere redshifts or colors. Linewidths are a measure of galaxy mass and should therefore be a more stable indicator of size than galaxy brightness, which can be badly affected by luminosity evolution. Masses may provide the best way to relate galaxies at different epochs, but for such a program to work, we must control systematic effects that could bias linewidth measurements at high redshift and skew comparisons with local Tully-Fisher calibrations. Potential sources of confusion in TF studies of galaxy structure and evolution include central or extended star bursts, infalling gas, turbulence and outflows, dust extinction, calibration of emission linewidths, and improper application of local TF calibrations to high redshift galaxies.

1 Introduction

Studies of galaxy structure and measurements of galaxy distances have often relied on the tight match between absolute luminosity and rotational linewidth, or Tully-Fisher relation (TFR). The best method for measuring spiral linewidths *locally* ($z \leq 0.05$) uses Hα rotation curves. For distant galaxies with $z > 0.4$, Hα is shifted out of the optical bandpass into a forest of night-sky lines, forcing reliance on bluer emission lines ([O II]λ3727, Hβ and [O III]$\lambda\lambda$4959,5007). Blue profiles may be affected by dust absorption more severely than Hα, and the radial distribution of [O II] and [O III] will depend on the abundance and temperature of the H II regions in a complicated way compared to Hα. For intermediate redshifts, spatially-resolved kinematics are no longer possible. Thus, one must first establish a calibration of blue and red, resolved[2D] and integrated [1D], galaxy linewidths in order to tie the local TFR to distant galaxies.

Courteau (1992; 1997; hereafter C92, C97) demonstrated the tight correlation between optical rotation curves[2D] and profiles[1D] and single-dish 21 cm profiles for a large sample of late-type spirals (see also Mathewson et al. 1992). Kobulnicky & Gebhardt (2000) extended this correlation to [O II] emission lines using a broad sample of Hubble types and showed that reliable linewidths could be obtained for distant galaxies. Blue linewidths for distant galaxies can therefore be compared to a local foil (Hα or H I) without introducing additional systematic effects. We confirm, and expand upon, the results of Kobulnicky & Gebhardt (2000) with a more extensive sample of Sc galaxies below. We also caution about important caveats in the application and interpretation of the TFR for distant galaxies.

1.1   Calibration of Blue Linewidths

Courteau (and Sandra Faber) collected optical long-slit spectra ([O II], Hβ, [O III], Hα, [N II]) of 20 Sc galaxies with the Lick 3-m telescope and Kast Double spectrograph (Courteau, Faber, & Sohn 2002; hereafter CFS02). Integration times were 1800s for each spectrum and R$\simeq$ 6000. Our observing strategy includes the measurement of narrow-slit major axis spectra and drift scans over the full size of the galaxy to simulate low-resolution integrated profiles as measured at large distances. The data were reduced according to standard procedures (C97); rotation curves were extracted by measuring intensity-weighted centroids at each spatial bins and accounting for instrumental broadening. The [O II] doublets were fitted with a double Gaussian function convolved with the instrumental profile. The extracted rotation curves were then modelled with a smooth function, e.g. an arctan, to obtain characteristic terminal or suitably chosen velocities. For late-type spirals, the velocity estimate, $V_{2.2}$, at 2.2 disk scale lengths (or equivalently $V_{1.3}$ at 1.3 effective radius) minimizes TF scatter (C97, Courteau & Rix 1999).

We verify that blue and red emission lines map the same velocity field (e.g. Fig. 1) though differences in spatial coverage, or flux ratios, exist. Thus, smooth fits to resolved rotation curves should yield matching ΔV's at all wavelengths. The rotation curve for UGC 11809 (Fig. 1) illustrates significant emission differences for Hα and blue linewidths at the center and outskirts of the galaxy. On average, blue and red emission lines trace the same extent (see Fig. 2), but for large systems ($R_{max} > 20$ kpc), Hα is systematically detected beyond the last point of blue emission.

We also construct 1-dimensional emission line profiles by collapsing the 2-dimensional rotation curves along the spatial axis (We account for the varying fraction of light at each radii covered by the long slit.) These profiles mimic the integrated linewidths of distant galaxies, and can be compared with lower resolution (and more realistic) drift scans. A suitable definition of integrated linewidth, measured at 20% of total flux, yields a good match to local 21cm linewidths (C97), both for collapsed profiles and drift scans. Note that it is *critical* that linewidths be measured similarly for calibrators and targets to prevent artificial biases.

We find a good correlation, with 15-20% scatter, between [O II], Hβ, and Hα *integrated* linewidths (CFS02). [O III] fluxes are comparatively low and [OIII] integrated profiles are generally too noisy for extragalactic investigations. For emission-line galaxies, the integrated flux at [OII] is on average four times greater than at Hβ; [O II] might therefore be favored as a kinematic tracer at high redshift. Hβ ΔV's are however more easily measured (the [O II] doublet requires careful deblending).

Also, in contrast with Kennicutt (1992), Jansen et al. (2001) show that Hβ is a considerably better tracer of star formation than [O II]. We suggest that Hβ and [O II] emission lines be secured whenever possible for linewidth confirmation and star formation estimates (see e.g. Charlot & Longhetti 2001).

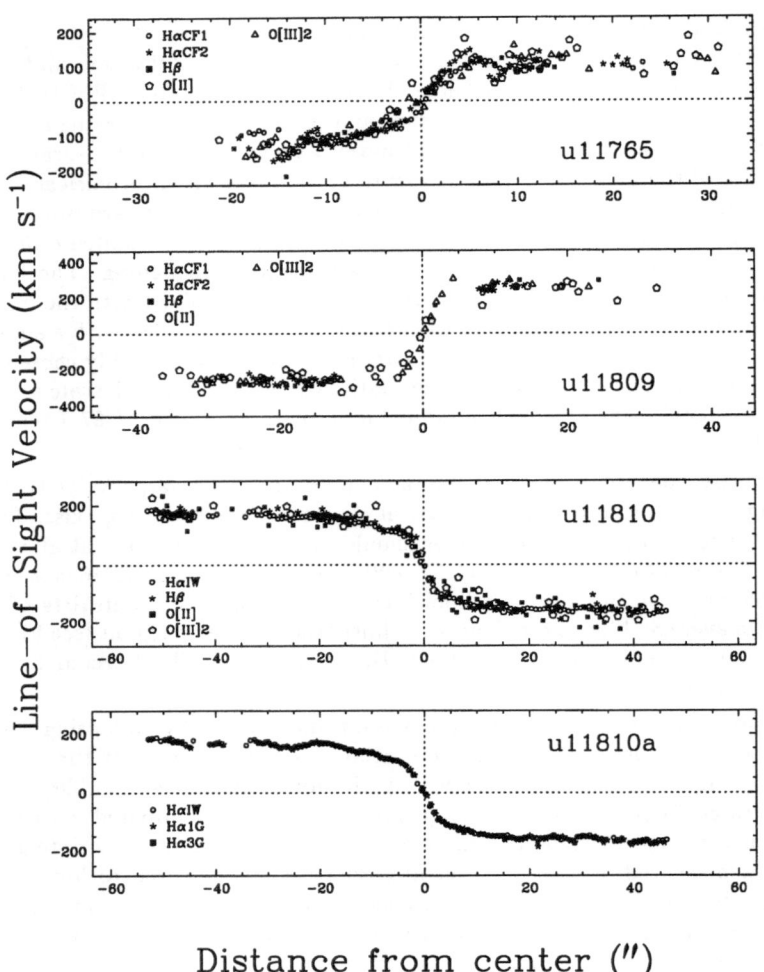

Fig. 1. Example of optical rotation curves for 3 UGC late-type spiral galaxies. HαCF refers to the compilation of Hα linewidths by Courteau (1992). The notation HαIW, Hα1G, and Hα3G refers to velocity centroids extracted using an intensity-weighted scheme, single-Gaussian fits (to the Hα line), and triple-Gaussian fits (to the Hα-[N II] complex). The different fitting techniques yield comparable results, as seen in the bottom panel. The match between different kinematic tracers (top three panels) is excellent

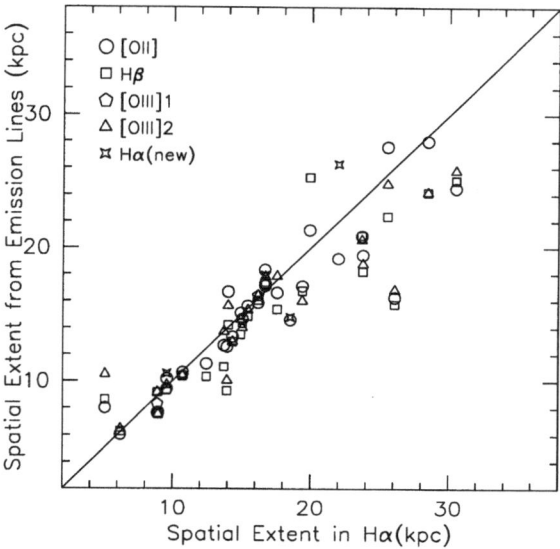

Fig. 2. Maximal extent of optical emission lines in late-type spirals. The $H\alpha$ measurements (x-axis) are taken from the CF sample (C92, C97)

1.2   Central Star Bursts

At intermediate redshifts, the rotation curves of spiral disks can no longer be resolved and studies of TF evolution must rely on integrated linewidths.

The major drawback of intensity-weighted profiles is that a flux integral favors the brightest H II and star-bursting regions (C92, C97). Unlike resolved rotation curves, integrated linewidths also suffer from seeing and instrumental broadening effects which further complicates their analysis.

Central bursts may cause important linewidth (mass) discrepancies beyond $z \sim 1$ where the population of bursting galaxies increases significantly (e.g. Steinmetz; this conference [tc]). Linewidths are usually measured relative to peak or total fluxes; either way, the effect of *central* starburst activity biases linewidths *low*. Galaxies with central bursts thus appear too bright (in a TF sense) for their linewidths, thereby mimicking luminosity evolution.

C92 compared linewidths of nearby galaxies with central starburst/AGN activity from resolved and integrated rotation profiles (see Fig. 3) and found that star-bursting bulges can yield linewidth measures that are 25-30% smaller than the true dynamical value. This corresponds to artificial offsets of $\sim$ 0.′′8 from the nominal TF relation, or a 75% increase in luminosity. Similarly, half-light radii can also be reduced by 40% or more by starburst activity. This *"starburst bias"* may explain the preponderance of blue compact (bursting) objects at intermediate redshifts (e.g. Guzman et al. 1997; Hammer et al. 2001; Koo [tc]) which would be the centers of massive, L^* or less, spiral galaxies (see e.g.

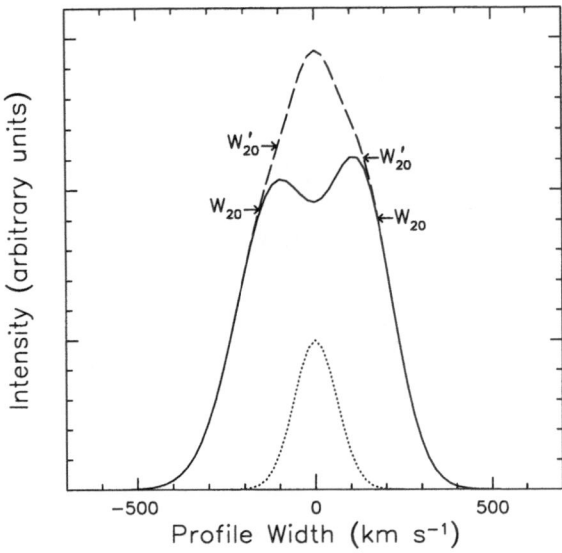

Fig. 3. Schematic representation of intensity-weighted rotation profiles with (dashed curve) and without (solid curve) a central starburst, based on an average of observations (C92). Linewidths measured at a fraction (e.g. 20%) of the total flux are biased *low* by the effect of a central starburst. The 20% linewidth baseline with and without starburst is measured between W'_{20} and W_{20}, respectively. The broad Gaussian-like emission profile from the star-bursting bulge is also shown as a dotted curve

Barton & van Zee 2001 [BvZ01]). BvZ01 have also considered the effect of star-bursting bulges on integrated linewidths and find somewhat milder trends than reported here but do not preclude more active bursts and thus larger offsets such as the ones we find. Effects which would bias integrated linewidths *high* include extended (non-central) star-bursts, infalling gas from satellites and dwarfs, turbulence, gas outflows, etc. Alternate definitions of linewidth measurements may help control some of these biases (CFS02).

AGN activity, from mergers or interactions, is likely to increase with redshift at a rate $\propto z^{2.3}$ (Patton et al. 2002), but we may be able to weed out bursting systems on the basis of their blue colors, especially if they are nuclear AGNs rather than starbursts (Kannappen et al. 2002), to the extent that a color-TF residual correlation can be identified at high z. Furthermore, little luminosity evolution has so far been detected up to $z \sim 1$ (e.g. Koo [tc]) and recent studies of TF evolution at intermediate redshifts based on integrated linewidths suggest only minor, if any, effects of central bursts (Vogt et al. 1997; Weiner [tc]).

1.3   Dust Extinction

The likely higher fraction of dust in the central parts of spiral galaxies at high redshifts is also a cause for concern. Dust extinction in the inner disk steepens the solid body part of the rotation curve, thus causing a central depression in the integrated profile. This, in turn, biases linewidths *high*. The comparison of rotation curves for blue and red emission lines (e.g. Fig. 1) shows no discernable effect due to opacity in the centers of these nearby, moderately inclined, disks (see also Prada 1996). Extinction effects, even at $H\alpha$, are most noticeable for tilted disks with $i > 84°$ and only 21-cm fluxes should be trusted in such cases (Courteau & Faber 1988). While our null test is reassuring for local TF studies, extinction effects should be revisited in TF studies of distant galaxies.

2   TF Applications at High Redshift

The greatest challenge to comparing local TFRs to distant galaxies is the matching of galaxy families. The nearby TF calibrators should be direct descendants of the more distant targets, or else structural/dynamical differences will be meaningless. Contrary to most local TF calibrations which are often heavily pruned (e.g. for cosmic flow studies), the range of Hubble types for distant galaxies is poorly controlled – in large part due to resolution and cosmological dimming effects.

A proper TF calibration sample must include a broad and unbiased range of morphologies to sample the general emission-line galaxy population – barred, starbursting, irregular, and all other types of spirals – as might be seen at high redshift. Only the "kitchen sink" calibrations (with all spiral types and morphologies), such as those of Barton et al. (2001) or Kannappan et al. (2002), should be used in cosmological TF studies (provided the calibration uses unambiguous, high-quality, magnitudes and linewidths, and is free of Hubble flow distortions). The intrinsic scatter of "all-inclusive" TF calibrations, $\sigma_{TF} \gtrsim 0.^{m}5$, is a full $0.^{m}1$-$0.^{m}2$ higher than values often quoted in studies of TF evolution and cosmic flows. Standardization of calibration and analysis techniques is also quite important, as scatter depends strongly on techniques. Renewed interest has also been given in B-band TF calibrations which match the targets' rest-frame luminosities. These calibrations have highest dispersions and are most sensitive to dust extinction corrections, thus complicating the work of TF practitioners.

Other important effects to luminosities and linewidths of distant galaxies which would mimic TF evolution include variations of the dust/gas and mass-to-light ratios, truncation of rotation curves due to interactions or $(1+z)^3$ surface brightness dimming, etc. See CFS02 for more details.

Luminosity-linewidth studies of distant galaxies, though far from trivial, hold the promise of very exciting advancements in our understanding of the evolution of galaxy populations.

References

1. Barton, E.J., Geller, M.J., Bromley, B.C., van Zee, L., & Kenyon, S.J. 2001, AJ, 121, 625
2. Barton, E.J., & van Zee, L. 2001, ApJ, 550, L35; see also these proceedings
3. Charlot, S., & Longhetti, M. 2001, MNRAS, 323, 887
4. Courteau, S. 1992, Ph.D. thesis, University of California, Santa Cruz [C92]
5. Courteau, S. 1997, AJ, 114, 2402 [C97]
6. Courteau, S. & Faber, S.M., 1988, in *The Extragalactic Distance Scale*, eds. S. van den Bergh & C. Pritchet, (San Francisco: ASP), Vol. 4, 366
7. Courteau, S., Faber, S.M., & Sohn, Y.-J. 2002, in preparation [CFS02]
8. Courteau, S. & Rix, H.-W. 1999, ApJ, 513, 561
9. Guzman, R., et al. 1997, 1997, ApJ, 489, 559
10. Hammer, F., Gruel, N., Thuan, T.X., Flores, H., & Infante, L. 2001, ApJ, 550, 570
11. Jansen, R. A., Franx, M., & Fabricant, D. 2001, ApJ, 551, 825
12. Kannappan, S. J., Fabricant, D. G., & Franx, M., 2002, AJ, in press
13. Kennicutt, R.C. 1992, ApJ, 388, 310
14. Kobulnicky, H.A. & Gebhardt, K. 2000, ApJ, 119, 1608
15. Mathewson, D.S, Ford, V.L., & Buchhorn, M. 1992, ApJS, 81, 413
16. Prada, F. 1996, PASP, 108, 549
17. Patton, D.R. et al. 2002, ApJ, in press
18. Vogt, N.P., et al. 1997, ApJ, 479, L121

Gas Minor-Axis Velocity Gradients in Early-Type Spiral Galaxies

E.M. Corsini, A. Pizzella, L. Coccato and F. Bertola

Dipartimento di Astronomia, Università di Padova,
vicolo dell'Osservatorio 2, I-35122 Padova, Italy

Abstract. We measured a remarkable gas velocity gradient along the minor axis of 5 early-type spiral galaxies. This phenomenon suggests the presence of a kinematically-decoupled component in orthogonal rotation with respect to the galaxy disk. If this is the case a second event has taken place in the history of the galaxy. Alternatively the gas velocity gradient is the result of non-circular motions induced by the potential of a triaxial bulge or of a bar.

Recently a velocity gradient has been measured along the minor axis of the disk of the two Sa galaxies NGC 4698 (Bertola et al. 1999) and NGC 4672 (Sarzi et al. 2000). In NGC 4698 the minor-axis velocity gradient is observed in both stars and ionized gas, while in NGC 4672 only in the stellar component. Such a stellar rotation along the disk minor axis indicates the presence of a kinematically isolated core, which is rotating perpendicularly with respect to the disk component. The analysis of the HST images of the nucleus of NGC 4698 shows that the isolated core is a nuclear stellar disk with a scalelength of few tens of pc (see Pizzella et al. in this volume). According to Bertola & Corsini (2000) the presence of these orthogonally-rotating isolated cores indicates that the entire galaxy disk could be the end result of a second event.

In order to look for possible links between morphology and decoupling in early-type spirals, we studied a sample of 10 S0/a and Sa galaxies selected to be morphologically similar to the few spirals known to host counterrotating components. By measuring their stellar and ionized-gas kinematics we found a new example of minor-axis velocity gradient which characterizes the gaseous component only. In fact in the innermost regions ($|r| < 1$ kpc) of 5 objects (Fig. 1) we observe a gas rotation along the galaxy minor axis. Since the rotation of the inner gas is confined into the bulge-dominated region and since bulges are generally triaxial (Bertola et al. 1991) we are left, with two alternative scenarios: *(a)* The kinematic decoupling between the inner and the outer gas is intrinsic (i.e. they have skewed angular momenta) and the two components are rotating around roughly orthogonal axes, which are the shortest and the longest axes of the triaxial bulge. If this is the case the two kinematically-decoupled gaseous components are settled onto the two orthogonal equilibrium planes of the bulge with the outermost gas on the bulge plane containing the disk. We propose this scenario for NGC 2855 (Corsini et al. 2002) due to the presence of a faint ring-like structure which surrounds the galaxy and is elongated in a direction close to the galaxy minor axis.

(b) The kinematic decoupling between the inner and the outer gas is apparent (i.e. they have parallel angular momenta) and the two components are both rotating in the equatorial plane of the triaxial bulge. In a non-axisymmetric potential gas in equilibrium moves onto closed elliptical orbits which become nearly circular as soon as we move at larger radii (de Zeeuw & Franx 1989). If this is the case the central velocity gradient measured along the minor axis is due to the gas non-circular motions. This seems to be the case of NGC 4586 where the peanut-shape bulge suggests the presence of a bar.

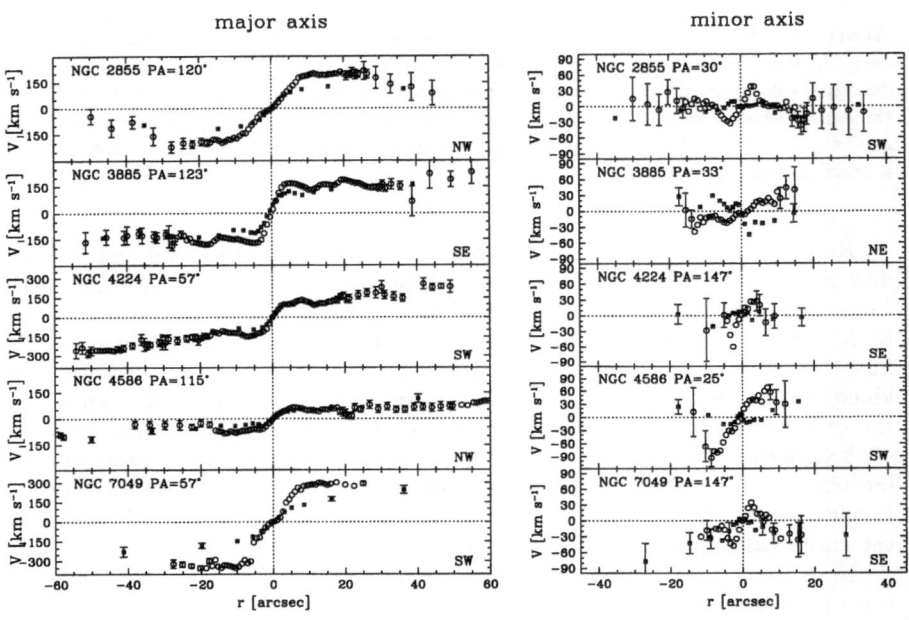

Fig. 1. Stellar (*filled squares*) and ionized-gas (*open circles*) kinematics along the major (*left panel*) and minor axis (*right panel*) of the Sa spirals NGC 2855, NGC 3885, NGC 4224, NGC 4586 and NGC 7049

References

1. Bertola, F., Corsini, E. M. 2000, in ASP Conf. Ser. 197, eds. F. Combes et al. (San Francisco: ASP), 115
2. Bertola, F., Vietri, M., Zeilinger, W. W. 1991, ApJ, 374, L13
3. Bertola, F., Corsini, E. M., Vega Beltrán, J. C., et al. 1999, ApJ, 519, L127
4. Corsini, E. M., Pizzella, A., Bertola, F. 2002, A&A, in press (astro-ph/0101592)
5. de Zeeuw, P. T., Franx, M. 1989, ApJ, 343, 617
6. Sarzi, M., Corsini, E. M., Pizzella, A., et al. 2000, A&A, 360, 439

Stellar M/L Ratios and Spiral Galaxy Dynamics

Roelof S. de Jong[1,2] and Eric F. Bell[2]

[1] STScI, 3700 San Martin Dr, Baltimore, MD 21218
[2] Steward Observatory, 933 N. Cherry Ave, Tucson, AZ 85721

Abstract. We present a simple technique to estimate mass-to-light (M/L) ratios of stellar populations in local universe galaxies based on two broadband photometry measurements, i.e. a color-M/L relation. The method is significantly better than using a fixed M/L, even in the near-IR. The main uncertainty stems from the assumed stellar IMF, which results in a zero-point uncertainty of the color-M/L relation. We constrain the zero-point using maximum disk rotation curve limits. We apply the color-M/L relation to the Tully-Fisher relation and show there is a universal stellar mass Tully-Fisher relation, independent of observed passband. We also apply the relation to galaxy rotation curves, but have to conclude that the current data is insufficient to constrain dark matter profile shapes across the full range of galaxy luminosities.

1 Galaxy Evolution Models and Mass-to-Light Ratios

We have used galaxy evolution models to investigate stellar M/L ratios of galaxies. Our models were tuned to fit the observed trends between the colors and the structural parameters of spiral galaxies [1,2]. Using a local gas density dependent star formation law, the photometric evolution is calculated, taking chemical evolution into account. As well as a closed box model we have models with gas infall and outflow, mass dependent formation epochs and star bursts. All models show large variations in M/L, amounting from a factor 8 in B to 2 in K, but in all models we find a strong correlation between M/L and optical color (e.g. Fig. 1a). The slope of the color-M/L relation is very robust against the particular stellar population synthesis model and against the exact details of the galaxy evolution model. The main uncertainty in the correlation is the zero-point, which is determined by the assumed stellar IMF.

A constraint on the color-M/L correlation zero-point can be obtained from galaxy rotation curves. The stellar disk in a galaxy cannot be more massive than allowed by its rotation curve, resulting in a maximum-disk M/L. Using the data from [5], we find that a standard Salpeter IMF over-predicts the maximum allowed mass for many galaxies, but a Salpeter IMF with a flat slope below $0.6\,M_\odot$ is consistent with all data (for $D_{\mathrm{UrsaMajor}} = 20$ Mpc).

2 Tully-Fisher Relations and Rotation Curves

Observed TF-relations are known to have a passband dependence, both in slope and in zero-point. After applying the extinction corrections of [4] and our color-M/L correlations, we can calculate stellar mass TF-relations from the observed

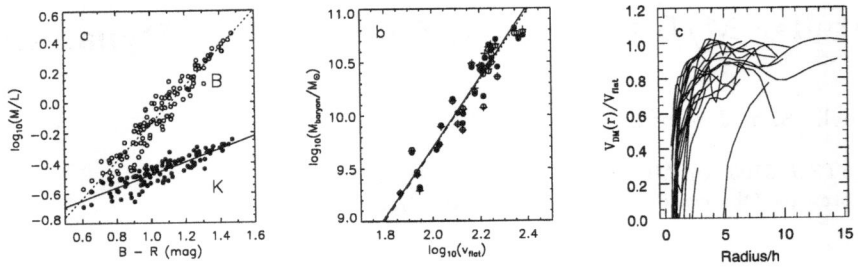

Fig. 1. a) M/L_B and M/L_K versus $B-R$ for our mass dependent formation epoch model with star bursts. **b)** Baryonic TF-relations derived from B (○), I (+) and K (•) observations. **c)** Dark Matter rotation curves derived by subtracting the stellar component from the observed rotation curves

TF-relations. We find that the stellar mass TF-relations derived from the different passbands are equal to within the uncertainties. By adding in the HI gas mass we can calculate the baryonic TF-relations (Fig. 1b). We find that the slope of the baryonic TF-relation must be less then 3.5 ± 0.2, significantly lower than found by [3], mainly due to our exclusion of low luminosity dwarfs with poorly determined inclinations and rotation velocities and a difference in assumed distance to the Ursa Major Cluster [see also McGaugh in these proceedings].

We have used our color-M/L relation to investigate the mass profiles of Dark Matter (DM) halos. We calculated radial stellar mass profiles for the galaxies of [5] from their luminosity and color profiles using our color-M/L relation. Subtracting the stellar and HI mass contributions from the observed rotation curves leaves in principle the DM rotation curves. We show stellar mass subtracted curves in Fig. 1c, normalized by disk scalelength and the velocity at the flat part of the rotation curve. Unfortunately, neither this nor any other form of normalization reveals a clear "universal" DM rotation curve. There are indications that this can be partly attributed to the lack of resolution in the HI rotation curves. We are currently processing high-resolution rotation curves for a sample of galaxies with deep near-IR data to address this issue in much more detail.

Support for RSdJ was provided by NASA through Hubble Fellowship grant #HF-01106.01-A from the Space Telescope Science Institute, which is operated by the AURA, Inc., under NASA contract NAS5-26555. Support for EFB was provided by NASA LTSA grant NAG5-8426 and NSF grant AST-9900789.

References

1. E.F. Bell, R.S. de Jong: MNRAS **312**, 497 (2000)
2. E.F. Bell, R.S. de Jong: ApJ **550**, 212 (2001)
3. S.S. McGaugh, J.M. Schombert, G.D. Bothun, W.J. de Blok.: ApJL **533**, L99 (2000)
4. R.B. Tully et al.: AJ **115**, 2264 (1998)
5. M.A.W. Verheijen, PhD thesis, Univ. of Groningen (1997)

Systematic Variations in Bulge and Disk Parameters

Lauren MacArthur[1], Stéphane Courteau[1], and Jon Holtzman[2]

[1] University of British Columbia, Vancouver BC, Canada
[2] New Mexico State University, Las Cruces NM 88003-8001, USA

1 Introduction

Disk scale lengths are a key ingredient to analyses of galactic structure and the modeling of stellar populations, dust content, and masses of spiral galaxies, yet they remain one of the most poorly constrained galaxy observables. Indeed, no universal definition of a scale length has ever been clearly defined. Published scale lengths often differ by 20% [4]. Errors in the sky and seeing estimates, choice of disk spatial extent and bulge shape parameter, and wavelength dependencies are some of the most important effects that must be considered in a rigorous determination of disk scale lengths. Additionally, many disk luminosity profiles show significant departures from a pure exponential and must be treated with care. We introduce below a new study aimed at refining our understanding of galaxy structural parameters.

2 Surface Brightness Profiles

The traditional description of the radial surface brightness (SB) profile of spiral galaxies is the sum of a bulge and exponential disk components. Two types of SB profiles are normally defined: Type I/II if the bulge-disk transition region lies above/below the inward disk extrapolation. We also introduce a third "transition" type for profiles that change from Type II at optical wavelengths to Type I in the infrared. A transition of this sort would be expected if the Type II dip is caused by dust extinction. In addition, we observe that many galaxies classified as Type II show a weakening of the inner profile dip at longer wavelengths. Transition galaxies are likely just a case of lesser dust content, whereas Type II systems remain optically thick, even at H-band.

Model profiles of spiral galaxies with different amounts and geometries of disk light absorption by dust have been used to characterize its effect on the disk scale length at different wavelengths [3]. Models with B-band central optical depths of 5 and a large range of dust-to-stars layering exhibit a clear Type II-like dip which indeed weakens at longer wavelengths. Stellar population effects cannot be ruled out as an explanation for the Type II dip.

3   The Data

We used a sub-sample of 123 galaxies from the multi-band (BVRH) catalog of UGC late-type high-SB (HSB) galaxies of Courteau & Holtzman with face-on and intermediate inclinations. Most galaxies have at least one set of BVRH images and multiple observations are used for 54 galaxies to estimate systematic errors. The profiles typically reach below 26 BVR and 22 H-mag arcsec^{-2}.

4   Bulge-to-Disk Decomposition Technique

One must derive bulge and disk parameters simultaneously in order to obtain reliable luminosity decompositions and disk scale length estimates. Inclusion of bulge light in disk fits can significantly bias disk scale lengths. Accordingly, we model bulge and disk components as generalized Sérsic and exponential profiles respectively, using 1D and 2D non-linear chi-square minimization routine. The bulge and disk model functions are convolved with a Gaussian PSF (with appropriate seeing FWHM). Surface brightnesses with random errors greater than 0.12 mag arcsec^{-2} are excluded; this effectively defines the spatial extent of our galaxy profiles. We have also tested the reliability of our B/D decompositions with extensive simulations.

5   Results

- Sky/seeing error estimates and the choice of disk spatial extent and bulge shape can cause significant errors (>50%) in the measurement of galaxy parameters. A rigorous analysis of Type I profiles based on B/D decompositions yields bulge and disk scale parameters with errors less than 20% and 4% respectively.
- Type I profiles appear to define the natural underlying stellar density distribution of all HSB spiral galaxies. Additional SB features are caused by dust and stellar populations.
- There is a range in the bulge shape parameter for late-type bulges with a mean value close to the exponential case (Sérsic $\langle n \rangle = 0.94 \pm 0.48$).
- There is a significant trend of decreasing scale length with increasing wavelength (also found in [2]).
- We confirm the finding by [1] that late-type spirals are best described, on average, by a double-exponential SB profile, with tight coupling between the bulge and disk parameters. This is best understood in a scenario where late-type bulges form via secular evolution of the disk.

References

1. Courteau, S., de Jong, R. S., & Broeils, A. H. 1996, ApJ, 457, L73
2. de Jong, R. S. 1996, A&A, 313, 45
3. Evans, R. 1994, MNRAS, 266, 511
4. Knapen, J. H. & van der Kruit, P. C. 1991, A&A, 248, 57

The Evolution of Distant Cluster Spirals

Bo Milvang-Jensen[1], Alfonso Aragón-Salamanca[1], George Hau[2], Inger Jørgensen[3], and Jens Hjorth[4]

[1] School of Physics and Astronomy, University of Nottingham, UK
[2] European Southern Observatory, Chile
[3] Gemini Observatory, USA
[4] University of Copenhagen, Denmark

Ground-based and HST observations indicate that the disk galaxy population in rich galaxy clusters has experienced remarkable evolution since $z = 1$. This contrasts with the mild evolution observed in the field spirals to $z \sim 1$ (cf. [2] and references therein). To quantify the evolution of the cluster spirals, we are studying the Tully–Fisher relation for morphologically-classified disk galaxies in rich galaxy clusters at $0.2 < z < 0.9$. The first cluster studied was MS1054−03 at $z = 0.83$. Multi-object spectroscopy was obtained at the VLT using two masks, with the slits aligned with the major axes of the galaxies. Rotation (tilted emission lines) was seen in ∼10 cluster spirals and 25 field spirals. Rotation curves were derived for the brighter emission lines in the sample from Gaussian fits at each spatial point along the slit (Fig. 1). For all the emission lines, $V_{rot} \sin i$ was also estimated visually from the 2-D spectra. The rotation curves in Fig. 1 have not been corrected for the effect of seeing and slit-width.

Fig. 2 shows a preliminary Tully–Fisher relation for a subsample of our galaxies. The magnitudes and ellipticities (inclinations) were measured in HST F814W images. The figure shows that at a fixed V_{rot} the high z cluster galaxies appear to be brighter on average than the high z field galaxies, and than the local relation. It is tempting to interpret this as the result of enhanced star formation on spiral galaxies falling onto the cluster. Our sample preferentially contains star forming spirals since we selected emission line galaxies from a rest frame B-magnitude limited sample. One could speculate that after this initial episode of enhanced star formation, the interaction with the intergalactic medium will remove much of the gas in these galaxies (cf. [1]), and the star formation will cease. After an E+A phase the spirals could turn into S0s. However, we must stress that a larger cluster sample, covering a range of redshifts, a more careful analysis of the data and detailed modelling are necessary to reach firm conclusions.

Acknowledgements – Generous financial support from the Danish Research Training Council (BM-J) and the Royal Society (AA-S) is acknowledged. This work is partially based on observations collected at ESO, Chile (N° 66.A-0376).

References

1. V. Quilis, B. Moore, R. Bower: Science **288**, 1617 (2000)
2. N. P. Vogt: ASP Conf. Ser. **197**, 435 (2000)

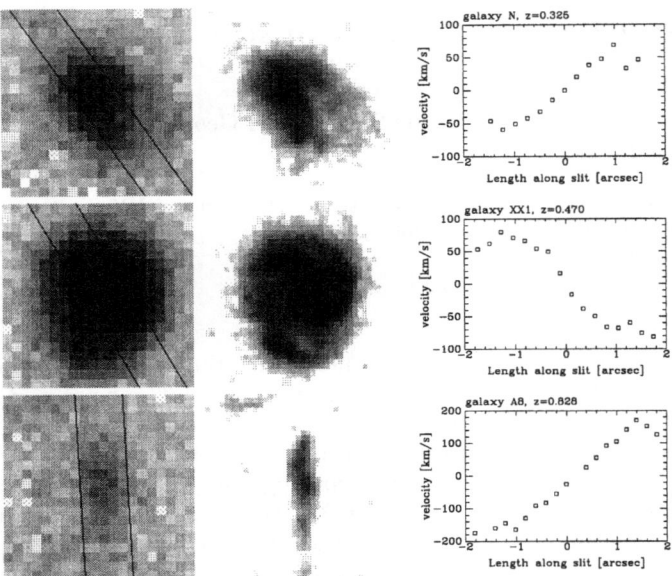

Fig. 1. Three examples of the galaxies for which we have obtained rotation curves. From left to right the panels show: FORS2 R-band images (4″ × 4″ section, 0.6″ seeing) with the 1″ slits overlayed; the WFPC2 F606W images; and the preliminary VLT rotation curves with no correction for the line-of-sight inclination of the galaxy, nor for the effect of the slit-width and seeing. The first two galaxies are field galaxies, and the third one is one of our faintest cluster galaxies. Their magnitudes are $R = 21.9$, 20.8 and 24.0 from top to bottom

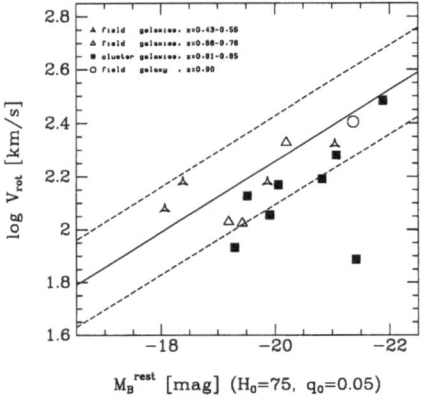

Fig. 2. Preliminary high redshift cluster and field Tully–Fisher relation. The solid line shows a local HI based TF relation for 32 cluster spirals (cf. [2]). The dashed lines show the 3 sigma limits. Filled symbols are cluster spirals, and open symbols field ones. The deviating cluster galaxy has uncertain morphological classification: edge-on disk with dust at the centre, or two almost overlapping edge-on disks

Nuclear Stellar Disks in Spiral Galaxies

A. Pizzella[1], E.M. Corsini[1], L. Morelli[1], M. Sarzi[1], C. Scarlata[2], M. Stiavelli[2], and F. Bertola[1]

[1] Dipartimento di Astronomia, Universitá di Padova, vicolo dell'Osservatorio 2, I-35122 Padova, Italy
[2] Space Telescope Science Institute, 3700 San Martin Dr., Baltimore, MD 21218, USA

Abstract. We report the discovery of a nuclear stellar disk in 3 early-type spirals, namely NGC 1425, NGC 3898 and NGC 4698, revealed by WFPC2/F606W images out of a sample of 38 spiral galaxies, selected from the HST Data Archive. We derived their central surface brightness and scalelength by assuming them to be infinitesimally thin exponential disks. No nuclear disk was found in barred galaxies or galaxies of Hubble type later than Sb. The disks have scalelength of about 20 to 30 pc and are the first of this kind found in spiral galaxies. The external origin of the disk in NGC 4698 is proved by its orthogonal geometrical decoupling with respect to the host galaxy.

1 Results

Our work is based on the application of the unsharp mask technique on the nucleus of the sample galaxies. With this technique we are able to efficiently identify galaxy images affected by dust absorption and discard them (74 galaxies out of 112 of our initial sample), and to identify the presence of nuclear stellar disks in the remaining sample of 38 galaxies not affected by dust. Successively, we measured the photometric radial profile (μ, ϵ, PA, a_4, a_6) in order confirm the presence of the nuclear disks and to measure their photometric parameters adopting the photometric method introduced by Scorza & Bender (1995). In Fig.1 we show the unsharp images of the nuclei of the 3 galaxies hosting a nuclear disks as well as 2 cases where the disk was not found or the presence of dust prevented further analysis.

From our work, it seems that the presence of nuclear disks is restricted to S0's and bulge-dominated unbarred spiral galaxies. In the framework of massive bulge formation through a process of hierarchical clustering merging, the nuclear disks may be the final result of dissipational and star-formation processes subsequent to a second acquisition event.

With NGC 4698 we showed for the first time that second events represent a viable mechanism to build a nuclear disk in the center of disk galaxies. Indeed the nuclear disk of NGC 4698 is geometrically (this paper) and kinematically (Bertola et al. 1999) decoupled in an orthogonal way with respect to the host galaxy. This phenomenon can be hardly explained without invoking the acquisition of external material from the galaxy outskirts (see Bertola & Corsini 2000).

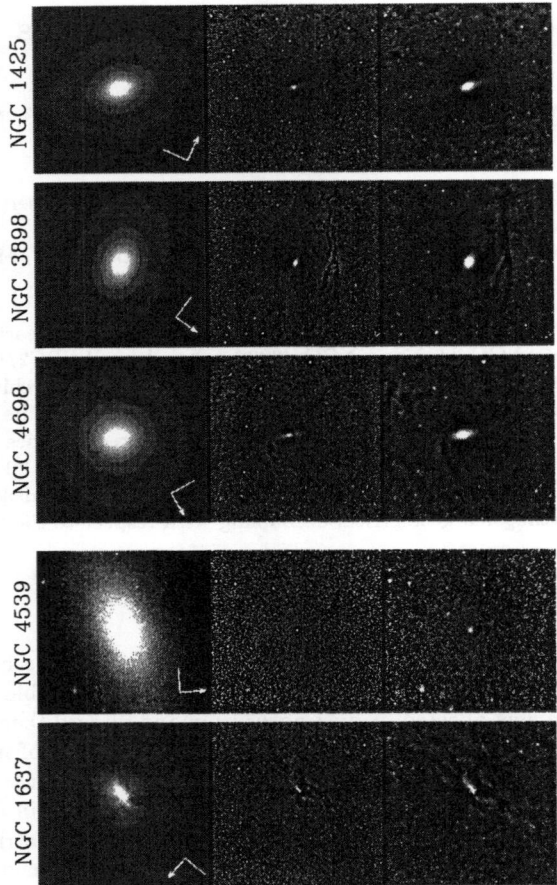

Fig. 1. *Left panels:* WFPC2/F606W images of NGC 1425, NGC 3898 and NGC 4698 (where we found a nuclear disk), NGC 4539 (where a nuclear disk is not present), and NGC 1637 (where dust patches prevent any further analysis). The size of the plotted region is $19''.3 \times 19''.3$. The orientation is specified by the arrow indicating north and the segment indicating east in the lower right corner of each panel. *Middle and right panels:* Unsharp masking of the WFPC2/F606W images obtained with $\sigma = 2$ and 6 pixels, respectively. Sizes and orientations are the same as in the left panels

References

1. Bertola, F. & Corsini, E. M. 2000, in ASP Conf. Ser. 197, eds. F. Combes et al. (San Francisco: ASP), 115
2. Bertola, F., Corsini, E. M., Beltrán, J. C. V., Pizzella, A., Sarzi, M., Cappellari, M., & Funes, S. J. 1999, ApJL, 519, L127
3. Scorza, C., & Bender, R., 1995, A&A, 293, 20

Measuring Galaxy Disk Mass with the SparsePak Integral Field Unit on WIYN

Marc Verheijen[1], Matthew Bershady[1], and David Andersen[2]

[1] University of Wisconsin, Madison WI 53706, USA
[2] Max Planck Institute for Astronomy, Heidelberg, Germany

We present first results from the commissioning data of the SparsePak Integral Field Unit on the WIYN telescope. SparsePak is a bundle of 82 fibers, arranged in a sparsely filled, 76×77 arcsec hexagonal grid. It pipes light from one of the Nysmith foci to the Bench Spectrograph. See http://www.astro.wisc.edu/~mab/research/sparsepak and Bershady et al. (2002a) for more details. The fibers are each 5 arcsec in diameter and are thus suitable to obtain spectroscopy on extended objects of lower surface brightness such as galaxy disks. In fact, SparsePak was purposely developed to measure stellar kinematics and velocity dispersions in disk galaxies with the ultimate goal of constraining the surface densities and masses of stellar disks. Such a measurement is crucial to improve our understanding of the principal dynamical components in disk galaxies from rotation curve decompositions.

As part of the SparsePak commissioning program we observed several galaxies, one of which was NGC 3982, a blue and very high surface brightness galaxy with $\mu_0^{obs}(B)=19.3$ mag/arcsec^2 and $B-K'=3.4$ (Tully et al. 1996). It was mapped in three spectral regions; around 5130Å(MgI), 6680Å(Hα) and 8660Å(CaII-triplet) with FWHM spectral resolutions of 24, 16 and 37 km/s respectively. About 3 dozen template stars with a range in T_{eff} and surface gravity were observed as well by drifting them across many fibers.

Three pointings were taken in the Hα line to fill the hexagonal grid. This yielded a contiguous Hα velocity field of very high signal-to-noise. Fitting a tilted-ring model to this velocity field yielded a kinematic inclination of 26 ± 2 degrees, consistent with the isophotal ellipticity, and an Hα rotation curve at 5 arcsec resolution which supplements a low resolution HI rotation curve.

Stellar absorption line observations were taken with a single SparsePak pointing, aimed at measuring the vertical velocity dispersion of the stars. Here we focus on the stellar velocity dispersions measured from the deepest CaII line. In the inner regions of the galaxy, the signal-to-noise in the spectra of individual fibers is high enough to determine the line centroids and to construct a stellar velocity field. Comparison of the Hα and stellar velocity fields clearly shows the effects of asymmetric drift inside 1.5 disk scale lengths h_R. Outside this radius the asymmetric drift approaches zero.

The layout of the fibers and the near face-on orientation of the galaxy allows for azimuthal averaging of 6, 6, 6, 12 and 18 fibers in 5 annuli to improve the signal-to-noise in the CaII absorption line in the outer regions of the galaxy.

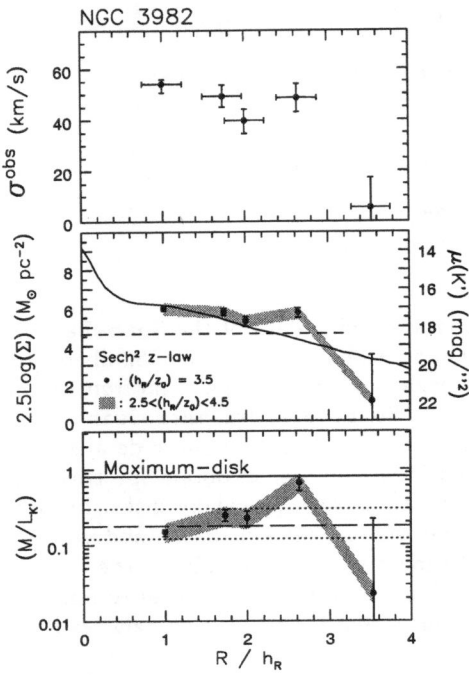

Fig. 1. Upper panel: Azimuthally averaged stellar velocity dispersions, corrected for instrumental and template broadening. Horizontal bars indicate the radial bin widths. Note the sudden drop beyond 3 disk scale lengths. Middle panel: Points indicate the inferred mass surface density for an isothermal vertical density profile and a range of scale heights. The solid line shows the radial K' surface brightness profile. Bottom panel: derived M/L of the disk in the K' band. The solid line shows the maximum disk value allowed by the rotation curve. The dashed line indicates the weighted radial average value and the dotted lines relate to the dotted lines in the right panel of Figure 2

The outermost annulus, with 18 fibers, has a radius of 3.5 h_R. Before averaging, the projected rotational velocities were taken out, using the centroids of the high signal-to-noise Hα lines. Stellar velocity dispersions were determined by convolving the spectrum of a K0.5-III template star with a Gaussian of varying FWHM and finding the best match to the azimuthally averaged galaxy spectra.

The results are summarized in Figure 1 and in Bershady et al. (2002b). The plotted velocity dispersions are the σ of the convolution Gaussian. No corrections were applied for contributions from the projected radial and tangential velocity dispersion components which are estimated to contribute less than 13%. The measured velocity dispersion is fairly constant with radius except for the outermost point where a significantly lower velocity dispersion is observed. Assuming an isothermal (sech^2) vertical density profile, a mass surface density can be derived for a variety of scale heights based on results from recent work by Kregel et al. (2002). This mass surface density includes any non-stellar component in the disk and is a factor ∼3.5 higher than the surface density of the Milky Way in the solar neighborhood (Kuijken & Gilmore 1991), indicated by the short dashed line in the middle panel. Given the K' luminosity profile, the mass-to-light ratio of the disk can be computed as a function of radius. A weighted radial average of $(M/L_{K'})=0.18$ is used in a rotation curve decomposition shown in Figure 2 which implies a substantially (∼ 25%) sub-maximum, yet very high SB disk.

This work was supported by NSF grant AST-9970780.

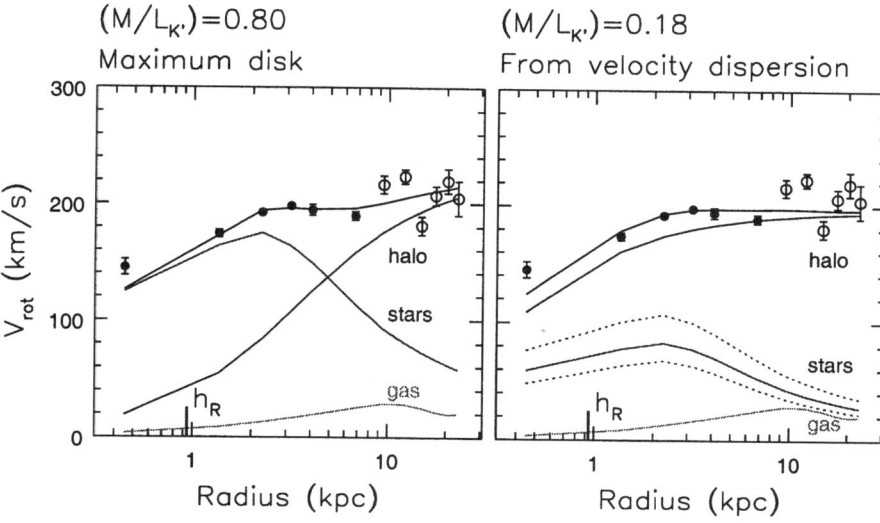

Fig. 2. Rotation curve decompositions for NGC 3982. Solid points are from SparsePak, open points are from HI measurements. Left panel: maximum-disk decomposition which implies a stellar mass-to-light ratio of 0.80 in the K' band. Right panel: rotation curve decomposition with a disk M/L of 0.18 based on the stellar velocity dispersion measurement of SparsePak. This suggests that NGC 3982 has a strongly sub-maximum disk. This is surprising given the fact that the stellar disk has a very short scale length and and extremely high surface brightness

References

1. M.A. Bershady, D.R. Andersen, J. Harker, M.A.W. Verheijen, L.W. Ramsey 2002a, PASP, in preparation
2. M.A. Bershady, D.R. Andersen and M.A.W. Verheijen 2002b, in *D*isks of Galaxies: Kinematics, Dynamics and Perturbations, eds. L. Athanassoula, I. Puerari, ASP Conference Series, in press
3. K. Kuijken and G. Gilmore 1991, Ap.J., 367, L9
4. M. Kregel, P.C. van der Kruit and R. de Grijs 2002, MNRAS, in press
5. R.B. Tully, M.A.W. Verheijen, M.J. Pierce, J.-S. Huang, R.J. Wainscoat 1996, A.J., 112, 2471

Evolution in the Linewidth-Magnitude Relation to $z > 1$ from the DEEP Groth Strip Survey

Benjamin J. Weiner and the DEEP Collaboration

UCO/Lick Observatory, 1156 High St. Santa Cruz, CA 95064, USA

Abstract. Galaxy kinematics are potentially a powerful tool to measure the evolution of Tully-Fisher-type relations and to see how properties of distant galaxies correlate with mass. We have measured velocity widths from emission lines for field galaxies to $z = 1.3$ in the HST Groth Survey Strip. At high redshifts, galaxies begin to deviate from the low-z relation: high redshift galaxies have lower linewidth or brighter magnitude, with $dM_B/dz \sim -2$. Possible explanations include: galaxies are brighter overall; there is a different high-z population; or that emission linewidths at $z \sim 1$ do not probe the full rotation velocity.

The DEEP Project has collected moderate resolution spectra for field galaxies in the HST Groth Strip with Keck/LRIS. We analyzed a subsample of these galaxies to obtain velocity linewidths from the integrated line emission, extracting 1-D spectra and obtaining the Gaussian σ, and applying an inclination cut of $\sin i > 0.75$, from the HST images analyzed with **gim2d** (Simard 1998).

Figure 1 shows the linewidth–absolute B magnitude distribution in four redshift bins. The diagonal line is a fit to the $0 < z < 0.6$ bin, and is plotted in the other bins for comparison. The low-z fit is consistent with the local TF relation, when the offset $\sigma_{corr} = 0.6 V_c$ is accounted for (as found from 2-D velocity fields of local galaxies, Rix et al. 1997). In the high-z bins, galaxies fall below the low-z relation, indicating that they have brighter M_B at given σ. The evolution indicated is $dM_B/dz \simeq -2$. The residual from the low-z relation is not strongly correlated with any single parameter, although there is a marginal tendency for large residuals to be in small, high [O II] EW, and blue galaxies.

Some possible explanations of the shift in $\sigma - M_B$ are: galaxies change little in mass but were brighter in the past; the $z \sim 1$ galaxies include a population which is not seen at low z; or linewidths at $z \sim 1$ may underestimate galaxy masses. Figure 2 shows good agreement for the 10 galaxies with both linewidth and circular velocity measurements. It is not yet clear why $\sigma - M_B$ indicates strong evolution while the sample analyzed for full rotation curves shows little evolution (Vogt et al. 1996, 1997). It is possibly because the linewidth sample probes higher z, smaller, and/or more peculiar galaxies.

References

1. Rix, H.-W., Guhathakurta, P., Colless, M., & Ing, K. 1997, MNRAS, 285, 779
2. Simard, L. 1998, ADASS VII, ed. R. Albrecht, R.N. Hook & H.A. Bushouse, 108

3. Vogt, N.P. et al. 1996, ApJL 465, L15
4. Vogt, N.P. et al. 1997, ApJL 479, L121

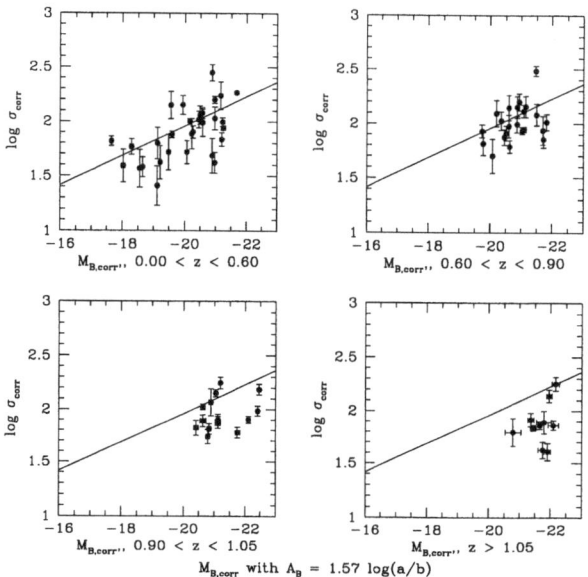

Fig. 1. Evolution in the linewidth-magnitude relation of field galaxies. Velocity linewidth log σ_{corr} versus rest-frame M_B for a sample of field galaxies from the DEEP 1 Survey, divided into four redshift ranges. The diagonal line is a fit to the low-z bin

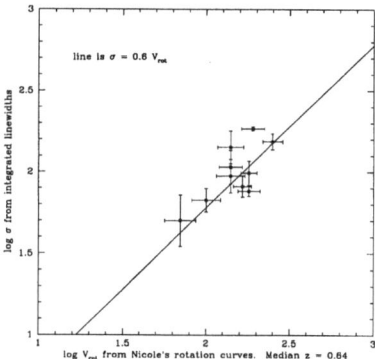

Fig. 2. Linewidth log σ_{corr} versus circular velocity log V_c measured from a full rotation curve analysis (Vogt et al. 1996, 1997). The diagonal line is $\sigma_{corr} = 0.6V_c$, not a fit, but as predicted from analysis of 2-D velocity fields of nearby galaxies (Rix et al. 1997)

The Masses of Distant Galaxies from Optical Emission Line Widths

Elizabeth Barton Gillespie[1] and Liese van Zee[2]

[1] Hubble fellow, University of Arizona, Steward Observatory,
933 N. Cherry St., Tucson, AZ, USA
[2] Indiana University, Department of Astronomy,
727 E. 3rd St., Bloomington, IN 47405-7105, USA

Abstract. Promising methods for studying galaxy evolution rely on optical emission line width measurements to compare intermediate-redshift objects to galaxies with equivalent masses at the present epoch. However, emission lines can be misleading. We show empirical examples of galaxies with concentrated central star formation from a survey of galaxies in pairs; HI observations of these galaxies indicate that the optical line emission fails to sample their full gravitational potentials. We use simple models of bulge-forming bursts of star formation to demonstrate that compact optical morphologies and small half-light radii can accompany these anomalously narrow emission lines; thus late-type bulges forming on rapid (0.5 – 1 Gyr) timescales at intermediate redshift would exhibit properties similar to those of heavily bursting dwarfs. We conclude that some of the luminous compact objects observed at intermediate and high redshift may be starbursts in the centers of massive galaxies and/or bulges in formation.

1 Introduction

Optical emission line widths are potentially important diagnostic tools for measuring the intrinsic gravitational masses of galaxies within their optical radii. Because of the sensitivity requirements for spatially resolved rotation curves, large surveys of galaxies at intermediate redshift and studies of galaxies at high redshift use unresolved or "integrated" emission line widths, computed from Gaussian fits to emission lines in the spectrum of the whole galaxy. However, the results are sometimes ambiguous in the case of compact star-forming galaxies.

The luminous compact blue galaxies observed at intermediate redshift (e.g., Koo et al. 1994; 1995) have small half-light radii ($R_e = 1 - 3.5$ kpc) and narrow emission-line velocity widths ($35 < \sigma < 126$ km s^{-1}). These properties suggest that although they are luminous galaxies, compact blue galaxies may be intrinsically faint galaxies undergoing a strong burst of star formation (Guzmán et al. 1996; 1997). However, Kobulnicky & Zaritsky (1999) measure high metallicities for these objects, appropriate only for massive galaxies, and HST images show possible evidence for surrounding older populations (Guzmán et al. 1998). Similarly, at higher redshifts ($z \sim 3$) the "Lyman break" galaxies also exhibit narrow integrated line widths that do not correlate with galaxy luminosity (Pettini et al. 2001). These observations taken together raise the question of whether emission line widths of compact objects accurately trace their potential wells (Kobulnicky & Zaritsky 1999).

We present evidence from observations of local galaxies and simple models of compact star formation that centrally concentrated star formation changes the measured emission line widths and half-light radii of galaxies (see also Kobulnicky & Gebhardt 2000; Pisano et al. 2001). This star formation can arise from major mergers (Mihos & Hernquist 1996), minor mergers (Mihos & Hernquist 1994), and secular evolution (Pfenniger & Norman 1990), processes that may be directly linked to bulge formation. The number of compact blue objects at intermediate redshift that are actually concentrations of star formation in larger galaxies remains unknown. If the luminous, compact blue galaxies are frequently bulges in formation, their number counts contain information about the timescales for evolution along the Hubble sequence.

2   Local Galaxies with Compact Central Star Formation

In a recent spectroscopic study of the centers of 502 galaxies in pairs, Barton, Geller, & Kenyon (2000) find evidence for correlations between the star-forming properties of interacting galaxies and the pair separations on the sky and in recession velocity. Their observations are broadly consistent with the Mihos & Hernquist (1996) picture of close galaxy-galaxy passes triggering gas infall and subsequent star formation in the central regions of some galaxies. Barton et al. (2001) examine the Tully-Fisher properties of a subset of the paired galaxies and find four galaxies that are apparently overluminous outliers to the relation. Barton & van Zee (2001) present VLA HI observations of two of the outliers; the radio line widths of the galaxies are substantially broader than their resolved optical emission line rotation curves. Thus, the observations support the possibility that centrally-concentrated star formation can give rise to anomalously narrow emission line widths that do not reflect the full gravitational potentials of the galaxies.

Fig. 1 shows one of the Barton et al. (2001) outliers, CGCG 132-062. The top panel shows the morphology of CGCG 132-062, including the disk surrounding the central, luminous region. The bottom panel shows the major-axis longslit spectra on the same scale. The emission is not spatially extended; it is confined to the central region of the galaxy and does not include the disky outskirts.

At higher redshifts, cosmological surface brightness dimming may render low surface brightness emission from the outskirts of a disk invisible. Thus, for distant galaxies, the observational bias against measuring full kinematic line widths likely extends to galaxies other than the 4 outliers in the Barton et al. (2001) study. Fig. 2 shows the H-alpha emission profile of a non-outlier spiral, NGC 470, from Barton et al. (2001). The H-alpha in the center of the galaxy is brighter than the outskirts by a factor of $\sim$100; this central part reflects only $\sim$50% of the kinematic width of the galaxy. An integrated line profile and perhaps even a 2-dimensional resolved rotation curve would miss this flux and therefore result in an anomalously small line width.

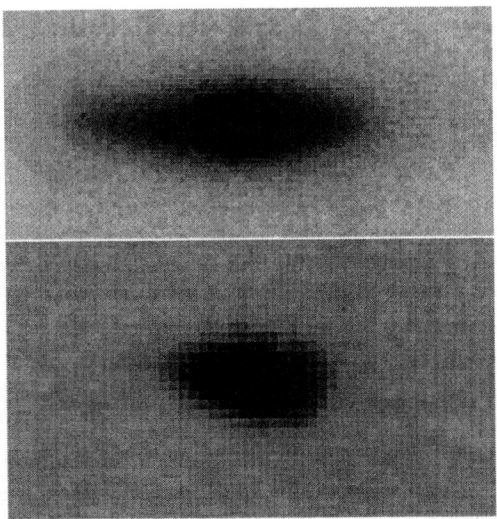

Fig. 1. CGCG 132-062, an interacting galaxy with centrally-concentrated optical emission line flux. We show *top* a B-band image and *bottom* a longslit spectrum on the same scale, where the horizontal direction is the spatial direction and the vertical direction is wavelength. The emission is largely confined to the central regions of the galaxy

3   The Effects of Centrally-Concentrated Star Formation

Large surveys frequently explore the intrinsic properties of galaxies using a limited set of structural parameters (e.g., half-light radius, R_e, or dispersion of fit to emission lines in the integrated spectrum, σ). However, if star formation rates vary in different components of the galaxies, the structural parameters of evolving galaxies will be affected by the differing mass-to-light ratios in the different components. In Fig. 3 we expand on the simple model of Barton & van Zee (2001) to show that a bulge-forming burst of star formation could profoundly effect the structural parameters of a galaxy during formation. Barton & van Zee (2001) describe the model in detail. We use the spectral synthesis models of Bruzual & Charlot (2001, in preparation) and typical "exponential bulge" parameters from Carollo (1999) and Courteau, de Jong, & Broeils (1996) [final $B/D = 0.1$; radius of disk is 12.5× radius of bulge]. After 7 Gyr, the previously bulge-less model spiral forms a bulge *in situ* in a brief period of time (instantaneous: *solid line*; $\tau = 0.5$ Gyr: *short-dashed line*; $\tau = 1$ Gyr: *long-dashed line*).

The top panels show that, depending on the formation timescale, the burst of star formation affects many of the basic structural parameters of the galaxy. During bulge formation, the total luminosity of the $\tau = 0.5$ or 1 Gyr models increases by < 1 magnitude, but the bulge-to-disk ratio can peak above 1 and the half-light ratio can dip to below 30% of its original value. Barton & van

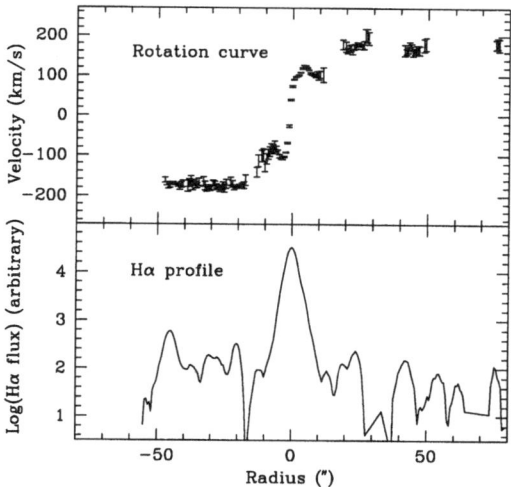

Fig. 2. High surface brightness star formation in the center of a paired galaxy. We plot the rotation curve of NGC 470 *top* and the profile of the Hα emission incident on the slit *bottom*. The center is ∼100× higher surface brightness than the disk but reflects only ∼half of the velocity width of the galaxy. Thus, although an accurate measurement at low redshift is possible, neither a high redshift spectrum nor an integrated spectrum would reflect the full velocity width of the galaxy

Zee (2001) use same model to show that σ for a maximal-disk rotation curve decreases to as little as 60% of its original value during bulge formation.

Although the model does not cover every possible evolutionary scenario, the results are generic and require only formation within a relatively short timescale – less than 1 Gyr – short enough to allow close galaxy-galaxy passes, minor mergers, and possibly single-episode secular evolution. With only ∼1 – 2 magnitudes or less of (transient) luminosity evolution, the temporary movement in σ – R_e space is enough to misjudge the nature of these galaxies during bulge formation (c.f., Fig. 2 in Barton & van Zee). Thus, some of the luminous compact blue galaxies observed at intermediate redshift may be intrinsically massive galaxies containing compact bursts of star formation.

4 Luminous Compact Blue Galaxies and Galaxy Evolution

The luminous compact blue objects, whether dwarfs or spirals, are clearly some of the most rapidly evolving galaxies observed at intermediate redshift. Although extreme examples of these objects are relatively rare locally, close galaxy-galaxy passes, minor mergers, and secular evolution may funnel gas into the centers of galaxies more efficiently at higher redshifts, where more gas is available.

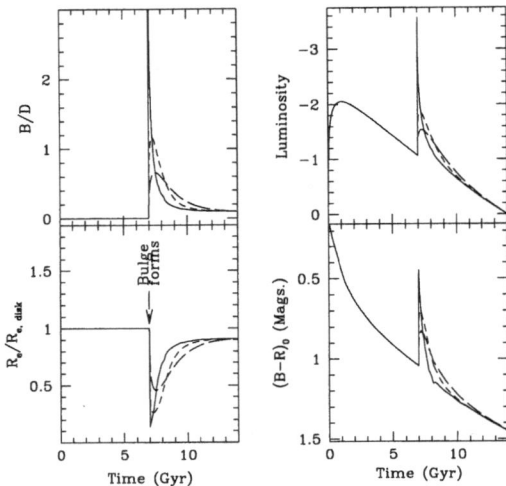

Fig. 3. A simple model for a bulge-forming burst of star formation. We plot the bulge-to-disk ratio (*upper left*), normalized half-light radius (*lower left*), normalized luminosity (*upper right*), and total $B - R$ color (*lower right*) for a model spiral galaxy that forms a bulge at intermediate redshift ($T = 7$ Gyr). See the text for more details

An understanding of both dwarf and luminous galaxy evolution at intermediate redshift requires measures of their intrinsic masses and sizes, hence their $z = 0$ morphologies and luminosities. "Exponential" bulges have distinctly different properties from $R^{1/4}$-law bulges (e.g., Andredakis & Sanders 1994); they are candidates for bulges formed via secular evolution (Pfenniger & Norman 1990) or any process that sends disk gas into the center of a galaxy. If the majority of the luminous, compact blue galaxies are actually spirals undergoing strong central bursts of star formation, they may be consistent with forming exponential bulges. Thus, the fraction of these objects that are actually bulges in formation may directly reveal the "exponential" bulge formation history of the Universe.

Although existing surveys are not deep enough, deep imaging holds promise for distinguishing star formation in intrinsically low- and high-mass galaxies. Fig. 4 shows results from simulations of face-on disks at $z = 1$ in galaxies with luminous bulges. The solid lines mark the approximate limits at which accurate structural parameters measurements are possible. The points are local disks from the de Jong (1996) sample. One to two orbits with WFPC2 are only enough to characterize the higher surface brightness disks. Although the Hubble deep fields detect the majority of even face-on disks at $z = 1$, their combined area is small, containing the progenitors of few late-type galaxies at $z \leq 1$. Only upcoming ACS surveys will probe deep enough over enough area on the sky to detect the majority of disks that may surround luminous, compact blue objects at $z = 1$.

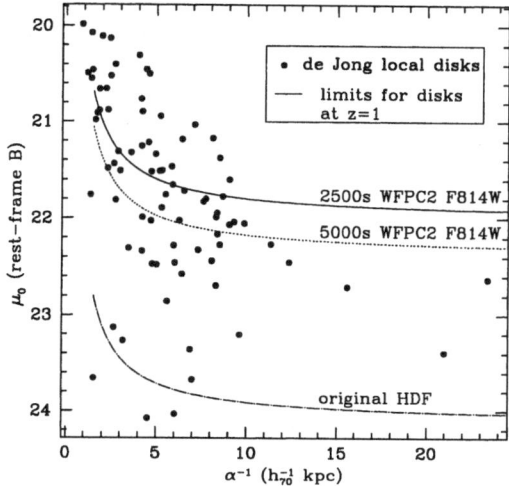

Fig. 4. Detection of face-on disks at $z = 1$. We plot the parameters of the de Jong (1996) spirals; solid lines show $z = 1$ limits with different WFPC2 exposures for accurate measurement of the parameters of face-on disks under luminous bulges

References

1. Andredakis, Y. C., & Sanders, R. H. 1994, MNRAS, 267, 283
2. Barton, E. J., & van Zee, L. 2001, ApJ, 550, L35
3. Barton, E. J., Geller, M. J., & Kenyon, S. J. 2000, ApJ, 530, 660
4. Barton, E. J., Geller, M. J., Bromley, B. C., van Zee, L., & Kenyon, S. J. 2001, AJ, 121, 625
5. Carollo, C. M. 1999, ApJ, 523, 566
6. Courteau, S., de Jong, R. S., & Broeils, A. H. 1996, ApJ, 457, L73
7. de Jong, R. S. 1996, A&AS, 118, 557
8. Guzmán, R., et al. 1996, ApJ, 460, L5
9. Guzmán, R., et al. 1997, ApJ, 489, 559
10. Guzmán, R., et al. 1998, ApJ, 495, L13
11. Kobulnicky, H. A., & Gebhardt, K. 2000, AJ, 119, 1608
12. Kobulnicky, H. A., & Zaritsky, D. 1999, ApJ, 511, 118
13. Koo, D. C., Bershady, M. A., Wirth, G. D., Stanford, S. A., & Majewski, S. R., 1994, ApJ, 427, L9
14. Koo, D. C., et al. 1995, ApJ, 440, L49
15. Mihos, J. C., & Hernquist, L. 1994, ApJ, 425, 12
16. Mihos, J. C., & Hernquist, L. 1996, ApJ, 464, 641
17. Pettini, M., et al. 2001, ApJ, 554, 981
18. Pisano, D. J., Kobulnicky, H. A., Guzmán, R., & Gallego, J. 2000, in preparation
19. Pfenniger, D., & Norman, C. 1990, ApJ, 363, 391

Kinematics of ISOCAM Selected Star-Forming Galaxies at z~1 in the Hubble Deep Field South

Dimitra Rigopoulou[1], Alberto Franceschini[2], Reinhard Genzel[1], and Niranjan Thatte[1]

[1] Max-Planck-Institut für extraterrestrische Physik, Postfach 1312, 85741 Garching, Germany
[2] Dipartimento di Astronomia, Universita' di Padova, Vicolo Osservatorio 5, I-35122 Padova, Italy

Abstract. The various deep ISOCAM surveys revealed a new class of infrared luminous galaxies which are characterized by a high rate of evolution and are found at redshifts of z~1. Based on our near-infrared low-resolution spectroscopy we find that these ISOCAM galaxies are dust-enshrouded star-forming galaxies. Here we report on the first spatially resolved H_α velocity profiles of ISOCAM galaxies in the Hubble Deep Field South. We find that some of these systems are in fact extremely massive galaxies. The galaxies show an offset of 1.6±0.3 magnitude in the rest frame B-band when compared to the local Tully-Fisher relation.

1 ISOCAM Observations of HDF-S and the VLT-ISAAC Sample

The ISOCAM observations of the Hubble Deep Field South (HDFS, Oliver et al., 2002, Aussel et al. 2002, in prep) resulted in the detection of 63 sources reaching down to a limiting flux of ~100 μJy. For our followup near-infrared spectroscopic observations we selected galaxies that had secure LW3 detections and I and/or K-band image counterparts (I-band data from Dennefeld et al. 2002, in prep, K-band from ESO-EIS Deep). We did not apply any selection based on colours. Our sample (hereafter VLT ISOHDFS sample) is thus a fair representation of the strongly evolving ISOCAM population near the peak of the differential source counts (Elbaz et al. 1999, Franceschini et al., 2001). Our VLT ISOHDFS sample contained about 30 galaxies. The LW3 flux ranges between 100–400 μJy. During 1999-2001 we have acquired secure spectroscopic redshifts and H_α spectra for ~25 galaxies (Rigopoulou et al., 2000, Franceschini et al. 2002, in prep). The strong H_α emission combined with the high Equivalent widths (EW) found in almost all VLT ISOHDFS galaxies imply that these objects are active star-forming galaxies.

2 Kinematics of ISOCAM HDFS Galaxies

2.1 The Sample

From our low-resolution sample we selected randomly a few objects as candidates for the high resolution spatially resolved kinematics studies. Figure 1 shows

HST/WFPC2 (F814 filter) and/or K-band ground based images of the galaxies studied: we chose an early type spiral (ISOHDFS S27), an interacting (double) system (ISOHDFS S25) and a late type spiral (ISOHDFS S55). The properties and morphological types of the candidate galaxies are representative of the properties of the entire sample of ISOCAM galaxies.

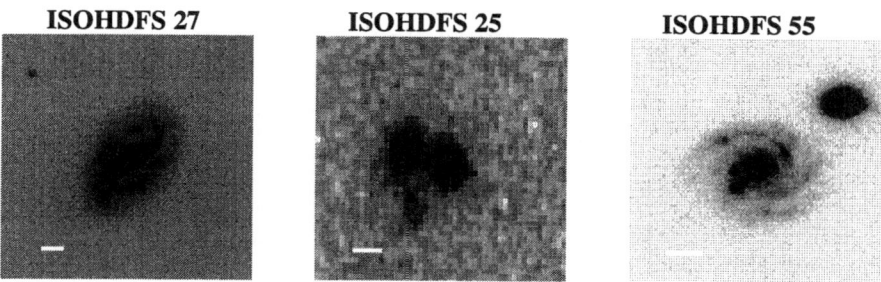

Fig. 1. HST/WFPC2 I_{814} -band images for ISOHDFS 27 and ISOHDFS 55. ESO-EIS K-band image for ISOHDFS 25 (outside of the WFPC2 HDFS image). The horizontal bar represents 1 arcsec

3   Results

3.1   The Spectra

The extracted spectra of the four galaxies are shown in Figure 2. The top panel shows the sky-subtracted 2-D spectra while in the middle panel the (observed) velocity profiles are shown. The positions of the strong nebular lines $H\alpha$ and [NII] are labelled.

ISOHDFS 27: The $H\alpha$ emission is clearly spatially extended over ~4 arcsec (FWHM) which at the distance of ISOHDFS 27 corresponds to about ~ 25 kpc. Two lobes are detected on each side of the $H\alpha$ continuum (same structure is seen in [NII]) which are originating in a rotating disk. The $H\alpha$ line velocity profile shows a double-horn feature, a clear signature of a rotating disk, as was already suggested by the 2-D spectra. The $H\alpha/$ [NII] line ratio is about 3 which is typical of HII regions. This line ratio combined with the extended morphology of the $H\alpha$ line rules out the presence of a central AGN as the prime ionizing source (as already suggested in Rigopoulou et al. 2000).

ISOHDFS 25: The two counter-rotating galaxies are clearly seen in the the sky-subtracted 2-D spectrum of ISOHDFS 25. $H\alpha$ emission is spatially extended in both galaxies with an extent of ~ 2 arcsec (FWHM) corresponding to about 15 kpc at the distance of the galaxies. The $H\alpha$ is clearly tilted in both galaxies which is attributed to ordered rotation. The velocity profiles are similar in shape

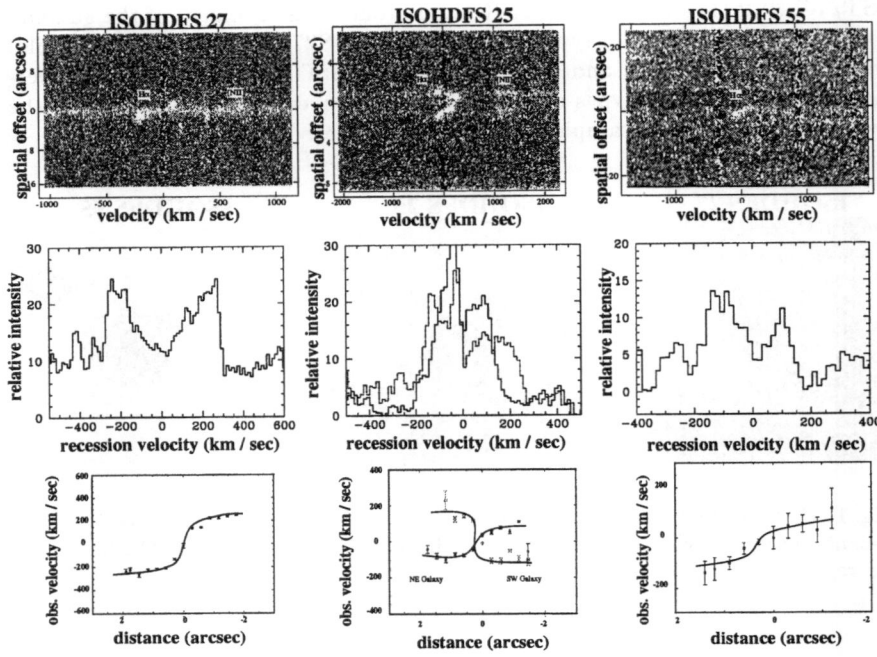

Fig. 2. top panel from left: 2D sub-images of the VLT-ISAAC spectra of ISOHDFS 27, ISOHDFS 25 and ISOHDFS 55. Positions of strong nebular lines are marked. The scale is 1.212 Å /pixel in the spectral direction. middle panel from left: Ha velocity profiles (integrated) for ISOHDFS 27, ISOHDFS 25 (two galaxies, NW straight line, SW dotted line), ISOHDFS 55. lower panel from left: Observed velocity curves for ISOHDFS 27, ISOHDFS 25 (two galaxies) and ISOHDFS 55. The points represent the observed velocities and the solid line is the model rotation curve

and have a p-p velocity of ~300 km s^{-1}. The shapes and linewidths are both typical of values found in nearby e.g. Rubin et al. 1985 and more distant systems.

ISOHDFS 55: The Ha emission is clearly extended although no continuum has been detected. The spatial extent of the Ha line over ~3 arcsec (smaller than the optical extent of the galaxy) implies that the Ha does not originate in centrally concentrated gas but most likely in a circumnuclear ring (similar to ISOHDFS 27). The double-horn velocity profile gives additional evidence for the emission originating in a ring-like structure. We see no nuclear Ha emission. This could be due either to excess extinction in the galaxy's nucleus or to the presence of a significant older stellar population component (Moy et al. 2002, in prep).

3.2   Measuring the True Rotation Velocity

To infer true rotational velocities V_{rot} we created a model where the galaxies were simulated with an exponential disk model. For the velocities of the model galaxies we followed the prescription of Persic & Salucci 1991. The model was then convolved with an appropriate Gaussian to account for the observing conditions. Emission-line fluxes were derived by using a model slit mask. The simulated line fluxes were analysed in the same way as the real data. The circular velocity of the model was adjusted by hand until the simulated and real (observed) emission lines matched at the velocity extremes. This model circular velocity was then adopted as the galaxy's true terminal rotation velocity V_{rot}.

3.3   Velocity Curves and Dynamical Masses

The observed velocity curves shown in Figure 2 (lower panel) have been derived using the higher S/N $H\alpha$ line. At each position along the slit we fit Gaussian profiles and derive amplitudes and widths based on the best-fit Gaussians. A velocity profile is then calculated.

The shape of the ISOHDFS 27 velocity curve is the one expected for a rotating disk. What is of real interest is the large rotational velocity of 540 ± 10 km s^{-1}. Rubin et al. 1985 presented a compilation of rotation curves for large samples of local Sa, Sb and Sc galaxies of various luminosities. They find that the rotational velocities decrease from Sa to Sc types. While the rotational velocities of the other three galaxies appear to be consistent with the median values for Sa or Sb (299–222 km s^{-1}), that of ISOHDFS 27 is considerably higher. There is only one other example of a local galaxy, UGC 12591 (an S0/Sa galaxy, Giovanelli et al. 1986) with a rotational velocity similar to that of ISOHDFS 27.

Finally, we estimated dynamical masses for the present ISOCAM galaxy sample (see Rigopoulou et al. 2002 in preparation for details). We find that ISOCAM galaxies are in fact very massive systems with masses of the order of a few $\times M \sim 10^{11} M_\odot$. ISOHDFS 27 is found to be an extremely massive galaxy containing mass of 10^{12} M$_\odot$ within the central 18 kpc area.

4   Tully-Fisher Relation

In Figure 3 we compare the present data of the four ISOCAM galaxies with the local Tully-Fisher (TF) in the rest-frame B-band, which corresponds to the observed I-band frame at z~0.6–0.8. Our ISOCAM galaxies are compared to the TF relation of Pierce & Tully 1992 (PT) based on data for 16 galaxies from the Ursa Major Cluster. The PT relationship is based on measurements of the HI velocity widths corrected for the effects of projection and internal turbulence. We have not applied a similar correction to our data since these corrections are rather small. However, since dust plays a crucial role in ISOCAM galaxies (Paper I, Franceschini et al., 2002, in preparation) we have applied extinction corrections to the four galaxies in the TF plot. A$_V$ values for each individual

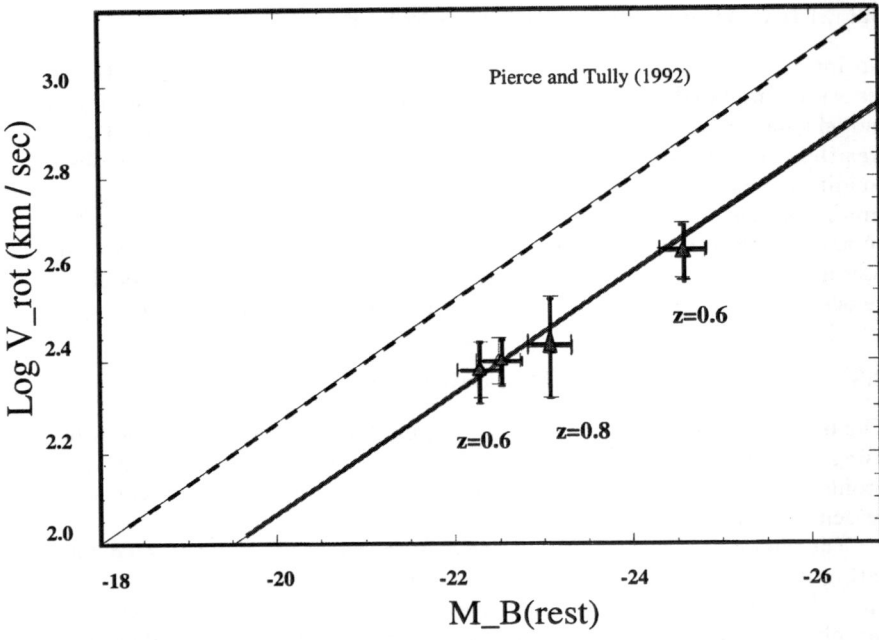

Fig. 3. The Tully–Fisher V_{rot} as a function of B_{rest} luminosity diagram. Our ISOCAM data are compared to the local TF by Pierce and Tully (1992, dotted line). The magnitudes have been corrected for internal extinction and the velocities for projection. The straight line is the fit to the ISOCAM points assuming same slope as the local TF. We find an offset of 1.6±0.3 from the local TF relation

galaxy can be found in Franceschini et al. (2002, in preparation). For the present data we have used a rather "conservative" A_V value of 1. The galaxies used in the local PT relationship have also been corrected for internal extinction following the prescription of Tully & Fouque 1985.

Using the data for the four galaxies we compute a weighted offset of 1.6±0.3 mag relative to the local B-band relation. At this stage our sample is obviously too small for statistically significant results however, a clear offset from the local TF is evident. We next discuss the importance of the various possible biases. First, as was shown in Paper I, ISOCAM galaxies are actively starforming, thus the sample is biased towards $H\alpha$ bright emission-line objects and overall higher luminosities. Second, the galaxies used to estimate the local TF do not include any mergers/barred/irregular systems, in other words, galaxies morphologically akin to the ones seen at higher redshifts. Corrections for these effects are bound to reduce the observed offsets. On the other hand dust extinction plays a crucial role: given the selection of ISOHDFS galaxies (dusty active starbursts) the true extinction is definitely higher than what has been assumed here and subsequently the offsets could increase. Although at this stage we cannot *quantify* the contributions of all these biases we can securely claim that an offset from the local TF of at *least 1 mag* is present in our data.

References

1. D. Elbaz et al., A&A **351**, L37 (1999)
2. Franceschini, A., Aussel, H., Cesarsky, C.J., Elbaz, D., Fadda, D., A&A **378**, 1, (2001)
3. Giovanelli, R., Haynes, M.P., Rubin, V.C., Kent, W.F., ApJ **301**, L7 (1986)
4. S. Oliver et al., MNRAS in press (2002)
5. Persic, M., & Salucci, P., ApJ, **368**, 60 (1991)
6. Pierce, M.J., & Tully, R.B., ApJ, **387**, 47 (1988)
7. Rigopoulou, D., et al, 2000, ApJ, **537**, 85 (2000)
8. Rubin, V.C., Burstein, D., Ford, W.K. Jr., and Thonnard, N., ApJ, **289**, 81, (1985)
9. Tully, R.B., & Fisher, J.R., A&A, **54**, 661, (1977)
10. Tully, R.B., & Fouqué, P., ApJS **58**, 67, (1985)
11. Vogt, P.N., et al., ApJ **479**, L121 (1997)

Stellar Mass of Faint Mid-Infrared Galaxies

Yasunori Sato

Institute of Astronomy, University of Tokyo,
2-21-1 Osawa, Mitaka, Tokyo, 181-0015 Japan

Abstract. Near-infrared emission is a good estimator of stellar masses of galaxies. For galaxies at high redshifts, mid-infrared observations are needed to trace the near-infrared emission at the rest-frame. A deep mid-infrared survey with the ISOCAM LW2 filter is sensitive enough to detect elliptical galaxies at high redshifts ($z > 1$). Stellar masses of the $6.7\,\mu$m galaxies are then derived, indicating a typical stellar mass of $\sim 10^{11}\,M_\odot$. Stellar mass density (SMD) is also estimated, which is almost comparable to the integration of the star formation rate density (SFRD) in the Universe.

1 Introduction

Stellar mass is one of the fundamental properties of galaxies such as star formation rate. At the near-infrared, a ratio of stellar mass to luminosity has a very small dispersion among galaxies with various ages and morphological types (star formation histories). This is due to the long lifetime of low mass stars contributing to the near-infrared emission and also to the negligible effect of dust extinction at the near-infrared. Thus, we can derive stellar masses of galaxies if we obtain their near-infrared luminosities. The accuracy of such photometrically-determined stellar masses depends on observing wavelength and redshifts of galaxies. Stellar mass estimated by the K band photometry is quite accurate at the local Universe. However, its accuracy becomes worse at higher redshift, as the K band shifts into the optical at the source's rest-frame. On the other hand, ISOCAM LW2 band photometry at $6.7\,\mu$m gives more accurate estimates (the errors should be less than $50\,\%$) for galaxies at $z > 1$. According to the GRASIL SED model [5], their E galaxy would be as bright as $20\,\mu$Jy at $z = 5$. It corresponds to a giant elliptical with $L_K \sim 4\,L^*$ and there is negative K correction for the $6.7\,\mu$m fluxes. A formation redshift of $z_f = 10$ and a cosmology with $\Omega_m = 0.3$, $\Omega_\Lambda = 0.7$, and $h = 0.65$ were assumed.

2 Results

Here I report initial results on stellar masses of faint mid-infrared galaxies. Full details and the updates will be reported in [4]. A sample of faint mid-infrared galaxies was obtained with a deep ISOCAM LW2 band survey in the SSA13 [2]. The survey reaches an $80\,\%$ completeness limit of $10\,\mu$Jy in the center of the survey area. The sample includes three SCUBA sources [3]. For all the $6.7\,\mu$m sources, their stellar masses were derived based on their photometric

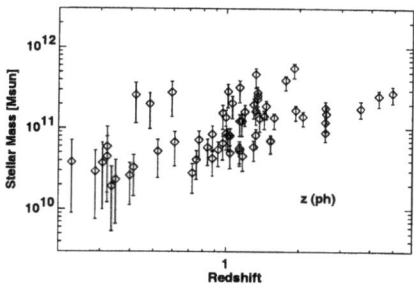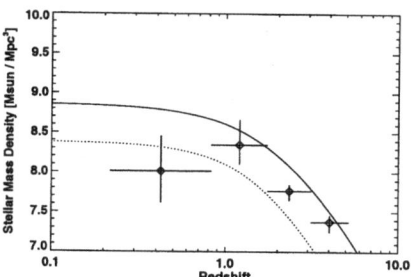

Fig. 1. (left) Stellar mass estimates of the 6.7 μm galaxies in the SSA13. Each estimate is a weighted mean of stellar masses based on the GRASIL SEDs [5] (E, Sa, Sb, and Sc). A weight of E : Sa : Sb : Sc = 0.4 : 0.2 : 0.2 : 0.2 is adopted. The error bars correspond to the range of the SEDs

Fig. 2. (right) Stellar mass density in the Universe derived from the 6.7 μm sample. Integrated contributions from the star formation densities parameterized in [1] are also shown: a case for $E(B - V) = 0.15$ (*solid curve*) and a case for no absorption by dust (*dotted curve*)

redshifts (Figure 1). The photometric redshifts were estimated via χ^2 minimization between the GRASIL SEDs and the available photometric data: mainly at the ISOCAM LW2, K, I, and B bands. For the sub-sample with spectroscopic redshifts, standard deviation of redshift differences between photometric and spectroscopic ones was 0.4. There is little correlation between stellar mass and redshift. A typical stellar mass for this sample is $\sim 10^{11} M_\odot$, almost comparable to M^*_{star} at the local Universe [1]. It should be noted that a deep ISOCAM LW2 imaging would preferentially select elliptical galaxies. Thus stellar masses of the 6.7 μm sources are likely at the upper ends of the error bars. Stellar mass density was then derived for this photometric redshift sample (Fig. 2). Contributions from each mid-infrared galaxies were summed up with the formula for stellar mass density as $\rho_{star} = \Sigma \, (M_{star}/V_{max})$. Incompleteness corrections has been applied. The stellar mass density for the 6.7 μm sample is almost comparable to the integrated contribution from the star formation rate density in the Universe. Because the 6.7 μm galaxies are not identified with UV galaxies, it would be important to distinguish whether the 6.7 μm galaxies are on the same evolutionary track as UV galaxies but at a different stage, or on a totally different track.

References

1. S. Cole, P. Norberg, C.M. Baugh, et al.: MNRAS **326**, 255 (2001)
2. Y. Sato, K. Kawara, L.L. Cowie, et al.: A&A submitted
3. Y. Sato, L.L. Cowie, K. Kawara, et al.: ApJL submitted
4. Y. Sato, et al.: in preparation
5. L. Silva, G.L. Granato, A. Bressan, L. Danese: ApJ **509**, 103 (1998)

M/L in Early-Type Galaxies to $z \sim 1$: Cluster vs Field

Garth Illingworth

UCO/Lick Observatory, Astronomy and Astrophysics Department,
University of California, Santa Cruz, CA 95064

Abstract. HST and Keck have revolutionized our understanding of distant galaxies, by enabling quantitative studies using length scales, surface brightnesses and spectroscopic velocity scales, as well as line strengths. Such measurements allow one to characterize galaxies on a mass scale, and to provide strong constraints on evolution from changes in M/L. Our program to obtain wide-field, multi-color WFPC2 mosaics with HST of intermediate redshift clusters, and spectroscopic membership and high S/N spectroscopy with LRIS on Keck, has provided new insights into the nature of elliptical and S0 galaxies in the cluster environment over a wide range of densities, from the core to the field. In particular, most ellipticals, and a significant fraction of the S0 population, have large luminosity-weighted ages, suggesting that their stellar populations were formed at redshifts beyond $z \sim 2$, though the existence of substantial numbers of major mergers in MS 1054-03 at $z = 0.83$ suggests that final assembly of such galaxies may not have occurred until much later for a significant fraction of early-type galaxies. Early-type galaxies in the field and in clusters differ less than expected.

1 Introduction

Keck LRIS multislit spectroscopy and HST WFPC2 imaging have been extensively used over the last few years to establish M/L variations for early-type E/S0 galaxies from the current day to redshifts near $z \sim 1$ in clusters and, more recently, to $z \sim 0.7$ in the field. Such data have been part of the recent revolution in our understanding of high redshift galaxies. This work enables us to begin to provide results on evolution that couple much more closely to the mass of galaxies and to remove some of the very large uncertainties associated with measures based solely on luminosity.

Three massive, X-ray selected clusters ranging in redshift from CL 1358+62 at $z = 0.33$, to MS 2053-04 at $z = 0.58$, and to MS 1054-03 at $z = 0.83$ have been the focus of our effort [1]. A key factor that distinguishes this work from other programs with HST is that we have used multiple pointings to cover a wide area around the cluster. The advantage of this approach for many issues relating to cluster and early-type galaxy evolution became clear as the project developed. The wide areal coverage also enabled us to select significant field early-type galaxy samples for comparison with the cluster galaxies.

The multiple pointings on HST with WFPC2 typically cover $5'-7' \times 5'-7'$ in size. This corresponds to a field size of roughly $1.5h_{65}^{-1} - 2h_{65}^{-1}$ Mpc, centered on the clusters. An example of the mosaic HST images for one of our cluster

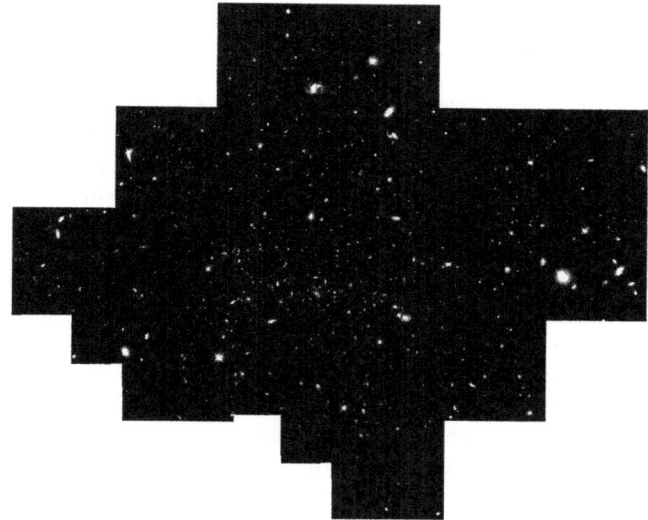

Fig. 1. The HST WFPC2 36-orbit, six-pointing two-color mosaic of the $z = 0.83$ X-ray cluster MS1054–03. This image covers $5' \times 7'$, i.e., out to $\sim 2h_{65}^{-1}$ Mpc radius

fields is shown in Fig. 1. The fields were imaged with two filters, F606W and F814W. Spectroscopy was then carried out using the Keck LRIS multi-object spectrograph, with a field comparable to that of the HST mosaics.

The scientific focus of this program is elucidating the timescales and processes surrounding early-type galaxy evolution in a range of environments. This involves, in particular, identifying the galaxy characteristics across the cluster as a function of environment and density, analysis of the cluster's spatial and velocity structure, and determining the mass distribution through weak lensing. The primary tools that we have used for characterizing the evolution of the early-type galaxy population have been the fundamental plane, the color-magnitude relations, and more recently, absorption line strengths.

2   Framework

The importance of the focus on early-type galaxies is that such galaxies carry the bulk of the baryonic mass in stars at the present epoch (more than 70% – [2]), and so the question "when did galaxies form?" is in large part "when/how did bulges/ellipticals form?" and "when were large disks assembled?". Late-type galaxy formation may provide key insights into the galaxy development process, but they carry but a tiny fraction of the mass in galaxies.

Just as very detailed studies of galaxies at $z = 0$ provides unique insights into galaxy evolution at early times, so it is that detailed studies at $z \sim 1$ have provided insights that are very difficult to realize through direct observations at higher redshifts where the galaxies are very faint. The considerable effort being expended on studies at redshifts $z \sim 3 - 6$ will be complemented by the more quantitative analyses possible at $z \sim 1$ where the universe, at a lookback time of 8 Gyr, is only $\sim 40\%$ of its present age. Even now, with the limited data available, it is striking how similar the universe at $z \sim 1$ is to the present-day universe, particularly for L* galaxies. For example, there is a recognizable Hubble sequence in place, large disks exist with a Tully-Fisher relation similar to that at $z \sim 0$ (with surprisingly modest passive evolution [3]), and an old elliptical population exists with comoving number densities within a factor 2 of that of the present day [4]. However, major differences do exist, namely the star formation rate (SFR) is substantially larger (5-10×), as is the merger rate and the blue galaxy fraction in a variety of environments. The $z \sim 1$ epoch is a time of fascinating contrasts. The HST Advanced Camera (ACS) and the new generation of spectrographs on Keck and the VLT (DEIMOS and VIMOS, to name but two), should enable a survey on a scale that will lead to datasets that are SDSS- or 2DF-like in their value for quantifying galaxy evolution.

One key issue which will require further careful analysis is that of "progenitor bias" [5], where those samples being observed at $z \sim 1$ are likely to be a subset of the present-day population since galaxies have undergone morphological transformation in the $z \sim 1$ to $z = 0$ timeframe.

A further issue will remain the biases introduced into optical-UV studies where the visible flux from galaxies is dominated by any recent (or not-so-recent) star formation. As the theme of this workshop emphasizes, measures which characterize mass are needed whenever possible.

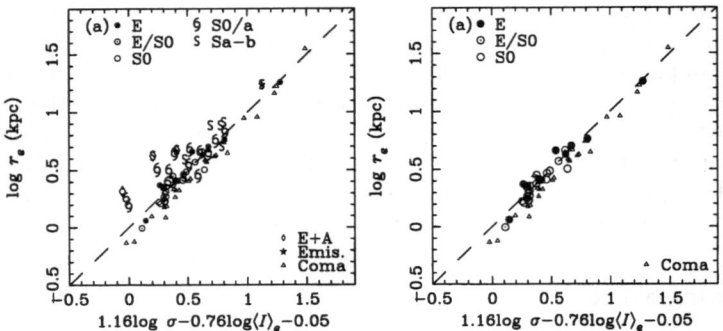

Fig. 2. The fundamental plane of the full sample (left) of 53 galaxies in CL 1358+62 at $z = 0.33$, including the early-type spirals and E+A galaxies. The fundamental plane of the 30 E/S0 galaxies (right) shows a very tight relation (comparable to that in Coma). The intrinsic scatter is only 14% in r_e. The (small) offset to Coma is evolutionary brightening, since the $(1+z)^4$ surface brightness dimming has been removed. The offsets for the early-type spirals and E+A galaxies are as expected for a younger luminosity-weighted age for their stellar populations (see [6])

3   Fundamental Plane

The fundamental plane is a key quantitative measure for studying distant galaxies. It is a tight relation in early-type galaxies between velocity dispersion, effective radius and mean surface brightness that is described by $r_e \propto \langle I_e \rangle^{-0.83} \sigma^{1.20}$. With reasonable assumptions about homology, this implies that $M/L \propto M^{0.25}$. The very small scatter, $\pm 23\%$ in Coma in M/L_V, makes it an ideal tool for establishing the evolution of M/L with redshift. Since age is related to M/L for early-type galaxy populations, constraints can be put on their (luminosity-weighted) ages and age distributions. This requires high S/N data, and attention to minimizing systematic errors. For $z \sim 0.3$–1 clusters, HST images are needed to derive r_e and I_e, while 8-10 m telescopes are needed for σ (6-8 hour integrations were used at Keck at $z = 0.83$).

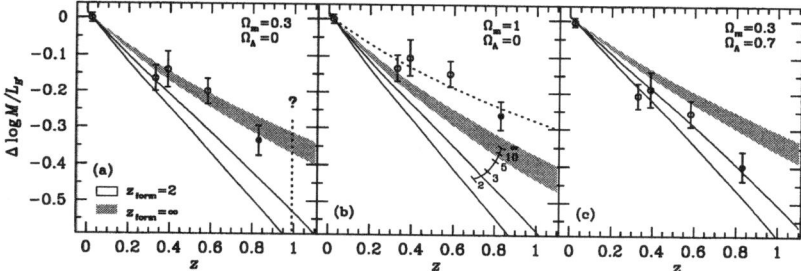

Fig. 3. Evolution of M/L_B (from [1]) with redshift from the fundamental plane measurements in Coma at $z = 0.02$, CL 1358+62 at $z = 0.33$, CL 0024+16 at $z = 0.39$, MS 2053-04 at $z = 0.58$, and MS 1054-03 at $z = 0.83$, for an (a) open $\Omega_m = 0.3$, a (b) flat, matter-dominated $\Omega_m = 1$, and (c) a lambda cosmology $\Omega_\Lambda = 0.7, \Omega_m = 0.3$. Single-burst model predictions are shown for a Salpeter IMF ($x = 1.35$) and a range of metallicities, as indicated by the spread in $\Delta \log M/L_{B'}$ for a given z_{form}. The sensitivity to the IMF is shown in (b) as a dotted line ($z_{form} = \infty$ and a steep IMF with $x = 2.35$), as is the effect of reducing z_{form} below $z_{form} = \infty$. The data favor high formation redshifts and a low q_0, but this result is sensitive to the IMF

For CL 1358+62, high S/N spectra were obtained of 53 galaxies in the HST mosaic, resulting in the fundamental plane in Fig. 2 [6]. These data made it possible to derive a fundamental plane at $z = 0.33$ that is comparable to Coma. The scatter in M/L in the E/S0 population in CL 1358+62 is just 16%. The offset from Coma constrains the luminosity-weighted ages to $z > 1$, and the consistency of the scatter in the fundamental plane and the color-magnitude relation constrains the luminosity-weighted age dispersion to be <15%.

Based on our fundamental plane measurements at this and higher redshift [7, 1], we have used the offsets in $\Delta M/L$ from Coma to set even stronger constraints on the luminosity-weighted ages, and on the cosmology, as shown in Fig. 3. For the popular lambda cosmology (right-hand panel) the luminosity-weighted star formation age is $z_f > 2$. These data give $\Delta \log M/L_B \propto -0.4z$. Tighter constraints would come from observing even slightly higher redshift clusters, though the spectroscopy becomes very time-consuming, even with Keck!

A key step recently has been the first measurements of the fundamental plane in the field [8] to $z \sim 0.5$, subsequently extended to $z \sim 0.7$ by [9]. Surprisingly, these data [8] suggest that field E/S0 galaxies are similar to cluster early-type galaxies, having an offset $\Delta(M/L) = -0.18 \pm 0.11$, suggesting that they are also quite old. They are only 21% younger, corresponding to a $z_f > 1.5$. Extension to higher redshifts is obviously the next step.

4   Major Mergers in MS 1054–03

The value of the large field coverage of our HST imaging was demonstrated again with our mosaic on MS 1054-03 at $z = 0.83$ (Fig. 1), when the morphological types of the 89 members that resulted from the Keck LRIS spectroscopic magnitude-limited sample ($I < 22.2$) were analyzed. The morphological types were split 22% E, 22% S0, 39% spiral and 17% *merger/peculiar*. To find such a high fraction of luminous merger/peculiar galaxies in such a rich cluster at such a late time was a surprise. The morphological types of the most luminous galaxies in MS 1054-03, plus the lower luminosity merger/peculiar objects, are depicted in Fig. 4.

These results are discussed in more detail in [10]. The color magnitude diagram shown in Fig. 4 demonstrates one of the remarkable aspects of the mergers in this cluster – they are quite red, with no detected [OII] 3727 Å emission, suggesting minimal ongoing star formation. The mergers are also found in the outer parts of the cluster, suggesting that they are in infalling clumps. Most of these mergers are likely to evolve into luminous ($\sim 2L^*$) early-type galaxies, presumably ellipticals, but also possibly S0s. If confirmed to be generally the case in other rich $z \sim 0.5$–1 clusters, this result suggests that the merging rate in the cluster environment is changing rapidly with redshift (possibly as fast as $(1+z)^6$). This important result is consistent with the predictions of hierarchical clustering models (see, e.g., [11]).

5   Summary

Quantitative approaches such as the use of the fundamental plane are allowing the use of mass-based (M/L) scales for characterizing galaxy evolution. Combined with other relations (e.g., color-magnitude and the forthcoming line strength studies), these provide complementary approaches to understanding the evolutionary history of early-type galaxies. Already our results show that the luminosity-weighted ages of E/S0 galaxies in intermediate redshift ($z \sim 0.3$–1) massive X-ray clusters are old, i.e., $z_f > 2$, with small $\Delta t/t \sim 0.15$–0.18. Tests on field galaxies indicate that they too also have quite old luminosity-weighted ages. Furthermore, our recent results on the major mergers in MS 1054-03 indicate that a substantial fraction of the E/S0 population could have undergone a final "assembly" phase at $z < 1$.

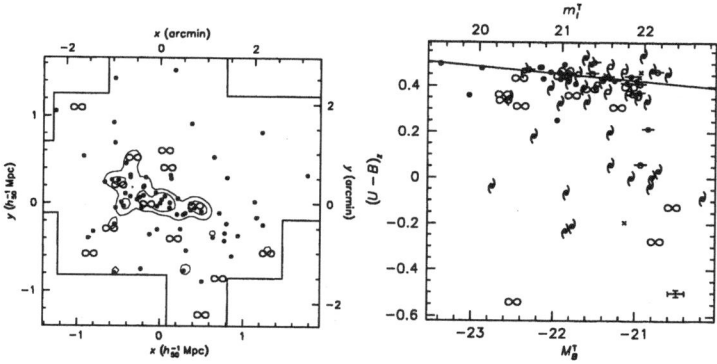

Fig. 4. Spatial distribution (left) of the confirmed cluster members in MS 1054-03, with contours of the the red galaxy distribution, and the mergers indicated as ∞ symbols. Color-magnitude relation (right) of the spectroscopically-confirmed members of MS 1054-03. Excluding the three very blue mergers, the bulk of the mergers are only ~ 0.07 mag bluer than the CM relation defined by the E/S0 early-type galaxies

Acknowledgements. I would like to thank the organizers for an excellent meeting at a great location. I would like also to thank my collaborators, Marijn Franx, Pieter van Dokkum, Dan Kelson, Dan Magee, and Kim-Vy Tran for making this such an excellent project. Support from STScI grants GO07372.01-A, GO07941.01-A, and GO08220.03-A, as well as NASA grant NAG5-7697 is gratefully acknowledged.

References

1. P.G. van Dokkum, M. Franx, D.D. Kelson, G.D. Illingworth: Ap. J. **504**, L17 (1998)
2. M. Fukugita, C.J. Hogan, P.J.E. Peebles: Ap. J. **503**, 518 (1998)
3. N.P. Vogt, A.C. Phillips, S.M. Faber, J. Gallego, C. Gronwall, R. Guzman, G.D. Illingworth, D.C. Koo, J.D. Lowenthal: Ap. J. **479**, L121 (1997)
4. M. Im, L. Simard, S.M. Faber, D.C. Koo, K. Gebhardt, C.N.A. Willmer, A. Phillips, G. Illingworth, N.P. Vogt, V.L. Sarajedini: Ap. J. in press (2002)
5. P.G. van Dokkum, M. Franx: Ap. J. **553**, 90 (2001)
6. D.D. Kelson, G.D. Illingworth, P.G. van Dokkum, M. Franx: Ap.J. **531**, 184 (2000)
7. D.D. Kelson, P.G. van Dokkum, M. Franx, G.D. Illingworth, D. Fabricant: Ap. J. **478**, L13 (1997)
8. P.G. van Dokkum, M. Franx, D.D. Kelson, G.D. Illingworth: Ap. J. **553**, L39 (2001)
9. T. Treu, M. Stiavelli, S. Casertano, P. Moller, G. Bertin: Ap. J. **564**, L13 (2002)
10. P.G. van Dokkum, M. Franx, D. Fabricant, D.D. Kelson, G.D. Illingworth: Ap. J. **520**, L95 (1999)
11. G. Kauffmann: MNRAS. **281**, 487 (1996)

The M/L Ratio of Distant Galaxies
from Faint Counts and the $K-z$ Diagram

Brigitte Rocca-Volmerange

Institut d'Astrophysique de Paris, 98bis Bd Arago, F-75014 Paris (rocca@iap.fr)

Abstract. Evolution scenarii of galaxies predict luminosity distributions and apparent ratios M/L as a function of z and type for various cosmogony and mass evolution schemes. Stellar emissions of elliptical to irregular galaxy populations, after k- and e-corrections, are then compared to observations. The best constraints on evolution scenarii are, at the present time, the deepest multispectral faint galaxy counts (B=29 to K=24). Another constraint is the $K-z$ diagram where distant galaxies from deep surveys delimitate a sharp linear limit up to $z > 4$. Compared to predictions of PÉGASE scenarii, a limit baryonic mass $M = 10^{12} M_\odot$ is then derived. Moreover this limit exists already at early epochs constraining formation models.

1   Galaxy Evolution Scenarii of the Model PÉGASE

In PÉGASE, the luminosity evolution scenarii follow conservative laws. Star formation rates are proportional to the current gas mass (M_{gas}^α with $\alpha =1$), except for starbursts and irregulars. The evolution of the gas content is followed from the beginning through ejecta, star formation, winds and cooling-flows. The initial mass function is chosen among a variety of laws, including Salpeter's and many others. In its new version PEGASE.2 (http://www.iap.fr/pegase), the model PÉGASE (Fioc & Rocca-Volmerange, 1997, 1999) is improved with an extinction factor depending on geometry and computed with a transfer code and Draine's dust grains. Moreover metallicity effects for gas and stars are coherently modelled. Seven spectral types (from starbursts to ellipticals) are proposed. In ellipticals, dust follows stars along a King's profile while an homogeneous distribution of stars, dust and gas is assumed in irregulars or spirals. The absolute luminosities are corrected by a distance modulus, cosmology dependent, which is the first parameter of the $K-z$ relation of distant galaxies (Rocca-Volmerange et al, 2002). The authors show the secund parameter is the baryonic mass.

2   Baryonic Masses of Distant Galaxies
from Apparent Magnitudes m_K

The galaxy evolution scenarii fit at best the deepest faint galaxy counts from the far-UV to the near-infrared in the classical cosmology $H_o = 70 km s^{-1} kpc^{-1}$, $\Omega_0 = 0.3$, $\Lambda_0 = 0.7$ (see figure 5 in Fioc & Rocca-Volmerange, 1999). With the same evolution scenarii, we are able to predict $K-z$ relations for various masses which significantly cover the area occupied by observed galaxies. We successively adopt

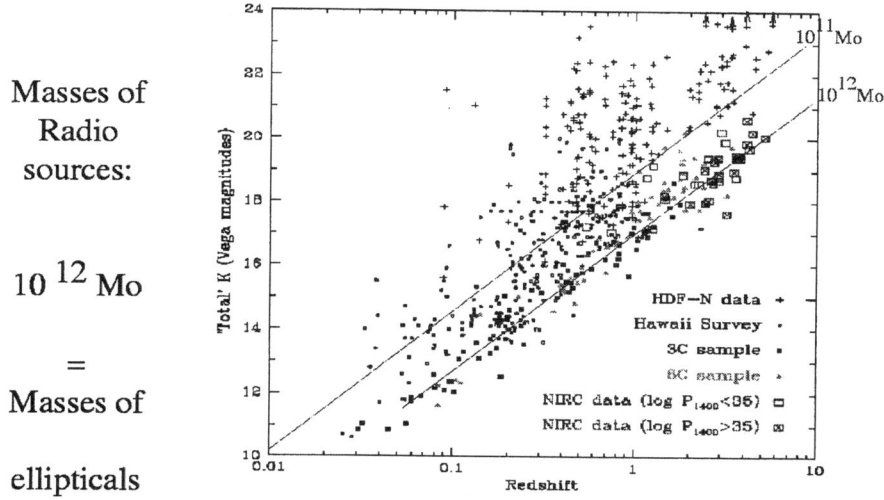

Masses of
Radio
sources:

10^{12} Mo

=

Masses of

ellipticals

Fig. 1. The $K-z$ diagram: radio galaxies and field galaxies are compared to iso-baryonic mass predictions with PÉGASE ($M = 10^{11} M_{\odot}$ and $10^{12} M_{\odot}$). The sharp $K-z$ limit corresponds to $10^{12} M_{\odot}$ ellipticals and radio galaxies, even at $z \geq 5$ (see also R.V. et al, 2002). Data are from de Breuck, 2000

the baryonic masses $M = 10^{11} M_{\odot}$ and $M = 10^{12} M_{\odot}$. Fig. 1 presents the two sequences of magnitudes K compared to observations of various catalogues of galaxies and radio galaxies compiled by de Breuck, 2000. The surprising correspondence of the sharp limit of mainly massive ellipticals and radio sources with the $M = 10^{12} M_{\odot}$ sequence, up to $z > 4$, is a strong result. Massive galaxies would form at the earliest epochs. Field galaxies of the HDF are better compared to sequences of $10^{11} M_{\odot}$ or less.

3   Conclusions

The evolution of the M/L ratio explains the apparent correlation of the $K-z$ diagram with only two parameters. An evolution in mass is only implied for the most massive objects, ellipticals and distant radio galaxies formed at $z > 5$, if it is active on a very short time scale (see RV et al, 2002, for details).

References

1. de Breuck, C., 2000, PhD thesis, Leiden University
2. Fioc M., Rocca-Volmerange B., 1997, Astron. & Astrophys. 326, 950
3. Fioc M., Rocca-Volmerange B., 1999, Astron. & Astrophys. 344, 393
4. Rocca-Volmerange B., de Breuck C., Le Borgne D., 2002, submitted

The Evolution of the Mass-to-Light Ratio of Field Early-Type Galaxies

Tommaso Treu

California Institute of Technology, Astronomy 105-24, Pasadena CA 91125

Abstract. I report on our recent measurement of the Fundamental Plane (FP) of *field* early-type galaxies (E/S0) in the redshift range $0.1 < z < 0.66$, and summarize our findings: *i)* the FP is defined and tight out to the highest redshift bin; *ii)* the intercept γ evolves as $d\gamma/dz = 0.58^{+0.09}_{-0.13}$ (for $\Omega = 0.3$, $\Omega_\Lambda = 0.7$), or, in terms of average effective mass to light ratio, as $d\log(M/L_B)/dz = -0.72^{+0.11}_{-0.16}$, i. e. faster than is observed for cluster E/S0 (-0.49 ± 0.05). In addition, we detect [OII] emission > 5Å in 22% of an enlarged sample of 42 *massive* E/S0 in the range $0.1 < z < 0.73$, in contrast with the quiescent population observed in clusters at similar z. We interpret these findings as evidence that a significant fraction of massive field E/S0 experiences secondary episodes of star-formation at $z < 1$.

1 The Evolution of Field E/S0 to $z \sim 0.7$

The Fundamental Plane ([3,4]; hereafter FP) is a tight correlation between the effective radius (R_e), the effective surface brightness (SB_e), and the central velocity dispersion (σ) of early-type galaxies (E/S0). Under the assumption that E/S0 are homologous dynamical systems the FP can be interpreted in terms of a power law relation between mass (M) and mass-to-light ratio (M/L) ([16] and references therein).

In recent years, the evolution of the FP with redshift has been used to measure the evolution of the M/L of E/S0 (e.g. [20,1,6]). In particular, our group ([9,10,16,17]) has been measuring the evolution of the FP of field E/S0, which – in a CDM scenario – are predicted to be assembled at later times than their cluster counterparts (e.g. [5]). Remarkably, we find that the FP is tight in the field out to $z \sim 0.7$, and the evolution of the intercept of the FP is faster in the field than in clusters ([17]). In addition, we find that 20% of field E/S0 at $z \sim 0.5$ have significant [OII] emission, in contrast with the quiescent stellar populations of E/S0 in the local Universe and in clusters at similar redshifts. These findings are consistent with a scenario where field E/S0 do not experience major structural changes between $z \sim 0.7$ and today, and the evolution of the M/L is driven by a small fraction of the stellar mass (1-10%) being formed in secondary bursts at $z < 1$.

2 Acknowledgments

This project is financially supported by STScI grant AR-09222.

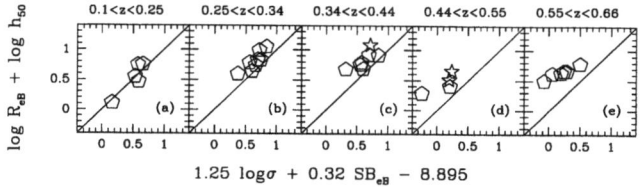

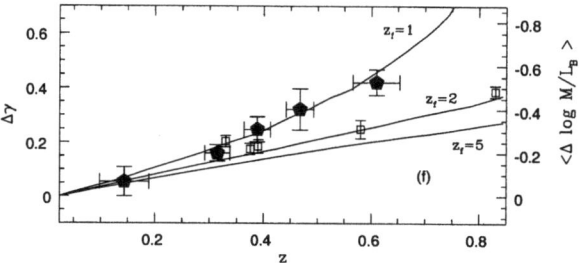

Fig. 1. FP in the rest-frame B band. In panels (a) to (e) we show the field E/S0 (open stars if [OII] emission is detected, open pentagons otherwise), binned in redshift and compared to the FP found in the Coma Cluster (Bender et al. 1998). Panel (f) shows the average offset of the intercept of field galaxies from the local FP relation of the Coma cluster ([1]) as a function of redshift (large filled pentagons) compared to the offset observed in clusters (open squares; [12,7,20,1,6]). The solid lines represent the evolution predicted for passively evolving stellar populations formed in a single burst at $z = 1, 2, 5$ (from top to bottom) computed using [2] models in the BC96 version ([16]). We assume $\Omega = 0.3$, $\Omega_\Lambda = 0.7$ and $H_0 = 65 \mathrm{kms}^{-1} \mathrm{Mpc}^{-1}$

References

1. R. Bender et al.: ApJ, **493**, 529 (1998)
2. G. Bruzual, S. Charlot: ApJ, **405**, 538 (1993)
3. S. G. Djorgovksi, M. Davis: ApJ, **313**, 59 (1987)
4. A. Dressler, D. Lynden-Bell, D. Burstein, R. L. Davies, S. M. Faber, R. Terlevich, G. Wegner: ApJ, **313**, 42 (1987)
5. G. Kauffmann: MNRAS, **281**, 478 (1996)
6. D. D. Kelson, G. D. Illingworth, P. G. van Dokkum, & M. Franx: ApJ, **531**, 184 (2000)
7. D. D. Kelson, P. G. van Dokkum, M. Franx, G. D. Illingworth, & D. Fabricant: ApJ, **478**, L13 (1997)
8. T. Treu, M. Stiavelli, G. Bertin, S. Casertano, P. Møller: MNRAS, **326**, 237 (2001)
9. T. Treu, M. Stiavelli, S. Casertano, P. Møller, G. Bertin: MNRAS, **308**, 1307 (1999)
10. T. Treu, M. Stiavelli, P. Møller, S. Casertano, G. Bertin: MNRAS, **326**, 221 (2001)
11. T. Treu, M. Stiavelli, S. Casertano, P. Møller, G. Bertin: ApJ, **564**, L00 (2002)
12. P. G. van Dokkum & M. Franx: MNRAS, **281**, 985
13. P. G. van Dokkum, M. Franx, D. D. Kelson, G. D. Illingworth, 1998, ApJ, **504**, L17

Measuring the Virial Masses of Disk Galaxies

Frank C. van den Bosch

Max-Planck Institut für Astrophysik, Postfach 1317, 85741 Garching, Germany

Abstract. I present detailed models for the formation of disk galaxies, and investigate which observables are best suited as virial mass estimators. Contrary to naive expectations, the luminosities and circular velocities of disk galaxies are extremely poor indicators of total virial mass. Instead, I show that the product of disk scale length and rotation velocity squared yields a much more robust estimate. Finally, I show how this estimator may be used to put limits on the efficiencies of cooling and feedback during the process of galaxy formation.

1 Introduction

Currently, the main uncertainties in our picture of galaxy formation are related to the intricate processes of cooling, star formation, and feedback. The cooling and feedback efficiencies are ultimately responsible for setting the galaxy mass fractions $f_{gal} = M_{gal}/M_{vir}$. Here M_{gal} is the total *baryonic* mass of the galaxy (stars plus gas, excluding the hot gas in the halo) and M_{vir} is the total virial mass.

Here I present new models for the formation of disk galaxies, which I use to investigate how well observables *extracted directly from these models* can be used to recover $f_{gal}(M_{vir})$. Even though the assumptions underlying the model are not necessarily correct, and the phenomenological descriptions of star formation and feedback are certainly oversimplified, this provides useful insights regarding the ability of actual observations to constrain the poorly understood astrophysical processes of galaxy formation.

2 Short Description of Models

The main assumptions that characterize the framework of the models are the following: (i) dark matter halos around disk galaxies grow by the smooth accretion of mass, (ii) in the absence of cooling the baryons have the same distribution of mass and angular momentum as the dark matter, and (iii) the baryons conserve their specific angular momentum when they cool. I follow Firmani & Avila-Reese (2000) and make the additional assumptions that (iv) the spin parameter of a given galaxy is constant with time, (v) each mass shell that virializes is in solid body rotation, and (vi) the rotation axes of all shells are aligned. Although neither of these assumptions is necessarily accurate, it was shown in van den

Bosch (2001) that they result in halo angular momentum profiles in excellent agreement with the high resolution N-body simulations of Bullock et al. (2001).

The main outline of the models is as follows. I set up a radial grid between $r = 0$ and the present day virial radius of the model galaxy and follow the formation and evolution of the disk galaxy using a few hundred time steps. Six mass components are considered: dark matter, hot gas, disk mass (both in stars and in cold gas), bulge mass, and mass ejected by outflows from the disk. The dark matter, hot gas, and bulge mass are assumed to be distributed in spherical shells, whereas the disk stars and cold gas are assumed to be in infinitesimally thin annuli. Each time step the changes in these various mass components in each radial bin are computed using the following prescriptions.

- The rate at which the total virial mass grows with time is given by the Universal mass accretion history (van den Bosch 2002a).
- At each redshift, the dark matter is assumed to follow an NFW density distribution (Navarro, Frenk & White 1997).
- The gas that enters the virial radius of a halo is added to the disk a time $t_c \equiv \max[t_{\mathrm{ff}}, t_{\mathrm{cool}}]$ later, where t_{ff} and t_{cool} correspond to the free-fall time and cooling time. The cooling time depends on the metallicity of the hot gas, which is taken to be a free parameter Z_{hot}.
- The radius at which the gas settles is governed by its specific angular momentum distribution, which follows from the assumptions (iv)–(vi) listed above.
- When the disk becomes unstable, part of the disk material is converted into a bulge component (cf. van den Bosch 1998).
- In the disk, only the cold gas with a surface density above the critical density given by Toomre's stability criterion is considered eligible for star formation. This gas is transformed into stars with a rate given by a simple Schmidt law.
- Part of the cold gas is expelled from the system by supernovae feedback. This is modeled as in van den Bosch (2000), and regulated by a free feedback efficiency parameter $\varepsilon_{\mathrm{fb}}$.

More details regarding these models can be found in van den Bosch (2001), where it is shown that these models yield disk galaxies in good agreement with observations. For instance, for the majority of the model galaxies the disk reveals an exponential surface brightness profile. Note that contrary to previous disk formation models (e.g., Mo, Mao & White 1998), this is not an *a priori* assumption of the model.

3   Galaxy Mass Fractions

In order to investigate how $f_{\mathrm{gal}}(M_{\mathrm{vir}})$ relates to the cooling and feedback efficiencies I discuss three models that only differ in the metallicity of the hot gas, Z_{hot}, and the feedback efficiency, $\varepsilon_{\mathrm{fb}}$: the 'Standard Model' with $Z_{\mathrm{hot}} = 0.3 Z_{\odot}$ (a typical value for the hot gas in clusters) and $\varepsilon_{\mathrm{fb}} = 0$ (i.e., no feedback), the 'Zero Metallicity Model' with $Z_{\mathrm{hot}} = 0.0$ and $\varepsilon_{\mathrm{fb}} = 0$, and the 'Feedback Model' with $Z_{\rm hot} = 0.3 Z_\odot$ and $\varepsilon_{\rm fb} = 0.02$ (i.e., two percent of the SN energy is converted to kinetic energy). All other model parameters are kept fixed at their fiducial values (see van den Bosch 2001). For each model a sample of 400 model galaxies is constructed. Present day virial masses are drawn from the Press-Schechter mass function with $10^{10} h^{-1}\,{\rm M}_\odot \leq M_{\rm vir}(0) \leq 10^{13} h^{-1}\,{\rm M}_\odot$, corresponding to $31\,{\rm km\,s^{-1}} \leq V_{\rm vir} \leq 312\,{\rm km\,s^{-1}}$, roughly the range expected for galaxies. Spin parameters, which parameterize the specific angular momentum, are drawn from a typical log-normal distribution.

Figure 1 plots the present day galaxy mass fractions $f_{\rm gal} = M_{\rm gal}/M_{\rm vir}$ as function of $M_{\rm vir}$ (crosses). In the models without feedback $f_{\rm gal}$ is virtually identical to the universal baryon fraction $f_{\rm bar}$ for low mass systems. For more massive systems, cooling becomes inefficient, causing $f_{\rm gal}$ to strongly decrease with increasing virial mass. This is more pronounced in the model with $Z_{\rm hot} = 0$, for which cooling is least efficient. In the Feedback Model (right panel), $f_{\rm gal} \ll f_{\rm bar}$ for the low mass systems, but with a large amount of scatter. This is a reflection of the scatter in halo spin parameters: systems with less angular momentum produce disks with higher surface densities, therefore have higher star formation rates, which induce a more efficient feedback. At the high mass end $f_{\rm gal}$ is fairly similar to the Standard Model without feedback. This owes to the fact that the mass ejection efficiency scales inversely with the square of the escape velocity, making feedback less efficient in more massive systems.

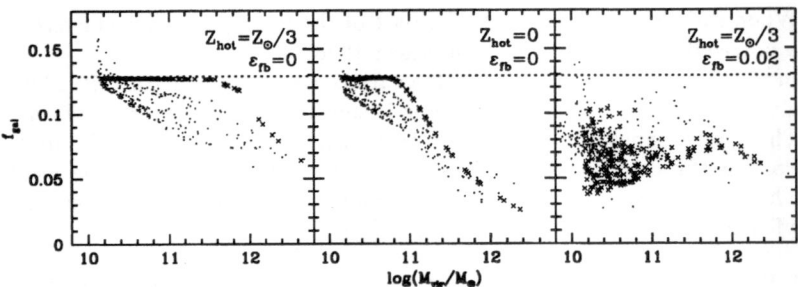

Fig. 1. A comparison of the true galaxy mass fraction $f_{\rm gal}$ as function of the true virial mass (crosses) with the same values estimated from the observables extracted from the models (dots). Results are plotted for all three models discussed in the text. The horizontal dotted line corresponds to the universal baryonic mass fraction $f_{\rm bar}$. In the model with feedback (right panel) the dots occupy the same parameter space as the crosses, indicating that the observables allow one to recover $f_{\rm gal}(M_{\rm vir})$, at least in a statistical sense. In the two models without feedback there are significant errors in the recovered $f_{\rm gal}(M_{\rm vir})$. Yet, the two $f_{\rm gal}(M_{\rm vir})$ are sufficiently different to discriminate between the two models; in particular, the estimated galaxy mass fractions nicely avoid the upper right regions of parameter space which contain information on the cooling efficiencies. Furthermore, the dots occupy different areas of parameter space in models with and without feedback, such that there is hope that the "observed" $f_{\rm gal}(M_{\rm vir})$ may be used to constrain the efficiency of feedback

Clearly $f_{gal}(M_{vir})$ depends strongly on both the cooling and feedback efficiencies. Therefore, if one could obtain a measure of $f_{gal}(M_{vir})$ observationally it would allow us to constrain the poorly understood physics of cooling and feedback. This requires one to be able to infer both the baryonic galaxy mass M_{gal} as well as the total virial mass M_{vir} from observations of the luminous (and gaseous) components. Using the models outlined above I now investigate which observables are best suited as the appropriate mass indicators.

For the galaxy mass one can write $M_{gal} = \Upsilon_B L_B + M_{gas}$. Here L_B is the total B-band luminosity, Υ_B is the corresponding stellar mass-to-light ratio, and M_{gas} is the galaxy's (cold) gas mass. L_B is easily obtained observationally, and I assume that M_{gas} is observationally accessible through HI measurements. The stellar mass-to-light ratio, however, is not directly accessible to an observer, but has been shown to correlate strongly with color. I therefore extract the $B - K$ color from the models from which I estimate Υ_B using the relation of Bell & de Jong (2001).

The galaxy virial mass is more difficult to obtain. In Figure 2 I plot three 'observables' as function of M_{vir} for galaxies in each of the three models. Note that both V_{max}, defined as the maximum rotation velocity inside the radial extent probed by the cold gas, and L_K are poor indicators of virial mass. First of all the slope and zero-points of the $V_{max}(M_{vir})$ and $L_K(M_{vir})$ relations depend on the input parameters of the model. This means that an observer trying to infer M_{vir} from either V_{max} or L_K needs to make assumptions about the efficiencies of cooling and feedback. However, it is exactly these efficiencies that we seek to constrain. Secondly, the scatter of both relations can be so large that even if the normalization of the relation were known, one could still not infer M_{vir} to better than an order of magnitude. In particular, the scatter in $V_{max}(M_{vir})$ can be very large. This owes entirely to the scatter in halo spin parameter, which sets the concentration of the baryonic mass component after cooling, therewith strongly influencing V_{max}.

As the lower panels in Figure 2 show, a much more reliable virial mass indicator is $R_d V_{max}^2 / G$. Here R_d is the disk scale length in the I-band, obtained from fitting an exponential to the I-band surface brightness distribution of the disk. Upon fitting all galaxies of all three models simultaneously I obtain

$$M_{vir} = 2.54 \times 10^{10} \, M_\odot \left(\frac{R_d}{kpc} \right) \left(\frac{V_{max}}{100 \, km \, s^{-1}} \right)^2 . \tag{1}$$

(rms scatter between 20 and 50 percent, depending on the amount of feedback). The fraction of model galaxies for which equation (1) yields an estimate of the true virial mass to better than a factor two is larger than 97 percent! It is remarkable that the zero-point for models with feedback is the same as for models without feedback. When matter is ejected it reduces V_{max} but at the same time increases the disk scale length such that $R_d V_{max}^2$ stays roughly constant. This is due to the star formation threshold criterion included in the models. If real galaxies follow a similar threshold criterion, equation (1) provides a fairly accurate estimate of the total virial mass of disk galaxies.

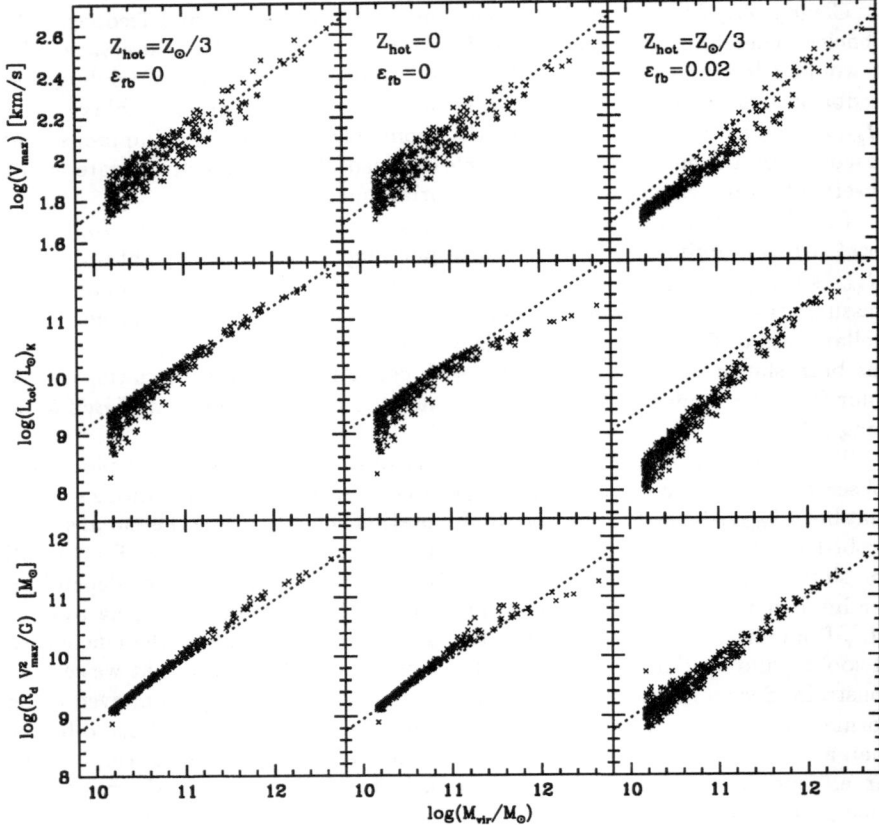

Fig. 2. The relation between various virial mass estimators and the actual virial mass for all three models. In the upper panels $\log(V_{\max})$ is plotted versus $\log(M_{\mathrm{vir}})$. The dotted line corresponds to $M_{\mathrm{vir}} \propto V_{\max}^{1/3}$, which is a reasonable description of the average relation. However, the scatter is large, and the zero-point depends on the actual model, which makes $V_{\max}$ unsuitable as virial mass indicator. The same goes for the K-band luminosity, which is plotted in the middle row of panels. Here the dotted line corresponds to $L_K \propto M_{\mathrm{vir}}$, which only yields a reasonable description for the brighter galaxies in models with $Z_{\mathrm{hot}} = Z_\odot/3$. For fainter galaxies, however, the scatter is large. Furthermore the slope of the $L_K(V_{\mathrm{vir}})$ relation depends strongly on the feedback efficiency. Thus total luminosity is also a poor indicator of total virial mass. The lower panels plot the virial mass estimator $R_d V_{\max}^2/G$ as function of M_{vir}. Here the dotted lines correspond to equation (1), which gives a reasonable fit to the model galaxies, independent of the cooling and/or feedback efficiencies. In addition, the scatter is relatively small, making the product of disk scale length and maximum rotation velocity squared a fairly accurate estimator of total virial mass

We now have all the tools in place to see whether we can recover $f_{gal}(M_{vir})$ from the "observables". The dots in Figure 1 correspond to the estimates of $f_{gal}(M_{vir})$ obtained using the method outlined above. In both models without feedback there are significant errors in the recovered $f_{gal}(M_{vir})$, which is dominated by errors in the estimate of M_{vir}. Yet, the recovered $f_{gal}(M_{vir})$ of the two models are sufficiently different to distinguish between them. In particular, in both models the dots avoid the regions in the upper right corner which contains information about the efficiency of cooling. In the feedback model the recovered values occupy roughly the same area of the $f_{gal} - M_{vir}$ plane as the intrinsic values. Although the one-to-one correspondence for individual model galaxies may be poor, statistically the method to recover $f_{gal}(M_{vir})$ explored here works reasonably well.

4 Conclusions

We have shown that the product of disk scale length and maximum rotation velocity squared can be used as a fairly accurate estimator of total virial mass. This in turn can be used to obtain estimates of the galaxy mass fractions as function of virial mass, which contains important information about the efficiencies of cooling and feedback. Another approach, that has been taken in the past, is to use published luminosity functions and luminosity-velocity relations to construct halo velocity functions (i.e., Newman & Davis 2000; Gonzales et al. 2000; Kochanek 2001). The main goal of these studies is similar to the work presented here, namely to circumvent the problems with poorly understood astrophysical processes when linking the observed properties of galaxies to those of their dark matter halos. Our results imply that great care is to be taken in linking an observable velocity such as V_{max} to the circular velocity of a dark matter halo. Based on our results, we suggest that the construction of a halo *mass* function using $M_{vir} \propto R_d V_{max}^2$ may proof more reliable. More details of the results presented here can be found in van den Bosch (2002b)

References

1. E. F. Bell, R. S. de Jong: ApJ, **550**, 212 (2001)
2. J. S. Bullock et al.: ApJ, **555**, 240 (2001)
3. C. Firmani, V. Avila-Reese: MNRAS, **315**, 457 (2000)
4. A. H. Gonzales et al.: ApJ, **528**, 145 (2000)
5. C. S. Kochanek: preprint, astro-ph/0108160 (2001)
6. H. J. Mo, S. Mao, S. D. M. White: MNRAS, **295**, 319 (1998)
7. J. F. Navarro, C. S. Frenk, S. D. W. White: ApJ, **490**, 493 (1997)
8. J. A. Newman, M. Davis: ApJ, **534**, L11 (2000)
9. F. C. van den Bosch: ApJ, **507**, 601 (1998)
10. F. C. van den Bosch: ApJ, **530**, 177 (2000)
11. F. C. van den Bosch: MNRAS, **327**, 1334 (2001)
12. F. C. van den Bosch: MNRAS, in press; astro-ph/0105158 (2002a)
13. F. C. van den Bosch: MNRAS, submitted; astro-ph/0112566 (2002b)

Ultradeep Near-Infrared ISAAC Observations of the Hubble Deep Field South: Selecting High-Redshift Galaxies in the Rest-Frame Optical

Ivo Labbé[1], Marijn Franx[1], Gregory Rudnick[2], Alan Moorwood[3], Natascha Förster Schreiber[1], Hans-Walter Rix[2], Lottie van Starkenburg[1], Peter van Dokkum[4], Paul van der Werf[1], Huub Röttgering[1], and Konrad Kuijken[5]

[1] Leiden Observatory, P.O. Box 9513, NL-2300 RA, Leiden, The Netherlands
[2] MPI für Astronomie, Königstuhl 17, D-69117, Heidelberg, Germany
[3] ESO, Karl-Schwarzschild-Str. 2, D-85748, Garching, Germany
[4] Caltech, 1200 East California Boulevard, Pasadena, CA, USA
[5] Kapteyn Institute, P.0. Box 800, 9700 AV, Groningen, The Netherlands

Abstract. We present very deep near-infrared imaging of the Hubble Deep Field South with ISAAC on the *Very Large Telescope*. We obtained more than 100 hours total integration time in J_s, H and K_s under the best seeing conditions; this resulted in the deepest ground-based infrared observations to date, and the deepest K_s-band in any field. We constructed a K_s-limited multicolor catalog, selecting high-redshift galaxies from their rest-frame optical properties. We detect luminous evolved galaxies, some of which are large in the rest-frame optical and have morphologies reminiscent of present-day spirals with red bulges and bluer disks. We also find a new population of optically faint galaxies at redshifts $2 < z_{phot} < 4$ with very red near-infrared colors ($J_s - K_s >$ 2.3), that would be missed by the standard U-dropout criteria. These galaxies are generally compact and many show pronounced breaks in the observed near-infrared, which we identify as the Balmer/4000 Å break; they may contain a significant fraction of the total stellar mass at this cosmic epoch. Surprisingly, both large U-dropouts and red galaxies were virtually absent in the HDF-North with its accompanying NICMOS data. In general, our results demonstrate the necessity of deep near-infrared imaging to obtain a more complete picture of the early universe.

1 Introduction

In the past decade, our ability to routinely identify and systematically study distant galaxies has dramatically advanced our knowledge of the high-redshift universe. In particular, the effective U-dropout technique [21,22] has enabled us to select distant galaxies using simple photometric criteria. Now more than 1000 of these Lyman break galaxies (LBGs) are spectroscopically confirmed at $z > 2$, and have been subject to targeted studies on spatial clustering [9], internal kinematics [16,17], dust properties [1], and stellar composition [20,15]. Although LBGs are among the best studied classes of distant galaxies to date, many of their properties are still unknown. More importantly, it is unclear whether the

U-dropout technique alone will give us a fair census of the galaxy population at $z \sim 3$, as it requires galaxies to be bright in the far-ultraviolet, blue and relatively compact, while we know that dust obscured systems, and most present-day galaxies do not satisfy these criteria.

In this context we initiated the Faint InfraRed Extragalactic Survey (FIRES) [6]; a large public program consisting of very deep near-infrared (NIR) imaging of two fields, carried out at the *Very Large Telescope* (VLT), to access the rest-frame optical emission of objects at redshifts $2 < z < 4$. We observed fields with existing deep optical WFPC2 imaging from the *Hubble Space Telescope* (HST): the WFPC2 main-field of Hubble Deep Field South, and the field around the distant cluster MS1054-03. By selecting sources in the near-infrared K_s-band, we hope to obtain a more complete view of the high-redshift universe. Compared to the rest-frame far-UV, rest-frame optical light is less sensitive to the effects of dust extinction and on-going star formation, and provides a better tracer of stellar mass. The extended wavelength coverage ($0.3 - 2.2\mu m$) is required not only to access the rest-frame optical, but also to determine the redshifts of faint galaxies from their broadband photometry alone.

We present here selected results of the HDF-S. The observations, reduction and analysis techniques are described in detail in [11] and [18]; the results on MS1054-03 will be presented in [7]. Throughout this paper, we assume a flat Λ-dominated cosmology ($\Omega_M = 0.3, \Lambda = 0.7, h = H_0/100 \text{ km s}^{-1} \text{ Mpc}^{-1} = 1.0$). Unless explicitly stated otherwise, all magnitudes are expressed in the Johnson [10] photometric system.

2   Observations and Photometry

The NIR observations of the HDF-S were obtained using the Infrared Spectrograph and Array camera (ISAAC) [14] on the VLT/Antu at Paranal. The ISAAC field of view is 2.5′ x 2.5′ with a pixel scale of 0.147″pixel^{-1}. We dithered many exposures in the J_s (1.24μm), H (1.65μm) and K_s (2.17μm) bands, with individual exposure times of 120s, 120s, and 60s, respectively. The total integration time in each passband is about 33 hours and the median seeing of all frames is about 0.45″. The sensitivity varies significantly over the field because of the dithered observations, with the formal limiting depths in J_s, H, and K_s in the deepest central 2′ x 2′ being 25.9, 24.9 and 24.4, respectively ($S/N \approx 5$ in a 0.7″ diameter aperture).

We complemented the ISAAC data with HST/WFPC2 imaging (version 2) [3] in filters close to the U, B, V and I. We constructed a K_s-band selected sample of galaxies and measured colors in all seven passbands within the K_s-detection isophote using the Sextractor software [2]. We accounted for differences in resolution and image quality between the HST and VLT dataset. Finally we determined the photometric redshifts of all extragalactic sources using the method detailed in [18].

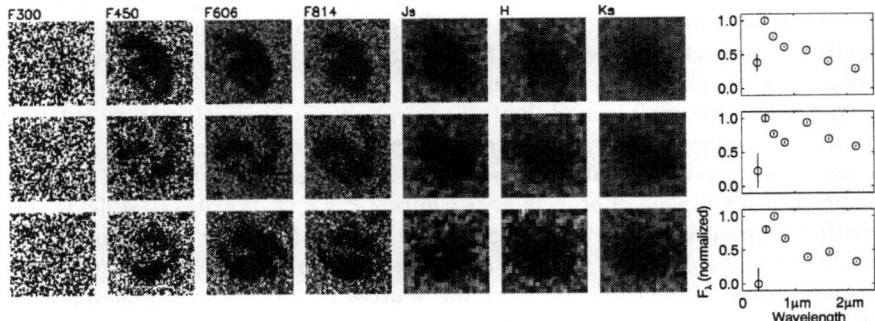

Fig. 1. Morphology of the most extended U-drop galaxies in the HDF-S. The left panels show the WFPC2 U, B, V, I data, and right hand panels show our VLT/ISAAC J_s, H, K_s data. The images are 3 x 3 arcsec on a side. The intensity is proportional to F_λ, with arbitrary normalization for each galaxy. The right hand column shows the spectral energy distributions. As can be seen, these U-drop galaxies are large in extent, and can show prominent Balmer breaks. The K-band infrared images (rest-frame optical) are generally more centrally concentrated than the WFPC2 images; this is a physical change and not caused by the difference in image quality. The galaxy in the bottom row is spectroscopically confirmed at $z_{spec} = 2.8$

3   NIR Properties of Lyman Break Galaxies

The LBGs identified in our K_s-band selected catalogue are different from those found in previous surveys, showing a large variation in both morphologies and spectral energy distributions (SEDs). We selected them by applying the criteria of Madau et al [13]. While most U-dropouts are compact at all wavelengths, some are large in the rest-frame optical and show profound differences between the rest-frame ultraviolet and optical morphologies, as shown in Fig. 1.

The I-band morphologies (rest-frame $2000-2700$ Å) are complex with knotty structure, whereas the NIR light is more centrally concentrated and extended, reminiscent of red bulges and bluer disks with star-forming clumps. In a few cases the rest-frame optical light is almost spatially distinct from the far-ultraviolet. The change in morphology when moving to the NIR is not dominated by the decreasing resolution of the ground-based data but reflects a change in the physical distribution of the light. These properties are in stark contrast with the compact and small sizes traditionally emphasized [8,12], and the segregated UV-optical morphology has not been seen in the HDF-N [4]. The LBGs with complex structures are amongst the brightest and reddest in the rest-frame optical, suggesting higher mass-to-light ratios, and they might be the most massive LBGs around. Their spectral energy distributions show a pronounced break in the rest-frame optical, which we identify as the age-sensitive Balmer/4000 Å break. Follow-up VLT/FORS spectroscopy [19] confirmed the redshift for one of the large galaxies, a disk-like U-dropout at $z_{spec} = 2.793 \pm 0.003$, implying a physical diameter of 10 h^{-1} kpc. While this is only one example, models of galaxy formation must

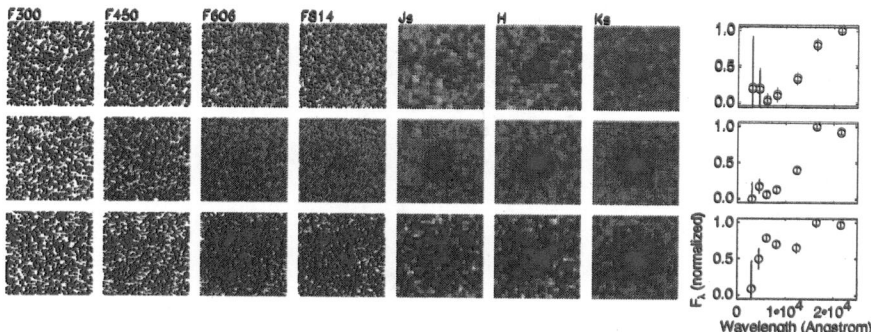

Fig. 2. same as Fig. 1, but for some of the red galaxies. Many of these galaxies are small, and show prominent breaks in the infrared (rest-frame optical). Note that the galaxy in the top row is barely visible even in J_s. The SED of the object in the middle is consistent with a strong Balmer break at $z \sim 2.5$. The galaxy in the bottom row is very extended in the rest-frame optical. It shows faint emission in the H-band out to $1''$, consistent with an exponential profile

produce objects of comparable sizes and in sufficient numbers to be consistent with our observations.

4 Balmer Break Galaxies

Evolved quiescent galaxies with a prominent Balmer discontinuity in the spectrum, like most present-day Hubble Type galaxies, would have very red observed NIR colors if placed at redshifts $z > 2$. In the HDF-S we find a substantial population of such red galaxies with $J_s - K_s > 2.3$ and photometric redshifts $2 < z_{phot} < 4$. These galaxies are generally compact, with exceptions, as can be seen in Fig. 2.

Most of them are extremely faint in the observed optical with red $V_{606} - H$ colors, as shown in Fig. 3, and would be missed by the standard U-dropout criteria. Whether the red NIR colors are due to age or dust has not been determined for the sample as a whole, but many spectral energy distributions exhibit a clear break between the J_s and H bands, more consistent with a Balmer/4000 Å break. One of the galaxies shows extremely red NIR colors similar to the curious "J-dropout" HDF-N JD1 [5], possibly a maximally old elliptical or highly reddened galaxy at $z \sim 3$. Another shows an H-band surface brightness profile which can be fit with an exponential out to $1''$ radius and this galaxy seems to host an arc of blue knots in the WFPC2 images; possibly star-forming sites embedded in a larger evolved system.

We find 13 red objects having $K_{J,tot} < 22$ compared to 37 U-dropouts to the same flux limit. Interestingly, the contribution of K_s-band light and the red rest-frame optical colors suggest this population contains a significant fraction

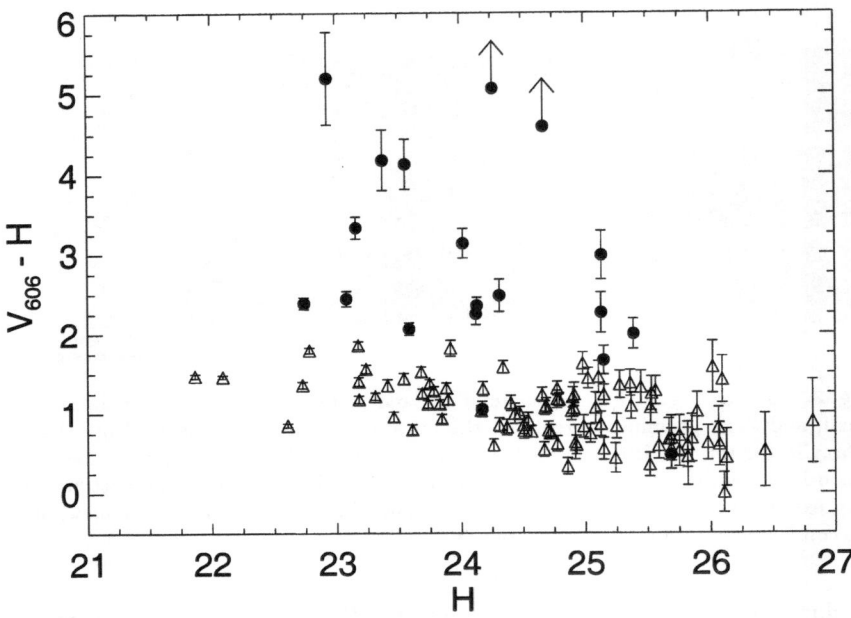

Fig. 3. Color-magnitude diagram for galaxies photometrically at $1.95 < z_{phot} < 3.5$ in the HDF-S. Filled circles show the $V_{606} - H$ colors of the red galaxies, while open triangles show the other objects. The number of candidates for red, non star-forming galaxies is much higher than in the HDF-N, see Fig. 1. of [15]. For easy comparison, we give the magnitudes and colors in the AB system, where $V_{606,AB} = V_{606} + 0.12$ and $H_{AB} = H + 1.38$. We have used 1 sigma upper limits for the fluxes in V_{606}. We find 7 galaxies redder than $V_{606,AB} - H_{AB} > 3$ and brighter than $H_{AB} < 25$, compared to one in the HDF-N

of the total stellar mass at this cosmic epoch. The surface density of these red sources is not well known; the HDF-N has far fewer red galaxies than the HDF-S, as is apparent from comparing Fig. 3 to Fig 1. of [15] which shows a similar plot of a NIR selected sample in the HDF-N. In the same diagram, we find 7 galaxies redder than $V_{606,AB} - H_{AB} > 3$ and brighter than $H_{AB} < 25$, compared to one in the HDF-N.

5   Conclusion

The first results of the HDF-S demonstrate the necessity of extending optical observations to near-IR wavelengths for a more complete census of the early universe. Our deepest K_s-band proves invaluable for it probes well into the rest-frame optical at $2 < z < 4$, where long-lived stars may dominate the light of galaxies. We find:

- high-redshift systems that are large in the rest-frame optical with complex UV morphologies. Some have spatially distinct rest-frame UV and optical light distributions, where in general the rest-frame optical is more centralized and extended. Some of them have disky structure with clumps of star formation, similar to local spiral galaxies.
- a previously undiscovered population of red galaxies with $(J_s - K_s > 2.3)$ at $z > 2$. Because most of them are barely detectable even in the deepest optical images, they would be missed by ultraviolet-optical color selection techniques, such as the U-dropout method. Many of these galaxies show a prominent jump in the NIR, which may be caused by the Balmer/4000 Å break. The red galaxies likely contribute significantly to the total stellar mass budget at redshifts $z \sim 3$.

We are pursuing our follow-up program to obtain more spectroscopic redshifts needed to confirm the above results. It must be stressed that while distant galaxies in the HDF-S look very different compared to what has been found in the HDF-N, both fields are rather small and results based on them may be seriously affected by large scale structure in the universe. This calls for more observations to similar limits with full optical-to-infrared coverage.

References

1. Adelberger, K. L. & Steidel, C. C. 2000, ApJ, 544, 218
2. Bertin, E. & Arnouts, S. 1996, A&AS, 117, 393
3. Casertano, S. et al. 2000, AJ, 120, 2747
4. Dickinson, M. 2000, Philos. Trans. R. Soc. London, A, 358, 2001
5. Dickinson, M. et al. 2000, ApJ, 531, 624
6. Franx, M. et al. 2000, The Messenger, 99, 20
7. Forster Schreiber, N.M. et al. 2002, in preparation
8. Giavalisco, M., Steidel, C. C., & Macchetto, F. D. 1996, ApJ, 470, 189
9. Giavalisco, M. & Dickinson, M. 2001, ApJ, 550, 177
10. Johnson, H. L. 1966, ARA&A, 4, 193
11. Labbe, I. F. L. et al. 2002, in preparation
12. Lowenthal, J. D. et al. 1997, ApJ, 481, 673
13. Madau, P., Ferguson, H. C., Dickinson, M. E., Giavalisco, M., Steidel, C. C., & Fruchter, A. 1996, MNRAS, 283, 1388
14. Moorwood, A. F. 1997, proc. SPIE, 2871, 1146
15. Papovich, C., Dickinson, M., & Ferguson, H. C. 2001, ApJ, 559, 620
16. Pettini, M., Kellogg, M., Steidel, C. C., Dickinson, M., Adelberger, K. L., & Giavalisco, M. 1998, ApJ, 508, 539
17. Pettini, M., Shapley, A. E., Steidel, C. C., Cuby, J., Dickinson, M., Moorwood, A. F. M., Adelberger, K. L., & Giavalisco, M. 2001, ApJ, 554, 981
18. Rudnick, G. et al. 2001, AJ, 122, 2205
19. Rudnick, G. et al. 2002, in preparation
20. Shapley, A. E., Steidel, C. C., Adelberger, K. L., Dickinson, M., Giavalisco, M., & Pettini, M. 2001, ApJ, 562, 95
21. Steidel, C. C., Giavalisco, M., Dickinson, M., & Adelberger, K. L. 1996, AJ, 112, 352
22. Steidel, C. C., Giavalisco, M., Pettini, M., Dickinson, M., & Adelberger, K. L. 1996, ApJ, 462, L17

HST Imaging of a $z = 1.55$ Old Galaxy Group

Andrew Bunker[1,2], Hyron Spinrad[2], Ross McLure[3,4], Arjun Dey[5],
James Dunlop[4], John Peacock[4], Daniel Stern[2,6], Rodger Thompson[7],
Ian Waddington[8,9], and Rogier Windhorst[8]

[1] Institute of Astronomy, Madingley Road, Cambridge CB3 0HA, UK
   email: bunker@ast.cam.ac.uk
[2] Department of Astronomy, 601 Campbell Hall, Berkeley CA 94720, USA
[3] Department of Astrophysics, Keble Road, Oxford OX1 3RH, UK
[4] Institute for Astronomy, Blackford Hill, Edinburgh EH9 3HJ, UK
[5] KPNO/NOAO, 950 N. Cherry Avenue, Tucson, AZ 85726, USA
[6] Jet Propulsion Laboratory/Caltech MS 169-327, Pasadena, CA 91109, USA
[7] Steward Obs., University of Arizona, N. Cherry Avenue, Tucson AZ 85721, USA
[8] Department of Physics, Arizona State University, Tempe, AZ 85287, USA
[9] Department of Physics, Bristol University, Tyndall Avenue, Bristol, BS8 1TL, UK

Abstract. We present high-resolution imaging in the rest-frame optical of the weak radio source LBDS53W091. Previous optical spectroscopy has shown that this object has an evolved stellar population of age $> 3\,\mathrm{Gyr}$ at $z = 1.55$, determined from the amplitude of rest-frame UV spectral breaks. We have obtained deep Hubble Space Telescope imaging over 10 orbits with NICMOS camera 2, using the F160W H-band filter ($1.6\,\mu$m) which is a good approximation to the rest-frame R-band. Our observations reveal a radial light profile which is well fit by a de Vaucouleurs' $r^{1/4}$ law, with a scale length of $r_e = 0.3''$ ($3\,h_{50}^{-1}$ kpc for $\Omega_M = 0.3$ and $\Omega_\Lambda = 0$). The elliptical morphology of the radio galaxy indicates a dynamically-evolved old system, consistent with the spectroscopic results. Some surrounding objects lie on the color:magnitude relation for a cluster at $z = 1.55$, and are likely to be associated. The group of galaxies are somewhat more luminous than the fundamental plane of ellipticals at $z = 0$ subject to $(1 + z)^4$ surface brightness dimming, but are consistent with estimates of the fundamental plane at high redshift subject to passive luminosity evolution from a formation epoch of $z > 3$.

1 Introduction

There has been a revolution in recent years in finding and studying star-forming galaxies at high redshift. An orthogonal approach to studying galaxy assembly is to focus on old stellar populations which formed at even earlier cosmic times. These old galaxies at intermediate redshift ($z \sim 1 - 2$) comprise some of the "Extremely Red Objects", a population with $(R - K) > 6$ mag. Here we present a morphological analysis of one of the oldest galaxies known at high redshift. The weak 1-mJy radio galaxy LBDS53W091 (1) has a stellar age of $> 3\,\mathrm{Gyr}$ at $z = 1.55$, based the amplitude of rest-frame UV spectral breaks seen in Keck/LRIS spectra (2,3) with a main-sequence turn-off around an F2-F6 star. If this spectral age-dating is correct, then the galaxy should be a dynamically-evolved elliptical. Ellipticals at the current epoch are remarkably homogeneous, with tight scaling relations - the Fundamental Plane (4,5). It is important to understand the true nature of this apparently old galaxy in the early Universe, and so trace the evolution of these scaling relations. Therefore, we have obtained deep, high-resolution imaging with the Hubble Space Telescope (HST) in the near-infrared and optical, allowing us to study its morphology and determine the structural parameters.

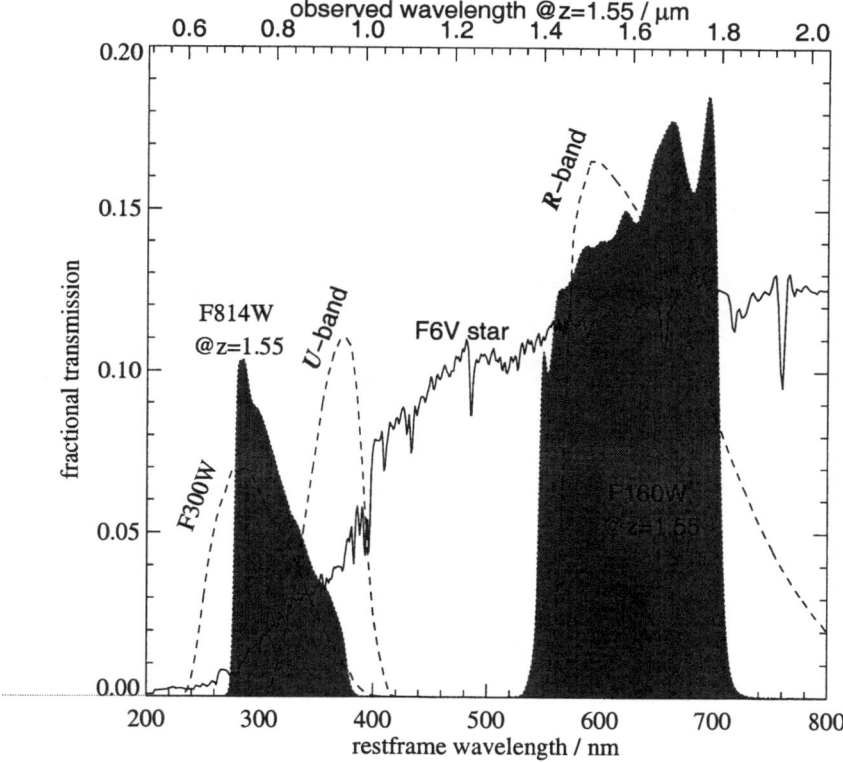

Fig. 1. HST I- and H-bands (shaded), showing location in the rest-frame of a galaxy at $z = 1.55$ with the spectral type of LBDS53W091 (solid line)

2   Observations with HST

We observed LBDS53W091 with HST over 10 orbits (19 ksec) using the Near Infrared Camera/Multi-Object Spectrograph, NICMOS (6). We used the diffraction-limited camera NIC2 (0.076 arcsec/pixel and a 19 arcsec field) and the F160W H-band filter (1.6 µm) which is a good approximation to the rest-frame R-band (see figure 1). Our near-infrared observations were complemented by a

2 orbit (5 ksec) WFPC2 image using the F814W "I-band" filter (0.8 μm), which at $z = 1.55$ is a good match to the rest-frame F300W filter. We used "drizzle" (7) to improve the resolution through sub-pixel sampling. Figure 2 shows the registered I- and H-band images. Prominent in the near-infrared is a group of extremely red galaxies around the location of the radio source.

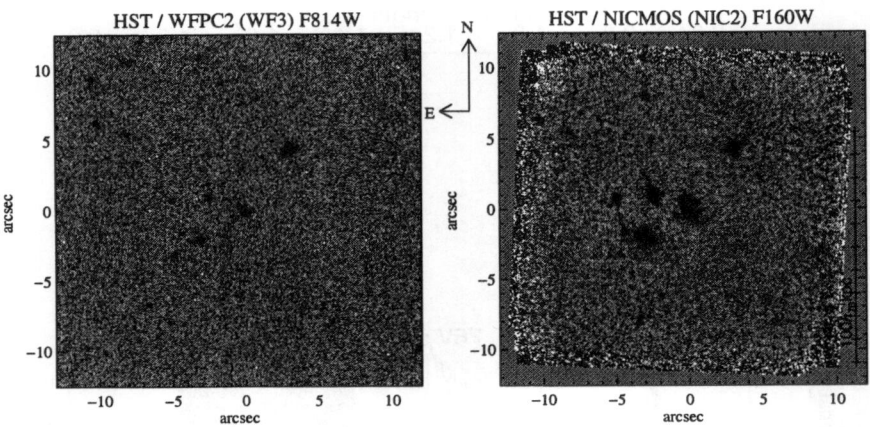

Fig. 2. The registered I- and H-band images. Prominent in the near-infrared is a group of extremely red galaxies around the location of the radio source (the coordinate zero-point)

3 Surface Brightness Profile

A range of two-dimensional models were convolved with the point spread function, then compared against the data to determine the best-fit parameters (figure 3). We fit the radial dependence of the surface brightness profile with various functional forms: an $r^{1/4}$ bulge (8); an exponential disk; a generic Sérsic $r^{1/n}$ profile (9); and a composition of disk and bulge (10). Our observations reveal a radial light profile which is well fit by a de Vaucouleurs' $r^{1/4}$ law, with a scale length of $r_e = 0.3$ arcsec ($3 h_{50}^{-1}$ kpc for $\Omega_M = 0.3$ and $\Omega_\Lambda = 0$). The optical I-band morphology (sampling the rest-UV and sensitive to younger, bluer stars) does show some evidence for a disk component, but the near-infrared H-band light is completely bulge-dominated. The elliptical morphology of the radio galaxy indicates a dynamically-evolved old system, consistent with the spectroscopic results. Down to our detection limits, we find no evidence for an unresolved central point source (11) due to AGN activity in this weak radio galaxy.

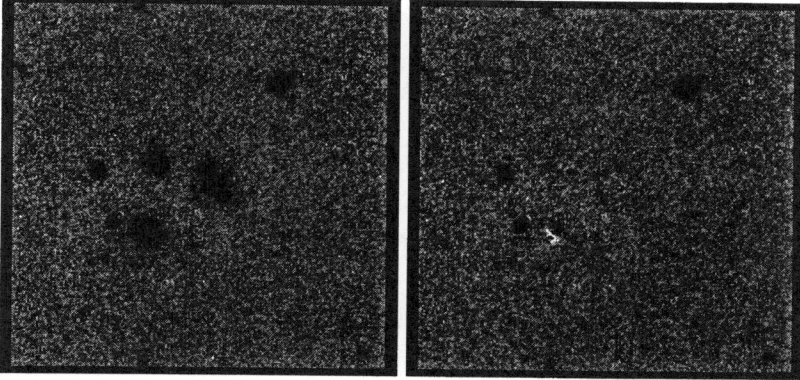

Fig. 3. The NICMOS H-band image (left), and with the best-fit two-dimensional de Vaucouleur $r^{1/4}$ models subtracted (right) for the three brightest galaxies. The small residuals indicate that these galaxies are bulge-dominated/elliptical in nature. The model fits have been produced using the GIM2D package of Simard et al. (1999)

4   The Fundamental Plane at $z = 1.55$

The "Fundamental Plane" (4,5) provides a tight correlation between the scale length (r_e), surface brightness (I_e) and central velocity dispersion (σ_c) of the form $r_e = k\,\sigma_c^{1.2}\,I_e^{-0.83}$; the precise physical origin of the fundamental plane is uncertain, but in essence it is a relation between galaxy mass and mass-to-light ratio ($M/L \propto M^{1/4}$, see 12). Recent work on clusters at moderate redshifts out to $z \approx 0.8$ has suggested an evolution in the zero-point of the fundamental plane (13,14), consistent with passive luminosity evolution of the elliptical stellar population which had formed at much higher redshifts. It is important to locate high redshift cluster ellipticals, such as those associated with LBDS53W091, on the fundamental plane as the inferred mass-to-light ratios may tell us how close these galaxies are to the formation epoch of ellipticals. In the case of LBDS53W091, no direct measurement of the central velocity dispersions exists - although our NICMOS imaging has yielded good determinations of the half-light radius and mean surface brightness. Plotting the edge-on projection of the fundamental plane for Coma (figure 4) shows that LBDS53W091 is more luminous than the fundamental plane at $z \approx 0$, correcting only for the $(1 + z)^4$ surface brightness dimming. However, if we also correct for passive luminosity evolution from a formation epoch of $z > 3$ then this $z = 1.55$ galaxy would lie on the fundamental plane if $\sigma \approx 200\,\text{km/s}$ (comparable with indirect velocity dispersion estimates – see 15).

5   An Old Group at $z = 1.55$?

Some surrounding objects lie on the color:magnitude relation for a cluster at $z = 1.55$, and are likely to be associated. The tightness of the color:magnitude

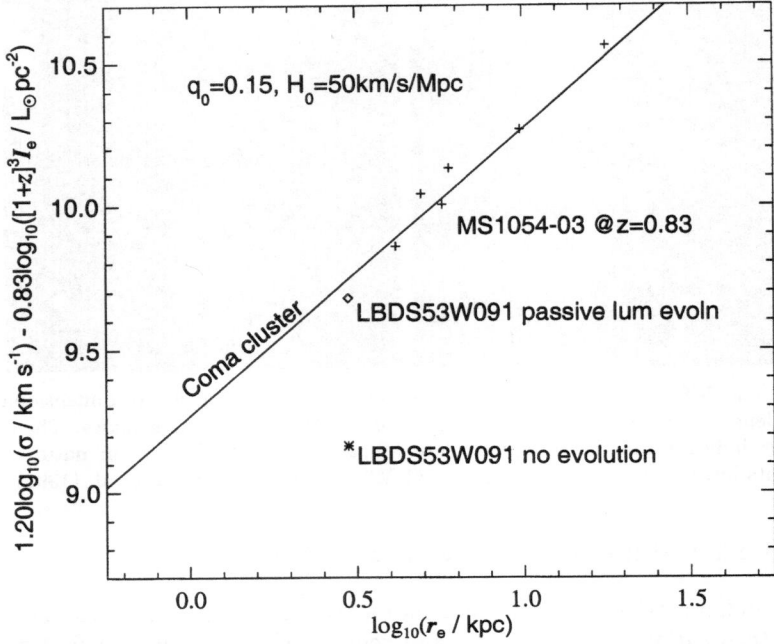

Fig. 4. The Coma fundamental plane viewed edge on (solid line). The crosses are ellipticals in the cluster MS1054-03 at $z = 0.83$ (see 14), corrected for passive luminosity evolution of $\Delta Mag \approx 0.4 \times z$. A cosmology of $\Omega_M = 0.3$, $\Omega_A = 0$, $H_0 = 50$ km/s/Mpc is used. The asterisk represents the galaxy LBDS53W091, assuming a velocity dispersion of $\sigma \sim 200$ km/s. This point lies well below the fundamental plane, implying the galaxy is brighter than present-day ellipticals. However, once correction for passive luminosity evolution since $z = 1.55$ has been made (diamond symbol), LBDS53W091 lies close to the fundamental plane for a reasonable range of velocity dispersions

diagram in clusters could be attributable either to a high formation redshift, or an extremely synchronized, coeval star-formation history of the member galaxies (16). The color:magnitude diagram determined at $z \approx 0$ from Coma is redder by $\Delta(U - B) \sim 0.5$ mag than the observed envelope for the field of LBDS53W091 (figure 5). Once again, this is entirely consistent with passive luminosity evolution and a high formation redshift.

References

1. Windhorst, van Heerde & Katgert (1984) A& A 58, 1
2. Dunlop et al. (1996) Nature 381, 581
3. Spinrad et al. (1997) ApJ 484, 581
4. Djorgovski & Davis (1987) ApJ 313, 59
5. Dressler et al. (1987) ApJ 313, 42

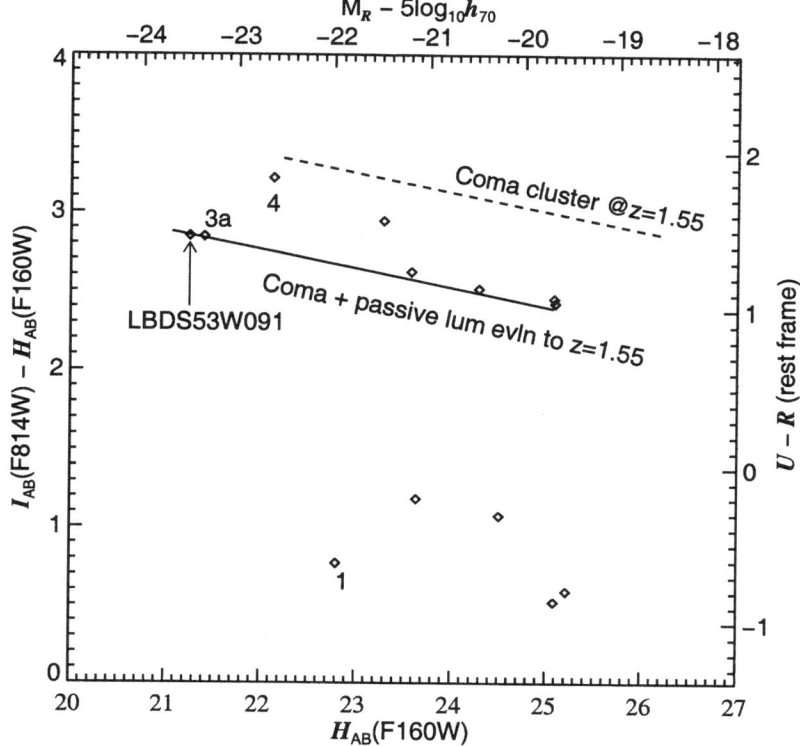

Fig. 5. The color:magnitude diagram for objects in the field of LBDS53W091. Some are significantly more blue than the radio galaxy, and are presumably foreground: "galaxy 1" has a spectroscopic redshift of $z = 1.01$ (see 3). However, there are half a dozen objects that lie close to the color:magnitude relation from Coma, corrected for passive luminosity evolution (solid line). This may be evidence for a cluster around LBDS53W091 at $z = 1.55$ – indeed, "galaxy 3a" has the same spectroscopic redshift (see 3)

6. Thompson et al. (1998) ApJLett 492, 95
7. Fruchter & Hook (1997) ASP Conf. Ser. Vol 125, p14
8. de Vaucouleurs (1948) Ann. d'Astroph., 11, 247
9. Sérsic (1968) "Atlas de galaxias australes"
10. Simard et al. (1999) 519, 563
11. McLure et al. (1999) 308, 377
12. Bender, Burstein & Faber (1992) ApJ 399, 462
13. Kelson et al. (2000) ApJ 531, 184
14. van Dokkum et al. (1998) ApJLett 504, 17
15. Peacock et al. (1998) 296, 1089
16. Bower, Lucey & Ellis (1992) MNRAS 254,589

The Masses of AGN Host Galaxies and the Origin of Radio Loudness

James S. Dunlop[1] and Ross J. McLure[2]

[1] Institute for Astronomy, Royal Observatory, Edinburgh, EH9 3HJ, UK
[2] Astrophysics, Department of Physics, Keble Road, Oxford, OX1 3RH, UK

Abstract. We highlight some of the principal results from our recent Hubble Space Telescope studies of quasars and radio galaxies. The hosts of these powerful AGN are normal massive ellipticals which lie on the region of the fundamental plane populated predominantly by massive ellipticals with boxy isophotes and distinct cores. The hosts of the radio-loud sources are on average $\simeq 1.5$ times brighter than their radio-quiet counterparts and appear to lie above a mass threshold $M_{sph} > 4 \times 10^{11} M_\odot$. This suggests that black holes more massive than $M_{bh} > 5 \times 10^8 M_\odot$ are required to produce a powerful radio source. However we show that this apparent threshold appears to be a consequence of an upper bound on radio output which is a strong function of black-hole mass, $L_{5GHz} \propto M_{bh}^{2.5}$. This steep mass dependence can explain why the hosts of the most powerful radio sources are good standard candles. Such objects were certainly fully assembled by $z \simeq 1$, and appear to have formed the bulk of their stars prior to $z \simeq 3$.

1 Introduction

Since the optical identification of Cygnus A (Baade & Minkowski 1954) it has been clear that the host galaxies of the most powerful radio sources in the nearby universe appear to be massive ellipticals. However, it is only in the last decade, since the repair of HST, that it has proved possible to perform a detailed comparison of the hosts of powerful radio-loud and radio-quiet AGN. In this brief article we summarize some of the main results from our recent HST studies of AGN hosts, with special emphasis on how their structures, sizes, luminosities and masses compare to those of 'normal' galaxies. We also explore what such studies can teach us about the physical difference between radio-loud and radio-quiet AGN, and about the formation history of massive elliptical galaxies.

2 Quasar Host Galaxies and 'Normal' Ellipticals

One of the most important new results from this work is the discovery not only that the hosts of powerful AGN (both radio-loud and radio-quiet) are almost exclusively ellipticals, but that these galaxies display a Kormendy relation indistinguishable in both slope and normalization from that displayed by normal massive ellipticals (Fig. 1; Dunlop et al. 2002). The Kormendy relation is the photometric projection of the Fundamental Plane (Djorgovski & Davis 1987; Dressler et al. 1987), but in the case of radio galaxies the third dimension (i.e.

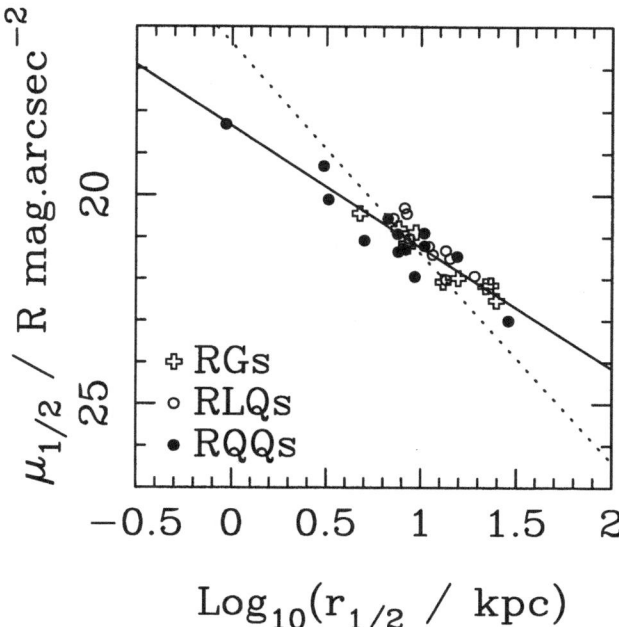

Fig. 1. The Kormendy relation followed by the hosts of all 33 powerful AGN imaged with the HST by Dunlop et al. (2002). The solid line is the least-squares fit to the data which has a slope of 2.90, in excellent agreement with the slope of 2.95 found by Kormendy (1977) for inactive ellipticals in the B-band. The dotted line has a slope of 5, indicative of what would be expected if the scale-lengths of the host galaxies had not been properly constrained

central stellar velocity dispersion) can be added with relative ease. This has recently been completed for a subset of 22 radio galaxies by Bettoni et al. (2002), who confirm that these objects lie towards the bright end of the same fundamental plane as defined by quiescent massive ellipticals.

The quasar hosts and radio galaxies are therefore all clearly large luminous galaxies with $L > L^\star$. However, the radio-loud hosts are more cleanly confined to a definite high mass regime, with 18/20 of the radio-loud hosts in the Dunlop et al. sample having spheroid masses $> 4 \times 10^{11} M_\odot$, compared with only 4/13 of the radio quiet hosts. The results of McLure & Dunlop (2002) demonstrate that this difference can reasonably be extrapolated to a difference in central black holes masses, with the radio-loud sources being confined to black-hole masses $M_{bh} > 5 \times 10^8 M_\odot$.

3   Radio-Power and Spheroid/Black-Hole Mass

At first sight, these results suggest the existence of a physical mass threshold above which galaxies (or their central black holes) are capable of producing powerful relativistic jets. This would also appear consistent with long-standing suggestions of a definite gap in the radio luminosity function of optically selected quasars. However, the recent study of Lacy et al. (2001) does not support the existence of any such gap or threshold. In fact Lacy et al. demonstrate the existence of a clear, albeit loose, correlation between radio power and black-hole/spheroid mass extending over 5 decades in radio power. However, the large scatter in the data, and the relatively gentle slope of the best-fitting relation ($L_{5GHz} \propto M_{bh}^{1.4}$) do not provide an obvious explanation of why the hosts of powerful radio sources should be such good standard candles.

Instead, Dunlop et al. (2002) have suggested that the distribution of AGN on the L_{5GHz}:M_{bh} plane is better described as being bounded by a lower and upper threshold for the radio output that can be produced by a black hole of given mass, and that these radio output thresholds are a much steeper function of mass, i.e. $L_{5GHz} \propto M_{bh}^{2.5}$. In Fig. 2 we demonstrate that the bounding relations deduced by Dunlop et al. also provide an excellent description of the data gathered by Lacy et al.. In fact, the lower boundary is essentially identical to the relation derived for nearby galaxies by Franceschini et al. (1998), who also concluded in favour of $L_{5GHz} \propto M_{bh}^{2.5}$. However, Fig. 2 makes the interesting (and perhaps surprising) point that the upper limit on black-hole radio output appears to be a similarly steep function of mass, simply offset by 5 decades in radio power.

This steep upper boundary on L_{5GHz} as a function of black-hole/spheroid mass provides a natural explanation for why the low-redshift radio-loud AGN hosts studied by Dunlop et al. lie above an apparently clean mass threshold. These objects have $L_{5GHz} > 10^{24}\mathrm{WHz}^{-1}\mathrm{sr}^{-1}$, and from Fig. 2 it can be seen that such radio powers can only be achieved by black holes with $M_{bh} > 2 \times 10^8 M_\odot$, and hence host spheroids with $M_{sph} > 2 \times 10^{11} M_\odot$.

Fig. 2 also provides a possible explanation for why the 3CR radio galaxies at $z \simeq 1$ appear to be even better standard candles that at low redshift; inclusion in the 3CR catalogue at $z \simeq 1$ requires $L_{5GHz} > 10^{26}\mathrm{WHz}^{-1}\mathrm{sr}^{-1}$, which Figure 3 indicates requires black holes with $M_{bh} > 10^9 M_\odot$, and hence host spheroids with $M_{sph} > 10^{12} M_\odot$. At such high masses the luminosity/mass function of elliptical galaxies is very steep (Kochanek et al. 2001), and so it is inevitable that any ellipticals which lie above this mass threshold will also lie very close to it.

4   The Origin of Radio Loudness

At the other end of the radio-power scale, Fig. 2 demonstrates the surely significant fact that many of the powerful optically-selected AGN produce a level of radio output which is indistinguishable from that produced by nearby quiescent ellipticals of comparable mass. In other words, the minimum radio power relation defined by the most radio-quiet quasars is the same as that defined by

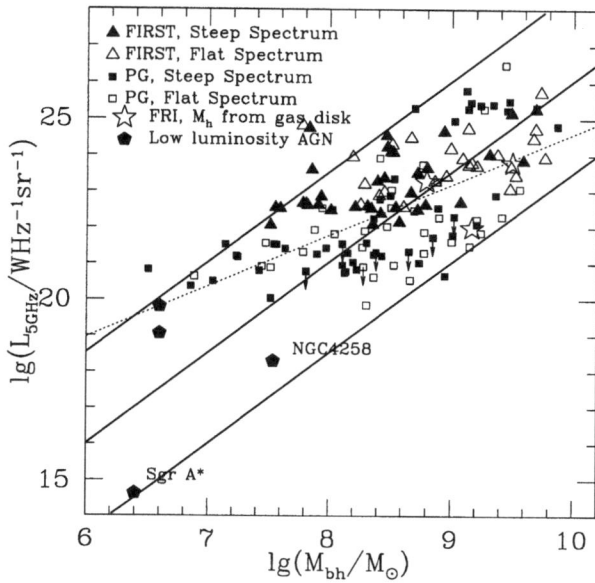

Fig. 2. A plot of $\lg L_{5Ghz}$ versus $\lg M_{bh}$ compiled from various samples of quasars by Lacy et al. (2001), with the bounding relations on minimum and maximum radio output suggested by Dunlop et al. (2002) superimposed (solid lines, $L_{5GHz} \propto M_{bh}^{2.5}$)

nearby 'quiescent' galaxies. This dramatically illustrates how very different the physical mechanisms for the production of optical and radio emission by a black hole must be, since the AGN are clearly in receipt of plenty of fuel.

These results therefore lead us to conclude that the difference between radio-loud and radio-quiet AGN cannot be explained as due to black-hole or host-galaxy mass, host-galaxy morphology, or indeed black-hole fueling rate. Rather there must be some other property of the central engine which determines whether a given object lies nearer to the upper or lower radio-power thresholds shown in Fig. 2. The only obvious remaining candidate is spin. This has been previously suggested and explored by many authors on the basis that angular momentum must surely be important for the definition of jet direction (e.g. Wilson & Colbert 1995; Blandford 2000). Here we have effectively arrived at the same conclusion by a process of elimination of the obvious alternatives.

5   The Assembly of Quasar Host Galaxies

A number of independent lines of evidence suggest that the hosts of powerful AGN formed at high redshift. This evidence is most convincing for the radio-loud population: allowing for the effects of passive evolution, radio galaxies at $z \simeq 1$

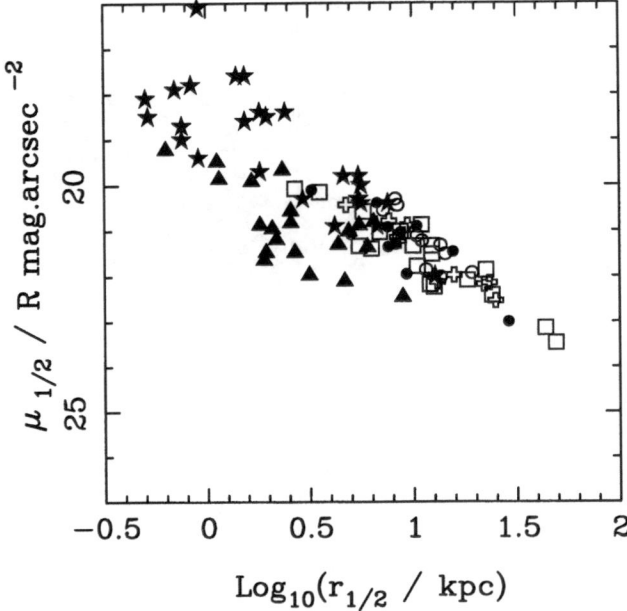

Fig. 3. A comparison of the properties of the AGN hosts with those displayed by various other types of spheroid on the photometric projection of the fundamental plane. Symbols for the quasar hosts and radio galaxies are as in Fig. 1. The stars are the data for ULIRGs and LIRGs from Genzel et al. (2001) transformed from the infrared to the R-band assuming $R - K = 2.5$. Triangles and squares indicate the positions of 'discy' and 'boxy' ellipticals from Faber et al. (1997) after conversion to $H_0 = 50 \, \mathrm{km s^{-1} \, Mpc^{-1}}$

lie on the same Kormendy relation as shown in Fig. 1 (McLure & Dunlop 2000; Waddington et al. 2002), the $K - z$ relation for powerful radio galaxies appears consistent with purely passive evolution out to $z > 3$ (van Breugel et al. 1998; Jarvis et al. 2002), and strong star-formation activity in powerful radio galaxies seems largely confined to $z > 2.5$ (Archibald et al. 2001).

The picture is currently somewhat less clear for the hosts of radio-quiet quasars. The colours and off-nuclear spectra of low-redshift quasar hosts indicate that their stellar populations are predominantly old (Dunlop et al. 2002; Nolan et al. 2000; McLure et al. 1999) but there is also some evidence that the hosts of radio-quiet quasars are significantly less massive by $z \simeq 2$ compared to the present day (Kukula et al. 2001; Ridgway et al. 2001). This raises the possibility that some of the low-redshift radio-quiet quasar population could be produced by the same sort of recent major mergers which power Ultra Luminous Infrared Galaxies (ULIRGS). In fact we can now begin to explore this possibility directly by combining our own results of quasar hosts with the results of near-infrared imaging and spectroscopy recently performed by Genzel et al. (2001).

While it is true that ULIRGs such as Arp220 have surface brightness profiles well-described by an $r^{1/4}$-law, Genzel et al. have shown that such remnants lie in a different region of the fundamental plane than that which we have found to be occupied by the quasar hosts. Specifically, the effective radii of the ULIRGs is typically an order of magnitude smaller than those of the quasar hosts. Indeed one can go further and conclude that whereas ULIRGs may well be the progenitors of the population of intermediate-mass ellipticals which display compact cores and cusps (Faber et al. 1997), the quasar hosts lie in a region of the $\mu_e - r_e$ plane which is occupied by boxy, giant, ellipticals with large cores. This comparison is illustrated in Fig. 3, where we have augmented the Kormendy diagram shown in Fig. 1 with the addition of the data from Genzel et al. on LIRGs and ULIRGs, and the data from Faber et al. (1997) on 'discy' and 'boxy' ellipticals. Thus, present evidence suggests that, at least at low redshift, any ULIRG $\rightarrow$ quasar evolutionary sequence can only apply to a fairly small subset of objects.

With the advent of the Advanced Camera on HST, and high-resolution near-infrared imaging on ground-based 8-m telescopes, the next few years should see some major advances in our understanding of the properties of quasar host galaxies as a function of redshift. Fig. 3 indicates that such studies should also shed light on the formation history of the high-mass end of the present-day quiescent elliptical galaxy population.

References

1. E.N. Archibald, et al.: MNRAS **323**, 417 (2001)
2. W. Baade & R. Minkowski: ApJ **119**, 206 (1954)
3. D. Bettoni, et al.: A&A in press, astro-ph/0110420
4. R.D. Blandford: astro-ph/0001499 (2000)
5. S. Djorgovski & M. Davis: ApJ **313**, 59 (1987)
6. A. Dressler, et al.: ApJ **313**, 42 (1987)
7. J.S. Dunlop, et al.: MNRAS in press, astro-ph/0108397
8. S.M. Faber, et al.: AJ **114**, 1771 (1997)
9. A. Franceschini, et al.: MNRAS **297**, 817 (1998)
10. R. Genzel, et al.: ApJ **536**, 527 (2001)
11. M. Jarvis, et al.: in: The mass of galaxies from low to high redshift, in press, astro-ph/0112341
12. C.S. Kochanek, et al.: ApJ **560**, 566 (2001)
13. J. Kormendy: ApJ **217**, 416 (1977)
14. M. Kukula, et al.: MNRAS **326**, 1533 (2001)
15. M. Lacy, et al: ApJ **551**, L17 (2001)
16. R.J. McLure & J.S. Dunlop: MNRAS **317**, 249 (2000)
17. R.J. McLure & J. Dunlop: MNRAS in press, astro-ph/0108417
18. R.J. McLure, et al.: MNRAS **308**, 377 (1999)
19. L. Nolan, et al.: MNRAS **323**, 308 (2001)
20. S. Ridgway, et al.: ApJ **550**, 122 (2001)
21. W. van Breugel, et al.: ApJ **502**, 614 (1998)
22. I. Waddington, et al.: MNRAS, submitted
23. A.S. Wilson & E.J.M. Colbert: ApJ **438**, 62 (1995)

The Evolution of Damped Lyman Alpha Systems

Sandhya Rao and David Turnshek

The University of Pittsburgh, Pittsburgh, PA 15260

1 Introduction

Surveys for damped Lyα (DLA) lines, historically defined to have $N_{HI} \geq 2 \times 10^{20}$ atoms cm^{-2}, have been used to study the distribution of neutral hydrogen in the Universe [1,2,4–6]. These surveys have shown that DLAs trace the bulk of the observable neutral gas mass at high redshift, and therefore, that they can be used to study the formation and evolution of galaxies. However, since the Lyα line falls in the UV for redshifts $z < 1.65$, HST is required to discover DLA systems at low redshift. With $z < 1.65$ spanning the most recent $\sim 75\%$ of the age of the Universe, the determination of the statistics and properties of DLA systems at these redshifts is crucial for understanding the evolution of galaxies. Here, we summarize results from an efficient, non-traditional HST-UV survey to discover DLA absorption-line systems at redshifts $z < 1.65$.

2 Summary of Results

For a detailed description of the survey the reader is referred to [2]. Here, we list the main results. (1) We uncovered a new selection criterion for the identification of DLAs, i.e., $\approx 50\%$ of the systems with $W_0^{\lambda 2796} > 0.5$ Å and $W_0^{\lambda 2600} > 0.5$ Å have DLA absorption (Figure 1a). (2) The incidence of DLAs per unit redshift, n_{DLA}, is observed to decrease with decreasing redshift (Figure 1b). The observed trend in n_{DLA} implies significant evolution only if the $z = 0$ data point is included in the analysis. (3) The cosmological mass density of neutral gas in DLA systems, Ω_{DLA}, at $z < 1.65$ is comparable to that observed at high redshift (Figure 2a). The error bars are large because $\Omega_{DLA}(z)$ is very sensitive to the small number of systems which have the highest column densities. $\Omega_{DLA}(z)$ is a factor of ~ 6.5 times larger than the value for $\Omega_{gas}(z = 0)$ inferred from local HI observations. (4) The HI column density distribution (CDD) of the low-redshift DLA absorber population is very different in comparison to high-redshift DLA absorbers, and in comparison to the column density distribution inferred from local spirals (Figure 2b). The low-redshift DLAs exhibit a significantly larger fraction of very high column density systems in comparison to determinations at both high redshift and locally. At no redshift does the CDD fall-off in proportion to $\sim N_{HI}^{-3}$. An $\sim N_{HI}^{-3}$ fall-off is theoretically predicted for disk-like systems and this is, in fact, observed locally in spiral samples [1,7].

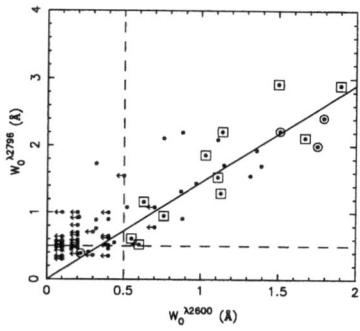
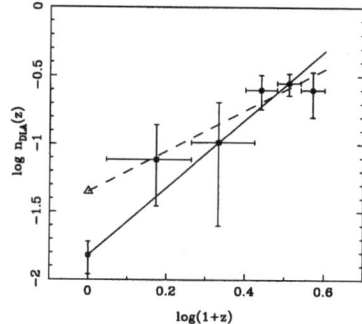

Fig. 1. (a) MgII $W_0^{\lambda 2796}$ versus FeII $W_0^{\lambda 2600}$. The DLAs are marked with open squares and previously known 21 cm absorbers are open circles. Half the systems for which $W_0^{\lambda 2796} > 0.5$ Å and $W_0^{\lambda 2600} > 0.5$ Å (shown by dashed lines) are DLAs. (b) The number density redshift distribution of DLAs. The solid squares are from [2] and the open circle is from HI in local spiral galaxies [1]. The solid line is forced to pass through the $z = 0$ data point and implies significant evolution. The dashed line does not include the $z = 0$ data point, but extrapolation to $z = 0$ (the open triangle) results in a value that is ~ 3 times larger than the observed incidence at $z = 0$

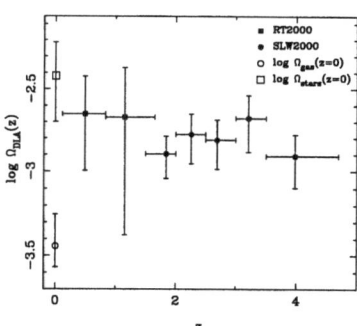
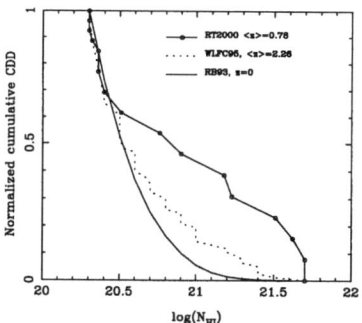

Fig. 2. (a) The cosmological mass density of neutral gas as a function of redshift for $H_0 = 65$ km s^{-1} Mpc^{-1}, $q_0 = 0.5$, and $\Lambda = 0$. (b) The normalized cumulative column density distributions for the low-redshift DLA sample [2], the high-redshift DLA sample [6], and local galaxies [1]

References

1. Lanzetta, K. M., et al. 1991, ApJS, 71, 1
2. Rao, S. M., & Briggs, F. H. 1993, ApJ, 419, 515
3. Rao, S. M., & Turnshek, D. A. 2000, ApJS, 130, 1
4. Storrie-Lombardi, L. J., & Wolfe, A. M. 2000, ApJ, 543, 552
5. Wolfe, A. M., Turnshek, D. A., Smith, H. E., & Cohen, R. D. 1986, ApJS, 61, 249
6. Wolfe, A. M., Lanzetta, K. M., Foltz, C. B., & Chaffee, F. H. 1995, ApJ, 454, 698
7. Zwaan, M. A., Verheijen, M. A. W., & Briggs, F. H. 1999, PASA, 16, 100

The Properties of Low-Redshift Damped Lyman Alpha Galaxies

David Turnshek, Sandhya Rao, and Daniel Nestor

The University of Pittsburgh, Pittsburgh, PA 15260

1 Introduction

Damped Lyman alpha (DLA) galaxies are those which give rise to the DLA lines seen in the spectra of background QSOs, indicative of neutral hydrogen column densities $N_{HI} \geq 2 \times 10^{20}$ atoms cm^{-2}. An understanding of this population of galaxies is important for studies of galaxy formation and evolution since DLA galaxies harbor the bulk of the neutral hydrogen in the Universe, and neutral gas is a necessary requirement for the subsequent formation of molecular clouds and the conversion of gas into stars.

2 Results

In a companion paper (this volume; see also [2]) Rao and Turnshek describe their survey for low-redshift ($z < 1.65$) DLA systems with HST. Thus far, optical/IR imaging of DLA fields at $z < 1$ with a variety of large telescopes has permitted us to search for and study the likely properties of about a dozen associated DLA galaxies. Figure 1 shows two identified DLA galaxies. Figures 2 and 3 show the properties of DLA galaxies at $< z >= 0.5$ in comparison to their expected properties if they were drawn from the sample of local ($z = 0$) gas-rich galaxies studied by Rao and Briggs [1]. It should be kept in mind that, in this identification work, it is generally impossible to rule out the possibility that a DLA absorber is in reality a fainter galaxy or a galaxy with a smaller impact parameter. Note that, in principle, an unseen DLA galaxy with a smaller impact parameter could be brighter if its light were sufficiently confused with the QSO's light. However, in all likelihood, the net result of these effects would lead to an over-estimate of either the DLA galaxy luminosity or impact parameter in some individual cases, which we believe would strengthen our conclusions below.

3 Conclusions

Figures 2 and 3 suggest that the local ($z = 0$) gas-rich galaxies are a subset of the low-redshift DLA galaxy population. This is evident from the spread in the distribution of DLA galaxy properties in comparison to the $z = 0$ simulation along 1000 sight lines. The comparison indicates that there may be an excess of low-redshift DLA galaxies in the form of dwarf and/or low surface brightness galaxies.

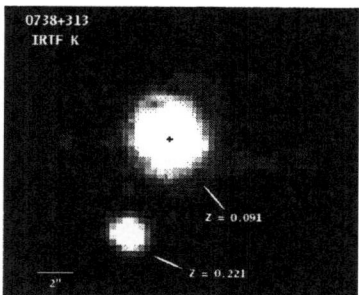

Fig. 1. Two examples of DLA galaxies along the sight-line towards the quasar 0738+313. The point spread function of the quasar, whose center is marked with a plus, has been subtracted. The $z = 0.09$ DLA galaxy is a low surface brightness dwarf galaxy with a luminosity of about $0.1L^*$. The $z = 0.22$ DLA galaxy is a compact dwarf galaxy, also with a luminosity of about $0.1L^*$.

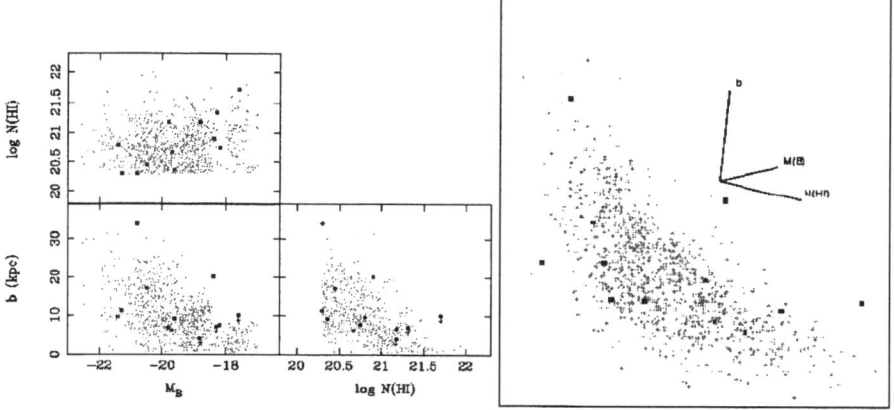

Fig. 2. (a) The solid squares are the 3-dimensional distribution of $< z >= 0.5$ DLA galaxy luminosities (M_B), impact parameters (b), and neutral hydrogen column densities (N_{HI}). The dots are a simulation of the expected properties of DLA galaxies for the case where the $< z >= 0.5$ DLA galaxies are drawn from the same population of local ($z = 0$) gas-rich galaxies studied by [1]. One thousand sight-lines were simulated. The possible effects of luminosity evolution were not considered. (b) This is a true 3-dimensional rendition of the results shown in Figure 2a. The orientation of the axes is adjusted such that the outlying nature of many of the squares representing the $< z >= 0.5$ DLA galaxies is apparent. This suggests that the set of local ($z = 0$) gas-rich galaxies studied by Rao and Briggs [1] is a subset of the low-redshift DLA galaxy population.

References

1. Rao, S. M., & Briggs, F. H. 1993, ApJ, 419, 515
2. Rao, S. M., & Turnshek, D. A. 2000, ApJS, 130, 1

The Assembly of Galaxies from a Wide/Deep IR Survey

Adriano Fontana

INAF – Osservatorio Astronomico di Roma, Via Frascati 33,
00040 Monte Porzio Catone, Rome, ITALY

Abstract. We show how deep multicolor imaging surveys may be used to trace the assembly of stellar mass in galaxies at different redshifts. Deep multicolor imaging from U to K bands are used to recover the overall spectral shape of a large K-selected sample of galaxies. We target fields where spectroscopic redshifts are available, to calibrate photometric redshifts for each sample. I will first discuss the uncertainties involved in the estimate of stellar masses in high z galaxies from broad band photometry, showing that the well known model degeneracies result in a minimum uncertainty of about 50%. When applied to the galaxies in the HDF–N and HDF–S, this analysis shows a clear decrease in the mass of the more massive galaxies observed for increasing redshift. This result is shown to be in agreement with the prediction of CDM models of galaxy formation.

1 Introduction

The efforts aimed at constructing a global picture of galaxy formation and evolution are hampered by converging difficulties. On the one hand, the baryonic processes involved are poorly understood, even in local galaxies, and can only be schematically parametrized. On the other hand, the direct observation of galaxies in the high redshift Universe is still difficult, in spite of the exciting progresses of recent years, and plagued by possible systematic effects. In this context, it is important to extend the observations and the interpretation of the data in order to collect the widest set of observables at different redshifts, that can be compared with the model prediction.

Following this strategy, we have analyzed in recent years the data coming from the Hubble Deep Fields and other surveys to obtain a set of observables, ranging from redshift distribution to luminosity functions to size distributions, that have been compared with the prediction of our rendition of current CDM models (e.g. Fontana et al 1999, Poli et al 2001).

Most of these observables are tied to the observed luminosities, that may be a biased tracer of the underlying mass. Much interest are therefore attracting the recent techniques to estimate the *stellar* mass content of local and high–z galaxies (Giallongo et al 1998, Brinchmann and Ellis 2000, Papovich et al 2001). These techniques rely on multicolor broad band imaging - extended into the near–IR range - to estimate the stellar content by a comparison with spectral synthesis models. In this contribution, I will provide the first results that this technique yields when applied to the HDF-N/S data at $0 < z < 3.2$.

2   The Data: A Deep Multicolor Imaging Survey

The data come from a multiwavelength deep imaging survey that we have been conducting over the last years. We have focussed in particular on the collection of K-selected deep samples, assuring that the UV and optical coverage and depth are adequate to follow the spectral shape of the reddest galaxies in the sample.

This has implied the use of NTT and VLT for the optical–nearIR imaging of the $K_{AB} \leq 22$ fields, and of the Hubble Deep Field South-WFPC3 data to complement the deepest VLT-ISAAC images that will eventually reach $K_{AB} \leq 25$. A distinctive feature of our survey is the combination with extended spectroscopic surveys: this way, we may use the data to better estimate the rest–frame parameters of the observed sample, and to obtain well calibrated "photometric" redshifts for the remaining sample.

An example of this approach is shown in fig. 1, where we present the calibration of photometric redshifts obtained in the framework of the "K20" spectroscopic redshift survey (A. Cimatti, this volume), toward the field of the $z = 3.65$ QSO Q0055-269. The imaging observations used here to reach $K \leq 20$ with adequate S/N extend over 10 bands ($UBGVRR_wIZJK$) obtained with NTT and VLT for a total of about 45 hours of integration. A technique to obtain an "optimal" aperture photometry has been developed to cope with the inhomogeneous image quality over the different bands. Photometric redshifts have been obtained through template fitting using the PEGASE 2.0 spectral package (Fioc & Rocca–Volmerange 1999, see Fontana et al 2002 for the details), and reach an accuracy of $\sigma_{\Delta z} = 0.08$ over the redshift range shown in Fig 1. The comparison,

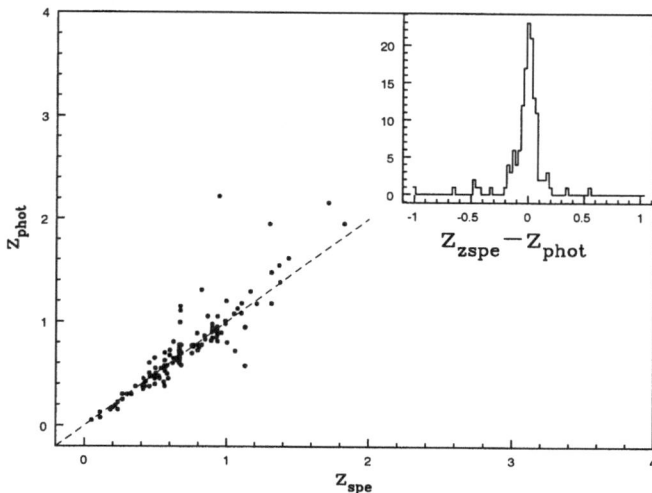

Fig. 1. Comparison between spectroscopic and photometric redshifts for 140 galaxies in the field of Q0055-269. Spectra come from the K20 survey, photometric redshifts from ground–based multicolor imaging

carried over the 140 galaxies with secure spectroscopic redshift, shows that it is possible to achieve a high reliability and accuracy of photometric redshifts even with ground–based data sets, provided that the imaging data are of adequate quality.

3   Estimating Galaxy Stellar Mass from Multicolor Samples

Our recipe for photometric redshifts is based on the use of spectral synthesis models, that allow - at least in principle - to take into account the spectral evolution of galaxies at high redshift. In practice, model input parameters (age, metallicity, dust content..) are varied over a large grid at each redshift to bracket the wide range of spectral properties possible at each redshift. Best-fit models to the observed colors are chosen from this dataset for each observed galaxy (see Fontana et al 2000 for details). Beside the good accuracy of photometric redshifts based on this method, this approach provides physical information on the properties of high redshift galaxies. From the input parameters used to draw the best fitting solution for each galaxy, we can indeed obtain estimates of the main physical quantities of each galaxy in the sample, like e.g. age of the last major starburst, mass already assembled in stars and dust content. More important, the statistical analysis may be used to evaluate the uncertainty on these best–fitting values induced by the degeneracies among the input parameters.

I focus here on the results concerning the stellar mass, as obtained from the best fitting spectrum in each galaxy. Fig 2 shows one example drawn from the HDFS sample. The IR observations and the multicolor catalog are taken from our VLT–ISAAC observations (Vanzella et al 2001). It is shown the spectral

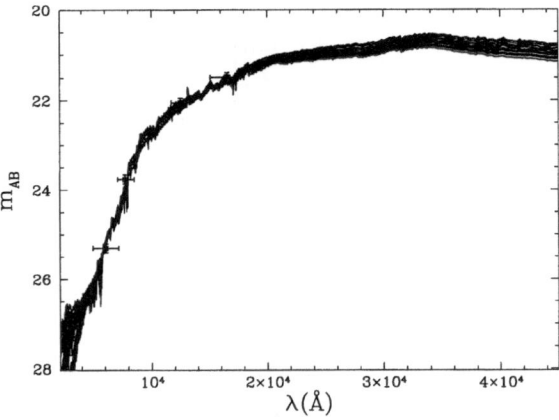

Fig. 2. Observed broad band spectral distribution of a $z = 1$ galaxy in the HDFS. Curves represent the Bruzual & Charlot spectra that can produce acceptable fits to the observed magnitudes

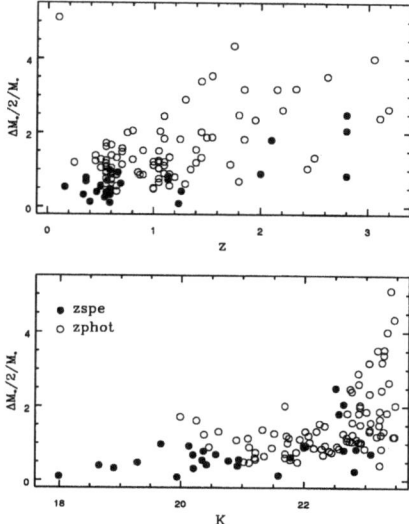

Fig. 3. Global uncertainties in the stellar mass values as obtained with best-fitting technique, as a function of redshift and observed K flux. Filled points represent objects with spectroscopic redshifts, empty with photometric redshifts

distribution of a $z \simeq 1$ galaxy, as estimated from broad–band photometry, along with the best fitting spectrum obtained from the Bruzual and Charlot 2000 library (for consistency with other works and with CDM theoretical modelling, we will adopt this library with Salpeter IMF hereafter). The best-fitting spectrum corresponds to a galaxy with about $20 M_\odot$, after normalization to the observed average flux level. However, we are well aware that nearly identical spectra may be drawn from the same library, by slightly changing the input parameters. To evaluate this effect, we select from the library all the spectra that produce an acceptable fit. They are overplotted in Fig 2. In this case, the resulting value for the stellar mass range from $\simeq 10 M_\odot$ to $\simeq 40 M_\odot$. The first conclusions that can be drawn from this single case is that even in the case of intermediate redshift galaxies observed at high S/N, an intrinsic uncertainty of about 50% in the stellar mass results from the degeneracies among some of the galaxy physical parameters. As shown in fig.2., this uncertainty cannot be significantly reduced by extending the observations at longer wavelengths.

To put this first statement on more quantitative grounds, we have applied the same analysis to the whole $K_{AB} \leq 23.5$ sample in the HDFS. For each object we have obtained the best fit values and total range of acceptable values $\Delta M_\odot = M_{MAX} - M_{min}$. We plot in fig. 3 the results, as a function of redshift and observed K-band, to show the different influence of progressively poor sampling of the near–IR spectrum at higher redshifts, and of progressively poorer S/N at fainter K flux levels. For object without spectroscopic redshift, we have

allowed the best–fitting spectra to shift in redshift around the best fit photometric redshift, to include this further uncertainty. The resulting picture is that
stellar masses may be constrained to within a factor of two with U-to-K broad
band imaging over a wide redshift range ($z \leq 2$) provided that IR observations
are correspondingly deep. At higher redshifts, however, the poor constraints on
the rest–frame near–IR spectrum result in a typical uncertainty of an order of
magnitude (see also Papovich et al 2001). Deep space–based observations will
reduce this value appreciably (Dickinson 2002, this volume).

4   The Hierarchical Growth of the Galaxy Stellar Mass

With these cautions in mind, one can compare the evolution of the stellar mass
in the HDF galaxies - obtained with the best- fitting technique described above
- with the prediction of the CDM models. For the latter, we use our analytic
rendition of the CDM models, that include the physical description of the Cole
et al 2000 models, with a refined treatment of the merging between satellite
galaxies (Menci et al, 2002). Given the uncertainties involved and the poor statistics of the HDF sample, we focus here on a first specific statement concerning
the growth of the most massive galaxies, and defer more refined treatment of
other integral quantities - like mass densities and mass functions -to forthcoming works. Fig. 4 shows the stellar mass derived for each galaxy in the HDFS
and HDFN. Both derive from the analysis described above, and based on our
catalogs already published (Fontana et al 1999, Vanzella et al 2001). The only

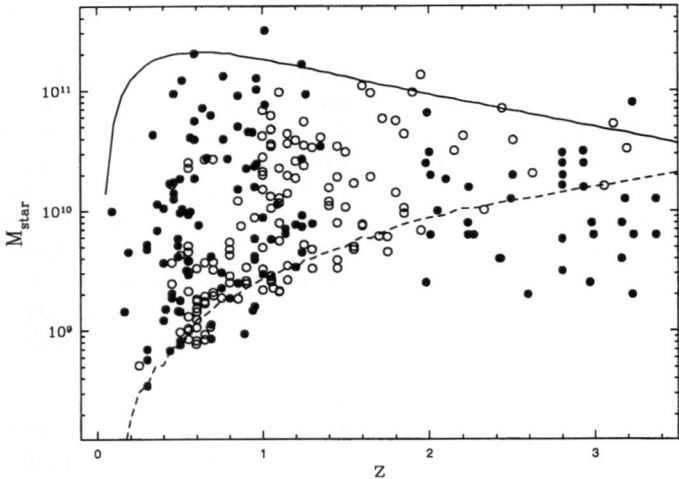

Fig. 4. Stellar mass for the galaxies in the HDFN and HDFS, obtained as described
in the text. Filled points represent objects with spectroscopic redshifts, empty with
photometric redshifts

exceptions are the data concerning the HDF–N galaxies at $z > 2$, that have been taken from Papovich et al 2001, where a similar analysis has been done using deeper NICMOS observations.

The lower envelope in the distribution is a result of the K selection criteria adopted here. The statement that we make here is about the *upper* envelope of the distribution, since there is no a priori selection against the more massive galaxies in the sample. *Fig. 4 shows that the mass of the most massive galaxies in the HDF samples increases with decreasing redshift.* This behaviour is also expected according to the hierarchical growth predicted by CDM models. To make this point quantitative, we have computed from the theoretical mass function a threshold defined such that one expects one galaxy above the threshold (in the area sampled by the HDFs observations) per unit redshift. In practice, our CDM rendition predicts about three galaxies to lie above the line shown in Fig. 4, in agreement with the observed quantities.

What can be learned from this comparison? By looking at Fig 4, it appears that galaxies with $simeq M_* = 10^{11} M_\odot$ start to be found at $z \leq 2$ in fields like the HDFs, and their number is in rough agreement with those expected from the high mass tail of theoretical CDM mass functions. To measure and describe the overall mass distribution will be a task for the next evolution of these survey.

Acknowledgments

The work presented here is the result of the coordinated efforts of several people, that I warmly thank here: E. Giallongo, N. Menci, F. Poli, A. Cimatti, S. Cristiani, S. D'Odorico, E. Daddi, M. Mignoli, L. Pozzetti, A. Renzini, P. Saracco, E. Vanzella, G. Zamorani.

References

1. Brichmann & Ellis 2000, ApJ 536, 77L
2. Cole et al 2000, MNRAS, 319, 168
3. Fioc, M. Rocca–Volmerange 1997, A&A 326, 950
4. Fontana et al 1999, MNRAS, 310, L27
5. Fontana et al 2000, AJ, 120, 2206
6. Fontana et al 2002, in prep.
7. Giallongo et al 1998, AJ, 115, 2169
8. Menci et al 2002, ApJ subm
9. Papovich, C. Dickinson, M., 2001, ApJ
10. Poli et al 2001, APJ, 551, 45L
11. Vanzella et al 2001, AJ, 122, 2190

The Masses of Lyman Break Galaxies

Joel R. Primack[1], Risa H. Wechsler[2], and Rachel S. Somerville[3]

[1] Physics Department, University of California, Santa Cruz, CA 95064, USA
[2] Department of Physics, University of Michigan, Ann Arbor, MI 48109, USA
[3] Department of Astronomy, University of Michigan, Ann Arbor, MI 48109, USA

Abstract. Data on galaxies at high redshift, identified by the Lyman-break photometric technique, can teach us about how galaxies form and evolve. The stellar masses and other properties of such Lyman break galaxies (LBGs) depend sensitively on the details of star formation. In this paper we consider three different star formation prescriptions, and use semi-analytic methods applied to the now-standard ΛCDM theory of hierarchical structure formation to show how these assumptions about star formation affect the predicted masses of the stars in these galaxies and the masses of the dark matter halos that host them. We find that, within the rather large uncertainties, recent estimates of the stellar masses of LBGs from multi-color photometry are consistent with the predictions of all three models. However, the estimated stellar masses are more consistent with the predictions of two of the models in which star formation is accelerated at high redshifts $z \gtrsim 3$, and of these models the one in which many of the LBGs are merger-driven starbursts is also more consistent with indications that many high redshift galaxies are gas rich. The clustering properties of LBGs have put some constraints on the masses of their host halos, but due to similarities in the halo occupation of the three models we consider and degeneracies between model parameters, current constraints are not yet sufficient to distinguish between realistic models.

1   Introduction

A great deal of effort devoted to determining the cosmological parameters has recently paid off. But, although there is good evidence that the cosmological parameters are roughly $\Omega_m = 0.3$, $\Omega_\Lambda = 0.7$, and $h = 0.7$, and that ΛCDM with these parameters is a good fit to the observed universe [1], this theory does not make unique predictions regarding the masses and other properties of galaxies at high redshift. Galaxy properties in cosmological theories also depend on assumptions about uncertain aspects of star formation, supernova feedback, and dust obscuration. Here we will focus on star formation.

We consider three different models of star formation, differing in the way that the star formation rate depends on galaxy properties, and discuss the implications for masses, clustering, and other properties of Lyman break galaxies (LBGs) in semi-analytic models. These models [2] all assume exactly the same underlying ΛCDM model with the parameters above, so the properties of the dark matter halos at any given redshift and the halo merging histories are the same. We also make the same assumptions in each model regarding the initial mass function (IMF), which we assume to be Salpeter between 0.1 and 100 $M_\odot$. We use the GISSEL00 stellar population synthesis models of Bruzual & Charlot

[3] with solar metallicity, and a simple model for dust obscuration, in which the optical depth is a power-law function of the unobscured ultraviolet luminosity:

$$\tau_{UV} = \tau_{UV*}(L_{UV}/L_{UV*})^\beta \, , \tag{1}$$

with τ_{UV*} an adjustable parameter and $\beta = 0.5$ [4].

The modern approach to semi-analytic modeling was pioneered by White & Frenk [5], and further developed by them and their collaborators in [6] and [7]. In [8], we reviewed and extended this work, and applied it to high-redshift galaxies [2] using three simplified models of star formation. The three star formation models we consider span the range of models proposed for high redshift galaxy formation.

The simplest model, termed Constant Efficiency Quiescent, assumes that the quiescent star formation rate per unit mass of cold gas is constant:

$$\dot{m}_* = \frac{m_{cold}}{\tau_*} \qquad \text{(CEQ)} \, . \tag{2}$$

We showed in [2] that the resulting predictions of this model are similar to those of [7] and to more detailed treatment [9,10] from the same group that included starbursts from major mergers. An alternative Accelerated Quiescent model assumes that

$$\dot{m}_* = \frac{m_{cold}}{\tau_{dyn}} \qquad \text{(AQ)} \, . \tag{3}$$

The predictions of our AQ model are like those of [6] even though those authors included starbursts from major mergers, because the above star formation prescription (based on data on star formation in nearby galaxies [11]) converts gas to stars so efficiently at high redshifts (since typically $\tau_{dyn} \propto 1/(1+z)$). Finally, we consider a Collisional StarBurst model in which the quiescent star formation efficiency is constant, but where there is also a burst mode of star formation triggered by galaxy interactions:

$$\dot{m}_* = \frac{m_{cold}}{\tau_*} + (\dot{m}_*)_{\text{starbursts}} \qquad \text{(CSB)} \, , \tag{4}$$

where $(\dot{m}_*)_{\text{starbursts}}$ is due to bursts in merging galaxies. The efficiency of star formation in these bursts is scaled according to a model based on hydrodynamical simulations [12], in which the efficiency scales as a power-law function of the mass ratio of the merger (see [2] for details). We find in our CSB semi-analytic model that most of the star formation at redshifts above unity occurs in starbursts driven by minor mergers, in which the merging satellite has mass less than $\frac{1}{3}$ that of the central galaxy. As noted, our CEQ model is similar to the Durham group model of several years ago [9]. At this conference, Baugh presented preliminary results from an alternative model that is similar to our CSB model.

Figure 1a shows the star formation rate density as a function of redshift for these three models. The CEQ model does not produce as many stars at high redshifts $z \gtrsim 3$ as the extinction-corrected observations indicate (see [2] for references), while both AQ and CSB models are acceptable in this regard,

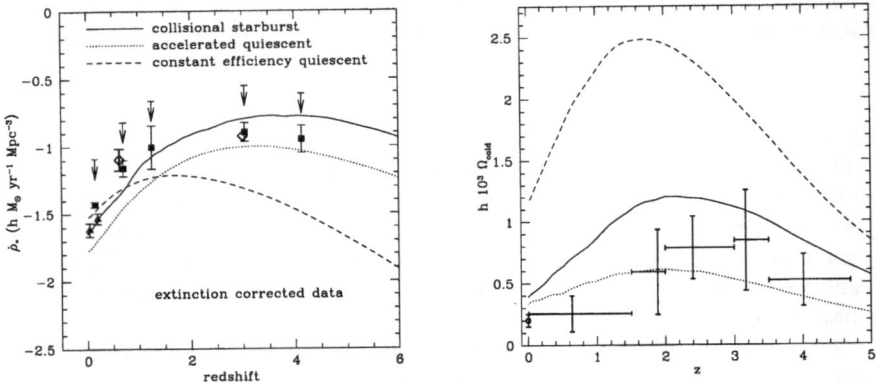

Fig. 1. Madau plot of star formation rate density (a) and cosmological density of neutral hydrogen (b) for our three models CSB, AQ, and CEQ. (From [2])

indicating that star formation at high redshift must be more efficient than locally. This argument is reinforced by the failure of the CEQ model to produce as many stars at $z > 2$ as are indicated by the fossil evidence (see Fig. 12 of [2]).

However, the AQ model converts gas to stars so efficiently that it may not have as much neutral hydrogen at $z > 2$ as is indicated by the data on damped Lyman-alpha systems – see Fig. 1b. The AQ model may also not have enough gas to fuel quasars at high redshifts [13], while the CSB model seems acceptable.

Further evidence that favors the CSB over the AQ model comes from the predicted LBG luminosity function. The CSB model predicts as many bright LBGs as are observed, although it slightly overpredicts the number of fainter ones, possibly because the dust obscuration prescription Eq. (1) is unrealistic in predicting very little extinction for lower-luminosity galaxies. But the AQ luminosity function predicts fewer bright LBGs than observed at $z = 3$, and far fewer than observed at $z = 4$ (see Figs. 4-7 of [2]). Bright LBGs only occur in massive halos in the CEQ and AQ models, and there are fewer such halos at higher redshifts.

2 LBG Masses

At this conference we have seen that analysis of the Hubble Deep Field data on LBGs [14] indicates that their stellar masses lie in the range $10^9 - 10^{11} M_\odot$, with a geometric mean of $6 \times 10^9 M_\odot$ (assuming a Salpeter IMF and solar metallicity, as we have done). Ground-based data on somewhat brighter LBGs indicate stellar masses in the same range, with a slightly higher median (see especially Fig. 10b of [15]). The predicted ranges of stellar masses and halo masses for all three of our models are shown in Fig. 2. (The results for the CSB model are similar to those represented by the histogram in the top right panel of Fig. 16 of [2], except that here they for $R_{AB} < 25.5$ while there they were for $V_{606} < 25.5$.) It is interesting that the distributions in the three models have similar medians, but very different widths. Perhaps counter-intuitively, the CSB model actually has *more* galaxies with large stellar masses than the other models! This is because there is more star formation activity at $z > 3$ in this model. However, in interpreting Fig. 2 it is important to keep in mind that the LBGs are predicted to be systematically fainter at a given stellar mass in the CEQ model.

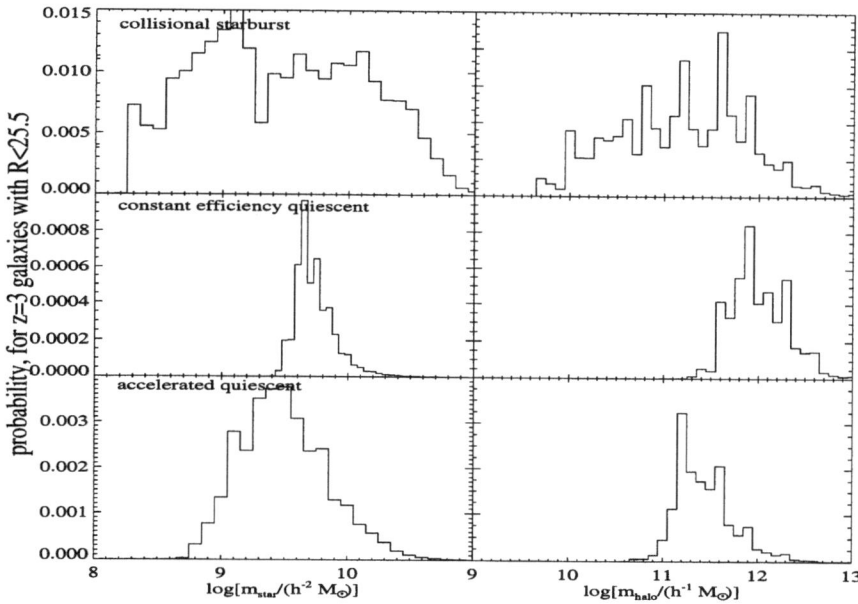

Fig. 2. Stellar and halo masses for galaxies brighter in R_{AB} than 25.5 at $z = 3$

It may therefore be more illuminating to look at the predicted relationship between stellar mass and rest-frame UV luminosity in all three models, shown in Fig. 3. The stellar masses deduced from the HDF-N data (see the lower right panel of Fig. 17 of [14]) agree well in zero point and slope with the predictions of the CSB and AQ models, but the stellar masses in the CEQ model are higher by about a factor of 2.5. However, the ground based data [15] do not show as clear a correlation of stellar mass with UV luminosity. Also, the deduced stellar masses are sensitive to the IMF and dust extinction, so there are large uncertainties making it impossible to rule out any of the models on this basis.

3   LBG Clustering

In principle, another way of estimating the masses of LBGs is via their clustering, since dark matter halos of higher mass are expected to be more correlated (e.g.

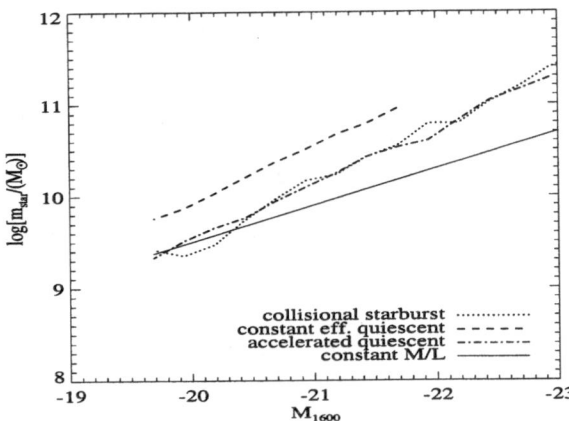

Fig. 3. Median stellar masses vs. rest-frame UV magnitude for CEQ, AQ, and CSB models. This relation is sensitive to the prescription for dust extinction; models that do not include dust extinction are closer to constant M/L

[16]). However, collisions between lower-mass halos are also more correlated than the halos themselves, since the collisions occur preferentially in denser regions [17] – thus knowledge of the LBG host halo masses from their clustering properties does not uniquely specify whether the LBGs are associated with galaxies in massive subhalos or with galaxy collisions. When we used N-body simulations combined with semi-analytic models to compute clustering properties of LBGs, we [18] found that the CEQ, AQ, and CSB models all predicted similar LBG clustering on both short and long scales, in general agreement with the available data – though CEQ produces the most clustered galaxies and is only marginally consistent. The similarity of LBG clustering properties in the three models reflects the larger similarity than might be expected in the dark matter halos that they occupy; we explored the physical reasons for this in [18]. However, the detailed clustering properties are affected by model ingredients that are still quite uncertain – for example, the efficiency of converting gas into stars in a galaxy collision – so further theoretical constraints on these parameters combined with recently improved observational constraints on LBG clustering may improve the potential to distinguish between modes of star formation.

A simple analytic model that constrains the dark matter halo masses hosting LBGs using data on their number density and clustering has shown that, if the typical number of galaxies in a halo of mass M is $N(M) = (M/M_1)^S$ for $M > M_{\min}$, current observational constraints on this model indicate that LBGs occupy halos greater than about $M \sim 10^{10} h^{-1} M_\odot$, with a power-law occupation of $S \sim 0.8$, so that typical LBGs reside in halos of a few $\times 10^{11} h^{-1} M_\odot$ [19]. Such a model can also be used to compare the host halo masses of different populations of objects, and has been used [20] to show that highly-clustered EROs inhabit halos that are an order of magnitude or two more massive than LBGs. The only possibly discrepant data on LBG clustering is the suggestion that the correlation length is a strong function of LBG brightness [21], in disagreement with each of our models [18]. However, this interpretation is controversial (cf. [18,1]).

4 Conclusions

While the masses of LBGs would seem to provide key information about their nature, due to the considerable uncertainties in our modeling and in deriving stellar or halo masses from the data, it is not currently possible to rule out any of the three very different recipes for star formation considered here on this basis. All three models are also roughly consistent with recent observational estimates of LBG clustering. However, the CEQ model predicts systematically higher stellar masses and also far too few bright LBGs especially at higher redshifts, and the AQ model may use up gas too efficiently to be consistent with other data. One way to check whether it is really true that many of the high-redshift bright galaxies are collision-driven starbursts is to see whether the morphologies of these objects resemble those produced in hydrodynamical simulations of interacting gas-rich galaxies, which are presently underway (see [23] for preliminary results).

References

1. J.R. Primack, 'The Nature of Dark Matter'. In: Proc. International School of Space Science 2001, ed. A. Morselli (Frascati Physics Series, 2002), astro-ph/0112255
2. R.S. Somerville, J.R. Primack, S.M. Faber: MNRAS **320**, 504 (2001)
3. G. Bruzual, S. Charlot: ApJ **405**, 538 (1993)
4. B. Wang, T. Heckman: ApJ **457**, 645 (1996)
5. S.D.M. White, C. Frenk: ApJ **379**, 52 (1991)
6. G. Kauffmann, S.D.M. White, B. Guiderdoni: MNRAS **264**, 201 (1993)
7. S. Cole et al.: MNRAS **271**, 781 (1994)
8. R.S. Somerville, J.R. Primack: MNRAS **310**, 1087 (1999)
9. C.M. Baugh, S. Cole, C.S. Frenk, C.G. Lacey: ApJ **498**, 504 (1998)
10. F. Governato et al.: Nature **392**, 359 (1998)
11. R. Kennicutt: ApJ **498**, 181 (1998)
12. J.C. Mihos, L. Hernquist: ApJ **425**, L13 (1994); **448**, 41 (1995); **464**, 641 (1996)
13. G. Kauffmann, M. Haehnelt: MNRAS **311**, 576 (2000)
14. C. Papovich, M. Dickinson, H.C. Ferguson: ApJ **559**, 620 (2001)
15. A.E. Shapley et al.: astro-ph/0107324 (2001)
16. H.-J. Mo, S.D.M. White: MNRAS **282**, 347 (1996)
17. T.S. Kolatt et al.: ApJ **523**, L109 (1999)
18. R.H. Wechsler, R.S. Somerville, J.S. Bullock, T.S. Kolatt, J.R. Primack, G.R. Blumenthal, A. Dekel: ApJ, **554**, 85 (2001)
19. J.S. Bullock, R.H. Wechsler, R.S. Somerville: MNRAS, **329**, 246 (2002)
20. L.A. Moustakas, R.S. Somerville: ApJ accepted, astro-ph/0110584
21. M. Giavalisco, M. Dickinson: ApJ **550**, 177 (2001)
22. S. Arnouts et al.: MNRAS **310**, 540 (1999)
23. R.S. Somerville: 'Disks at high redshift: interactions, mergers, and starbursts'. In: *Galaxy Disks and Disk Galaxies*, (ASP Conference Series, Vol. 230) eds. J.G. Funes, S.J., and E.M. Corsini, p. 477 (2001)

Deep Near-Infrared Imaging of Galaxies at z > 2 with Subaru

Toru Yamada and Masaru Kajisawa

National Astronomical Observatory of Japan, 2-21-1, Osawa, Mitaka, Tokyo 181-8588

Abstract. We present the results of the deep NIR imaging observations using the Subaru telescope. At the Hubble Deep Field North, we obtained the new and deepest K'-band image (10 hours) to study the rest-frame optical morphology of galaxies at $z > 3$ and the stellar mass distribution of galaxies at $2 < z < 4.5$. We also study the rest-frame optical properties of the 'building blocks' in the field of 53W002 and overall color distribution there. We then combined these data to compare the color and magnitude distribution in these two fields. The opt-NIR color-magnitude distributions in the two fields looks very similar and we report the conspicuous color change below $K_{AB} = 22$ and discuss the cause of the feature.

1 Extremely Deep NIR Imaging at the Hubble Deep Field North

While there exist quite deep and high-resolution WFPC2 and NICMOS images at the HDF-N, it has been desired to have the ground-base K-band data at the field with comparable depth and image quality in order to study more rest-frame optical properties of high-redshift galaxies. For the purpose, we obtained the very deep K'-band image using the Subaru telescope equipped with a NIR camera, CISCO. With about 10 hours of net integration under the condition of 0.3-0.6 arcsec seeing, we reached to the depth of $K_{AB} \sim 25.5$ as the peak of the counts of the detected galaxies.

Using the data, combined with the archived HST data taken in the WFPC2 $U_{300}B_{450}V_{606}I_{814}$ as well as NICMOS $J_{110}H_{160}$ bands, we studied the (i) rest-frame optical morphology of the galaxies at $z > 3$, (2) stellar mass distribution of high-redshift galaxies, and (3) correlation between galaxy properties with the stellar mass.

Figure 1 shows the Subaru K'-band image together with the HST I_{814}-band images. There is no clear Hubble sequence for galaxies at $z \sim 3$. The morphological feature seen in the HST WFPC2 images are well recognized in the Subaru K'-band image.

Figure 2 shows the obtained stellar mass distribution for the K'-band selected galaxies between $z = 1.9$ and 4.5. This is obtained by fitting the HST $UBVIJH$ and Subaru K'-band photometric data by the stellar evolutionary synthesis models (GISSEL96; Buruzual and Charlot 1993) changing the star-formation history, IMF, age, extinction, etc (Papovich et al. 2001, Shapley et al. 2001). The filled circles show the galaxies with spectroscopic redshift and those only with photometric redshift are by the open circles. We note that the stellar mass of the HDF-N K'-selected galaxies are typically very small, in the range between 10^9-10^{10} $M_\odot$.

In Figure 3, we present the possible correlation between stellar-mass and the rest-frame $U - V$ color for the galaxies. There are few massive (in stellar mass) galaxies with the bluest color and the massive galaxies tend to have redder $U - V$ color. If the rest $U - V$ color represents the age difference, then massive objects tend to have rather old average age, which may imply that the star-formation in these galaxies have been occurred rather successively, may be through the merging or assembly process.

z = 2.931

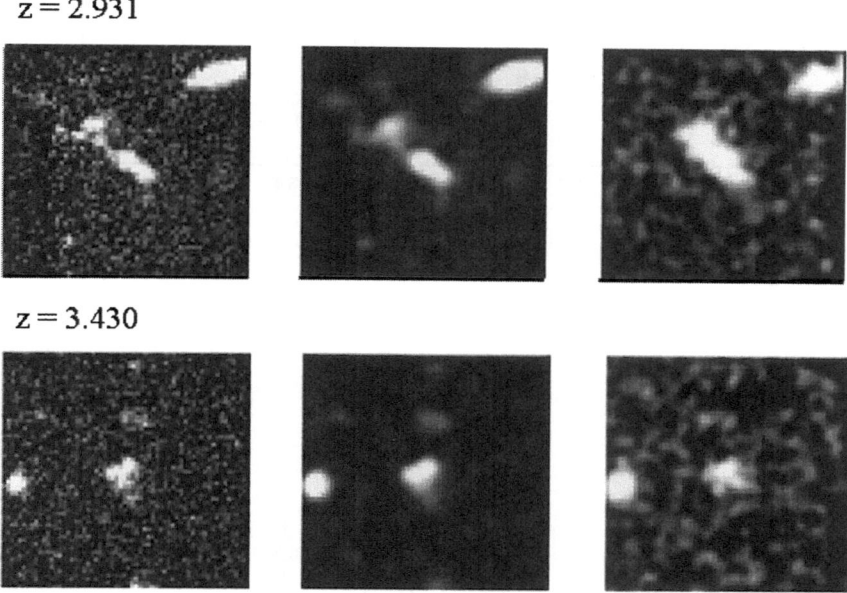

z = 3.430

Fig. 1. The rest-frame optical morphology of galaxies at z=3. We compare the WFPC2 raw (left) and seeing-convolved (middle) I_{814} images with CISCO K' ones

2   The Field of the Radio Galaxy 53W002 at z=2.4

We also obtained moderately deep J and K'-band images at the field of a radio galaxy 53W002 at z=2.4 where Pascarelle et al. (1996) discovered a dozen of emission-line objects and candidates using the HST WFPC2 intermediate-band image. Keel et al. (1999) also discovered that the field lies in the high-density region of strong emission line objects from the ground-base narrow-band images.

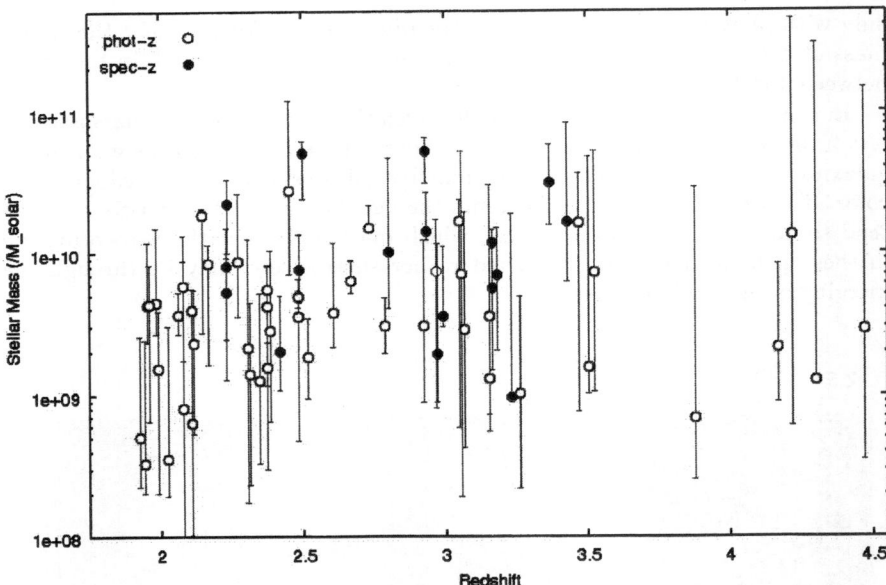

Fig. 2. Stellar mass distribution of the K'-selected galaxies at z=1.9-4.5

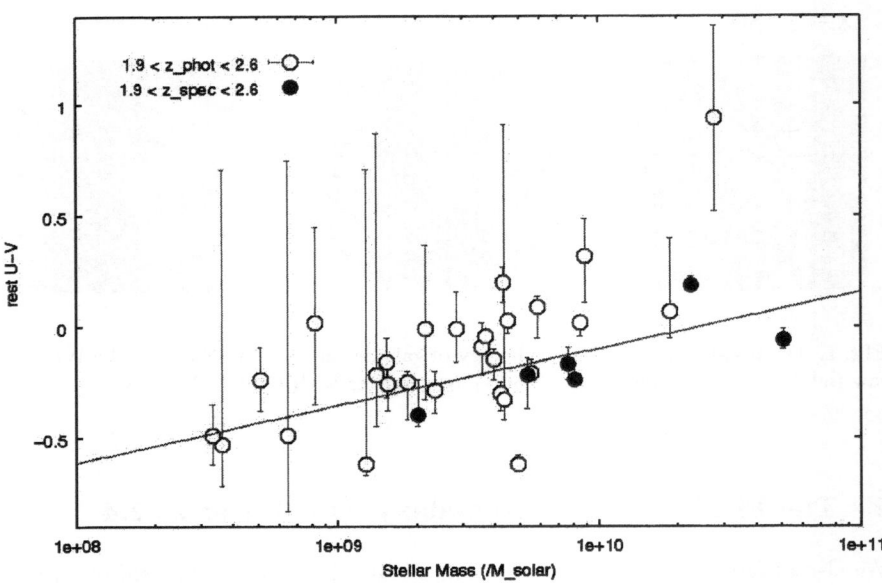

Fig. 3. Correlation between the obtained stellar mass and rest-frame $U - V$ color of the galaxies at z~2

We studied (i) the rest-frame optical properties of the emission-line selected objects and (ii) surface density of old quiescent galaxies at the redshift in the field using the Subaru data combined with the HST WFPC2 and NICMOS archive data. The detailed results are presented in Yamada et al. (2001). In brief summary, many of the emission-line objects are faint in NIR wavelength and so they are intrinsically compact and small objects dominated by the on-going star formation, and there are few developed quiescent galaxies that are older than ~ 1 Gyr in the field. The detailed discussion about the mass of the radio galaxy 53W002 itself based on the Subaru OHS and CISCO spectra is presented in Motohara et al. (2001).

Figure 4 shows the $I - K'$ color-magnitude diagram of the K'-selected galaxies in the field of 53W002 (filled circles). We also noticed a somewhat conspicuous change of the color distribution at $K'_{AB} \sim 22$. While the brighter objects show the colors distributed over the expected color rang for old and young galaxies at $z < 1$, below $K'_{AB} \sim 22$, most of the galaxies have blue colors ($I_{814} - K'_{AB} < 1.5$) and a small fraction have further red colors ($I_{814} - K'_{AB} > 2.5$). There are few galaxies with intermediate color range and thus there is a conspicuous color 'gap' or 'void' in the diagram. This trend is also seen in the $J - K'$ vs K diagram (Yamada et al. 2001).

3   Comparison of the Color-Magnitude Diagram in the Two Fields

Then we compare the color-magnitude distribution on these two fields. The data for the objects in HDF-N are shown by the open squares in Figure 4, and strikingly, the color-magnitude distributions in the two fields are quite similar, although faint red population is not seen in HDF-N. It suggests that this kind of color-magnitude distribution is a typical aspect of the universe if not an average.

What causes the conspicuous 'bluing' of the $I - K$ colors? Similar trend has in fact been reported by previous authors (Cowie et al. 1995, McCraken et al. 2000, Gardner 1995, Sarraco et al. 2001, but see Sarraco et al. 1999). We found that at least two phenomena are responsible for the distribution, namely, (i) deficit of intrinsically faint ($< L*$) red galaxies at intermediate redshift ($z \gtrsim 0.5$-1), and (ii) deficit of red galaxies including luminous ones at $z > 1$. This can be well recognized by comparing the data with the passive evolutionary models for the color-magnitude relation in the Coma cluster (Kodama et al. 1998) that are also shown if figure 4. Three lines represent the tracks of passively-evolving galaxies that are formed at $z = 4.6$ and have $M_V = -22, -20, -18$ at present epoch. The cosmology is $\Omega_0=1$, $\Omega_m=0.3$, $\Omega_\lambda=0.7$, and $H_0=70$ km s^{-1} Mpc^{-1}.

For the galaxies in HDF-N, we have the data of spectroscopic and photometric redshift. We can see clear deficit of faint red galaxies for those between $z = 0.5$ and $z = 1$ as shown in Figure 5. This is quite puzzling and interesting phenomena, since it cannot be explained by simple field-to-field variation effects because there are a certain number of luminous red galaxies at the same redshift. The number density of low-mass objects are expected to be larger than that of

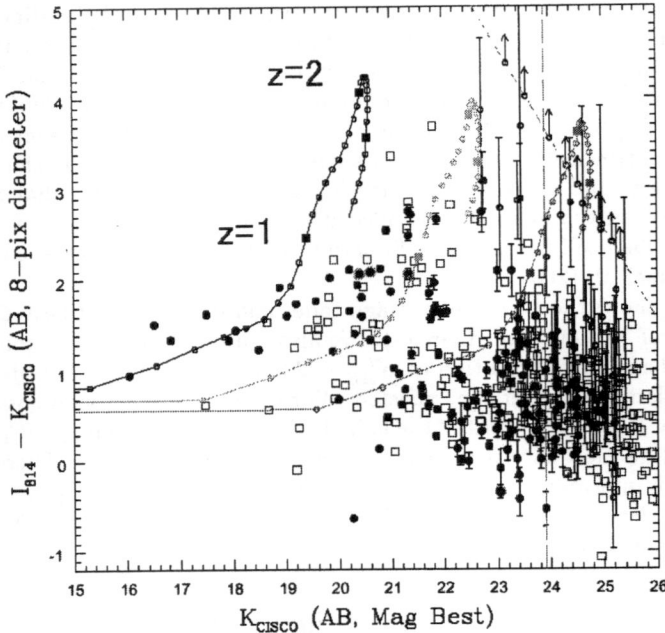

Fig. 4. Color-magnitude diagram for the K'-selected objects in the 53W002 field (filled dots) and HDF-N (open squares). Passive evolution track of the Coma metallicity sequence model in Kodama et al. (1998) is also plotted. The formation epoch is z=4.6 and their present absolute magnitude values are $M_V = -22, -20,$ and -18, respectively

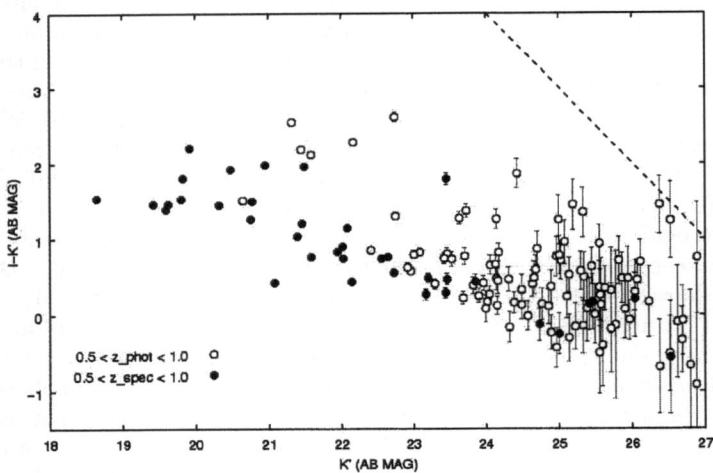

Fig. 5. Color-magnitude diagram for the K'-selected galaxies at z=0.5-1.0

the luminous objects since the local luminosity function have exponential cut-off at the luminous end at any band. Our result may imply that we rarely see the less massive galaxies that are dominated by old-stellar population at the intermediate redshift where old massive galaxies exist. It may be related to the negative faint-end of the local E/S0 luminosity function.

Kajisawa and Yamada (2001) investigated a complete volume-limited sample of galaxies with $M_V < -20$ up to z=2 in the HDF-N. Based on their careful morphological classification as well as evaluation of the possible uncertainty in photometric redshift, they concluded that the number density of early-type galaxies in HDF-N have very conspicuous decrease above z=1. So the deficit of red galaxies at $z > 1$ seen in Figure 4 may not be surprising although we still do not fully understand why such rapid change of the number density is observed at $z \sim 1$ after considering the possible effects of field-to-field variation and dust extinction, etc. We are planning to conduct more extended deep NIR and optical survey with Subaru to fully resolve these questions.

We thank Ichi Tanaka and Kentaro Motohara for useful discussions and help in observations. We also thank Taddy Kodama for useful suggestions and providing us the galaxy evolutionary synthesis models.

References

1. Bruzual A., G. & Charlot, S. 1993, ApJ, 405, 538
2. Cowie, L. L., Hu, E. M., & Songaila, A. 1995, AJ, 110, 1576
3. Dickinson, M. 2000, Royal Society of London Philosophical Transactions Series, 358, 2001
4. Gardner, J. P., Cowie, L. L., & Wainscoat, R. J. 1993, ApJL, 415, L9
5. Kajisawa, M. & Yamada, T. 2001, PASJ, 53, 833
6. Keel, W. C., Cohen, S. H., Windhorst, R. A., & Waddington, I. 1999, AJ, 118, 2547
7. Kodama, T., Arimoto, N., Barger, A. J., & Arag'on-Salamanca, A. 1998, A$A, 334, 99
8. McCracken, H. J., Metcalfe, N., Shanks, T., Campos, A., Gardner, J. P., & Fong, R. 2000, MNRAS, 311, 707
9. Motohara, K. et al. 2001, PASJ, 53, 459
10. Pascarelle, S. M., Windhorst, R. A., Keel, W. C., Scoville, N., & Armus, L. 1996, American Astronomical Society Meeting, 189, 8302
11. Papovich, C., Dickinson, M., & Ferguson, H. C. 2001, ApJ, 559, 620
12. Saracco, P., Giallongo, E., Cristiani, S., D'Odorico, S., Fontana, A., Iovino, A., Poli, F., & Vanzella, E. 2001, A&A, 375, 1
13. Saracco, P., D'Odorico, S., Moorwood, A., Buzzoni, A., Cuby, J.-G., & Lidman, C. 1999, A&A, 349, 751
14. Shapley, A. E., Steidel, C. C., Adelberger, K. L., Dickinson, M., Giavalisco, M., & Pettini, M. 2001, ApJ, 562, 95
15. Yamada, T. et al. 2001, PASJ, 53, in press.

Stellar Masses of High-Redshift Galaxies

Casey Papovich[1,2], Mark Dickinson[2], and Henry C. Ferguson[2]

[1] Steward Observatory, Univ. of Arizona, Tucson AZ 85721, USA
[2] Space Telescope Science Inst., Baltimore MD 21218, USA

Abstract. We present constraints on the stellar–mass distribution of distant galaxies. These stellar mass estimates derive from fitting population–synthesis models to the galaxies' observed multiband spectrophotometry. We discuss the complex uncertainties (both statistical and systematic) that are inherent to this method, and offer future prospects to improve the constraints. Typical uncertainties for galaxies at $z \sim 2.5$ are $\delta(\mathcal{M}) \sim 0.3$ dex (statistical), and factors of $\gtrsim 3$ (systematic). By applying this method to a catalog of NICMOS–selected galaxies in the Hubble Deep Field North, we generally find a lack of high–redshift galaxies ($z \gtrsim 2$) with masses comparable to those of present–day "L^*" galaxies. At $z \lesssim 1.8$, galaxies with L^*–sized masses do emerge, but with a number–density below that at the present epoch. Thus, it seems massive, present–day galaxies were not fully assembled by $z \sim 2.5$, and that further star formation and/or merging are required to assemble them from these high–redshift progenitors. Future progress on this subject will greatly benefit from upcoming surveys such as those planned with *HST*/ACS and *SIRTF*.

1 Introduction and Motivation

With current observations and those of the near future, we are able to observe distant galaxies ($z \gtrsim 2$) in their primeval stages, i.e., at an era when they are vigorously assembling their stellar content. However, no conclusive picture has yet emerged to describe how these high–redshift galaxies fit into the ancestral history of the present–day galaxy population. By measuring the stellar–mass distribution (which contains a complete historical record of star formation) for galaxies as a function of redshift, one can directly probe the global mass assembly history. This provides a stringent test for cosmological models that recount how high–redshift galaxies evolve into the present–day galaxy population.

However, a galaxy's stellar mass is *not* a directly measurable quantity: it must be inferred from models of the galaxy's mass–to–light ratios and the observed multiband photometry. In this contribution, we discuss the method used to obtain stellar mass estimates of distant galaxies and some the underlying caveats inherent in the process. We then present results from applying this method to a NICMOS–selected sample of galaxies in the *Hubble Deep Field North* (HDF–N).

2 Methodology

As the sample for our study, we have investigated the stellar–mass content of galaxies in the HDF–N using the multiband photometry from the *HST*/WFPC2

$(U_{300}B_{450}V_{606}I_{814})$, HST/NIC3 $(J_{110}H_{160})$, and ground–based K_s [1]. We initially focused on a sample of 31 "Lyman–break galaxies" (LBGs) with $2 \lesssim z \lesssim$ 3.5 [11] and fit their spectrophotometry with a suite of stellar–population–synthesis models [3,4], varying the age, SFR "e–folding" timescale (τ_{SF}), extinction (A_λ), and stellar mass; and also considered a range of metallicities ($0.001 - 3 \, Z_\odot$), and IMF (Salpeter; Scalo; Miller & Scalo). In general, we found only loose constraints on the parameters of the galaxies' stellar populations (i.e., age, τ_{SF}, A_λ, Z, and IMF). However, we derive fairly robust constraints for galaxy stellar masses (typical *statistical* uncertainty is ~ 0.3 dex).

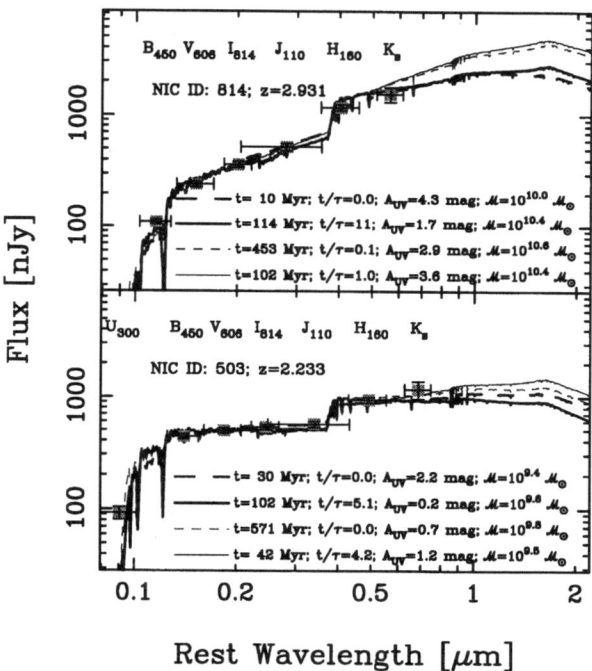

Fig. 1. Stellar population synthesis model fits to the observed photometry for two LBGs in the HDF–N. Each panel shows four models (with parameters inset), all of which fit the observed photometry at the 95% confidence level

Figure 1 illustrates the range of model parameters capable of fitting the observed photometry for two of the galaxies in the sample. Note that for each galaxy, there exist acceptable model fits with a wide range (i.e., more than an order of magnitude) of population age, τ_{SF}, and extinction. However, the stellar mass fits remain roughly constant [$\delta(\log\mathcal{M}) \sim 0.3$ dex]. This is also depicted in fig. 2. Note also that while the models fit the observed photometry (out to rest–frame ~ 6000 Å), they diverge strongly for $\lambda \gtrsim 1 \, \mu$m. This is generally true for the entire galaxy sample: there are large degeneracies in model age and extinction, which translates to a statistical uncertainty on the stellar mass estimates. Improved constraints would be possible with the incorporation of independent measurements of the instantaneous star–formation rate (e.g., nebular emission lines, FIR flux measures, etc.).

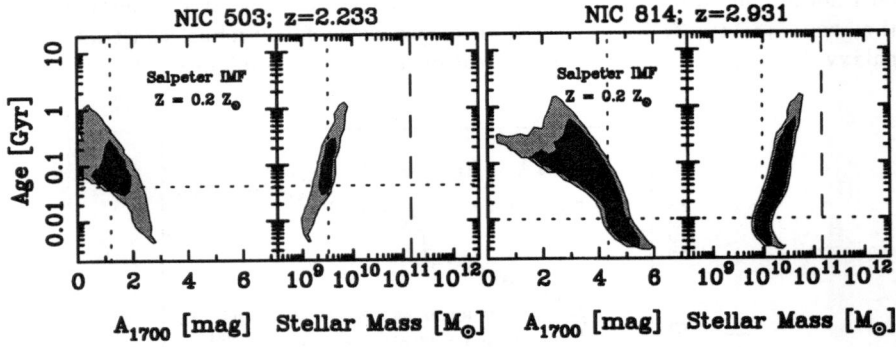

Fig. 2. Confidence intervals on the fitted parameters for the two galaxies in fig. 1. The equivalent 68% and 95% confidence intervals of extinction (at 1700 Å) and stellar mass are plotted versus the population age. Dotted lines indicate the best–fit solutions. The dashed line shows the characteristic stellar mass of a present–day "L^*" galaxy [2]

Although the statistical uncertainties on the galaxies' stellar–mass estimates are generally low, there remain inherent systematic uncertainties that must be considered when interpreting the results. Some of this arises from assumptions in the population–synthesis models (e.g., metallicity, IMF; see [11]). The metallicities of high–redshift galaxies are only weakly constrained; optical and near–IR spectra suggest $\sim 1/4 - 1/3\ Z_\odot$ (e.g., [6]). Varying the metallicity assumed in the synthesis models causes systematic shifts in the distribution of best–fit-model parameters, including ~ 0.3 dex in the inferred stellar masses. Similarly, the IMF at high–redshifts is essentially unconstrained. We find that a gamut of IMF models (Salpeter; Scalo; or Miller–Scalo) are all statistically consistent with the data; no one model is more preferred. The Scalo and Miller & Scalo IMFs systematically favor younger ages, lower extinctions, and somewhat higher stellar–masses. The lack of knowledge of the low–mass end of the IMF is also problematic. E.g., for a Salpeter IMF with a low–mass cutoff of 1 $\mathcal{M}_\odot$, the total stellar mass would be 39% that derived with a cutoff at 0.1 $\mathcal{M}_\odot$.

Systematics also arise from the assumptions of the galaxies' star–formation histories. All results presented thus far have used a monotonic, exponentially decaying (or constant) star–formation history. Such models likely only pertain to the youngest (and most dominant) stellar populations and as such neglect the contribution from any underlying, older stellar population. Because the single–component star–formation histories pertain to the youngest stellar populations,

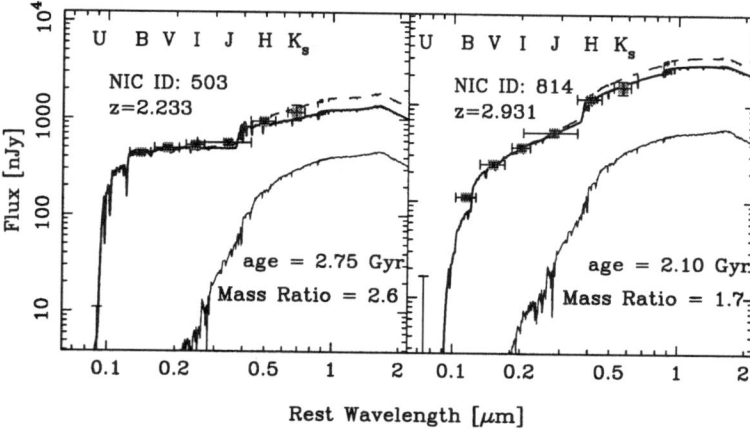

Fig. 3. Schematic illustration showing the effect on the stellar–mass estimates when adding a component corresponding to a maximally old stellar population. The thick line shows the best–fitting "young" model, and the thin line shows the maximum allowable contribution from an additional, old stellar population whose age is that of the universe at the observed redshift. The mass ratios of the old–to–young stellar populations are given in the panels. The dashed lines show the superposition of the two models

they arguably provide a *minimal* inferred $\mathcal{M}/L$ – and thus mass – for the galaxy. One can consider the flux contribution of an old stellar population from previous star–formation that is hidden "beneath the glare" of the young stars. We have investigated this effect by considering the sum of the fluxes from a maximally–old stellar component to that from the single–component models. The old–stellar component predominantly contributes to the flux longward of $\sim$ 6000 Å (see fig. 3). These two–component models yield a scenario where some fraction of the galaxies' stellar populations formed in a "burst" in the distant past. Such a scenario produces a maximal inferred $\mathcal{M}/L$, and thus translates to an upper bound to the galaxies' total stellar mass. For the HDF–N LBGs, the two–component models on average provide stellar mass estimates $\sim$ 3 times those from the single–component fits. Such a scenario is somewhat nonphysical (it assumes that most of the galaxies' observed stellar mass formed at $z \approx \infty$ and has since evolved passively), and considering this population merely serves as a fiducial with which to constrain the upper bound on the galaxy stellar masses.

3   Discussion and Results

Although at present the stellar–mass estimates for high–redshift galaxies' have significant uncertainties, these constraints are interesting nevertheless. For LBGs with "L^*" UV luminosities [16], we infer stellar mass estimates of $\sim 10^{10}\ \mathcal{M}_\odot$ or $\sim 1/10$th that of a present–day L^* galaxy [2]. Extending this analysis to all galaxies in the NICMOS HDF–N catalog allows a comparison between the LBG

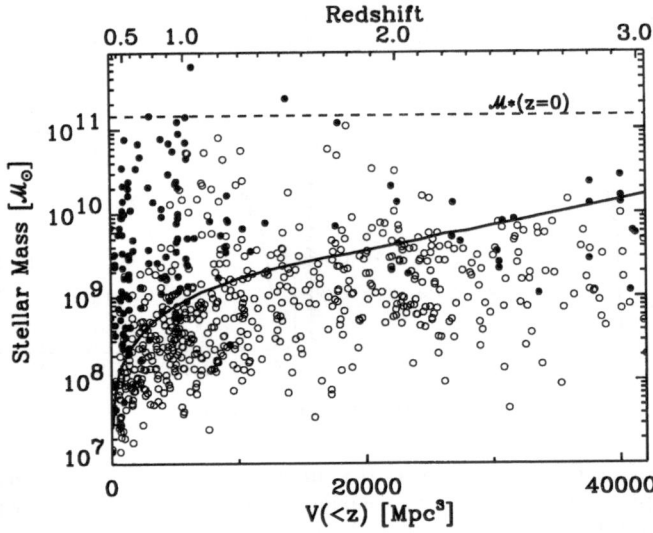

Fig. 4. Distribution of stellar mass for galaxies in the HDF–N as a function of comoving volume. The stellar mass estimates assume models with solar metallicity, Salpeter IMF, and single–component star–formation histories. Solid symbols denote galaxies with spectroscopically confirmed redshifts, and open symbols those galaxies where only photometric redshifts are available. The horizontal dashed line indicates the characteristic stellar mass of a present–day L^* galaxy [2]. The solid curve traces the "mass–limit" for a maximally old galaxy formed as a burst at $z \sim \infty$ with passive evolution, and normalized to the flux of the NICMOS detection limit ($H_{\rm AB} \approx 26.5$)

population and galaxies down to more modest redshifts ($z \gtrsim 0.5$). In fig. 4, we show the distribution of galaxy stellar mass in the HDF–N as a function of comoving volume. Here, all stellar masses assume solar metallicity, a Salpeter IMF, and use only the single–component star–formation histories. As such, they are nominally strict lower limits. Also shown in the figure is a fiducial curve denoting the minimal detectable stellar mass of a maximally old galaxy as a function of redshift and the NICMOS detection limit. Old galaxies would be detectable with masses *above* this curve. This, however, does not limit the minimal detectable masses of galaxies with lower mass–to–light ratios.

There are several interesting implications from fig. 4. Firstly, the HDF–N exhibits a lack of of $\mathcal{M} \gtrsim \mathcal{M}^*(z = 0)$ galaxies at $z \gtrsim 2$. Such galaxies should be detected (if present) in the deep NICMOS data, even to $z \sim 3$ (beyond which the NICMOS H band shifts below the 4000 Å/Balmer break and the stellar mass estimates are less secure). However, as shown in fig. 5, there are few (if any) galaxies in this redshift range with $V_{606} - H_{160}$ colors indicative of a galaxy dominated by old stellar populations. Thus, it is unlikely that we are missing such them if they were present (however, see recent results from the HDF–S, e.g.,

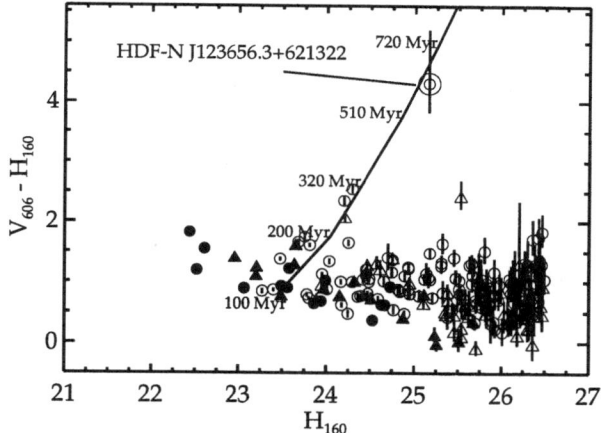

Fig. 5. Color–magnitude diagram for the HDF–N galaxies with $1.9 \leq z \leq 3.5$. Solid symbols denote galaxies with spectroscopically confirmed redshifts, and the open symbols those galaxies with only photometric redshifts. The solid curves denotes the evolution of a $10^{10} \, \mathcal{M}_\odot$ galaxy at $z = 2.7$ formed in a δ–function star–formation history. Note that the "J"–dropout, HDF–N J123656.3+621322 [7], is the only candidate for an old, red galaxy in the HDF–N in this redshift range

Labbé et al., this volume). It is a possibility that we have underestimated their stellar masses due to the uncertainties described above. Secondly, by $z \lesssim 1.8$, the upper envelope of stellar mass in the HDF–N increases to include massive, "L^*"–sized galaxies. Thus, it seems that the stellar populations of the progenitors to the massive galaxy population do not appear to be fully assembled in $z \gtrsim 2$ progenitors. This in turn suggests that more star–formation or merging (or both) are required for $z \lesssim 2$ to construct the large–galaxy population observed at $z \lesssim 1$ and at the present–epoch.

We wish to thank the conference organizers for arranging such a stimulating meeting in a beautiful setting. Support for this work was provided by NASA through grant GO–07817.01-96A.

References

1. M. Dickinson, et al.: *in preparation*
2. C. Papovich, M. Dickinson, & H. C. Ferguson: Astrophys. J. **559**, 620 (2001)
3. G. A. Bruzual, & S. Charlot: Astrophys. J. **405**, 538 (1993)
4. G. A. Bruzual: Astrophys. & Sp. Sci. Sup. **227**, 221 (2001)
5. S. Cole, et al.: Mon. Not. of the R. Astro. Soc. **326**, 255 (2001)
6. M. Pettini, et al.: Astrophys. J. **554**, 981 (2001)
7. M. Dickinson, et al.: Astrophys. J. **531**, 624, (2000)
8. C. C. Steidel, et al.: Astrophys. J. **519**, 1 (1999)

Rotation and Masses of Galaxies at z ≃ 3.2

Alan Moorwood[1], Paul van der Werf[2], Jean Gabriel Cuby[3], and Tino Oliva[4]

[1] European Southern Observatory, Karl-Schwarzschild-Str. 2, D-85748 Garching, Germany
[2] Leiden Observatory, P.O. Box 9513, 2300 RA Leiden, The Netherlands
[3] European Southern Observatory, Alonso de Cordova 3107, Santiago, Chile
[4] Osservatorio di Arcetri and Telescopio Nazionale Galileo, PO Box 565, 38700 S/C de la Palma, Tenerife, Spain

Abstract. We present infrared spectra at R ≃ 3000 around 2.1μm obtained recently with ISAAC at the VLT of a small sample of 5(6) high z galaxies discovered by narrowband infrared imaging and confirmed initially as emission line objects on the basis of relatively low s/n spectra. The primary aim of the new observations was to obtain improved velocity dispersions and, ideally, rotation curves in order to extend dynamical mass estimates to much higher redshifts than possible so far using visible spectroscopy. They were therefore conducted in service mode under excellent seeing conditions and with integration times of several hours. In most cases the emission 'line' can be unambiguously identified as the [OIII]5007/4959Å doublet at z ≃ 3.2. Measured velocity dispersions are typically ≃ 80 km s^{-1}. Of particular interest is the galaxy observed with the highest s/n and spatial resolution whose emission line exhibits a position versus velocity line tilt consistent both with ordered rotation and the velocity dispersion. The remainder exhibit either some or no rotation and with maximum velocities smaller than their velocity dispersions. This is most probably due to inclination and/or sensitivity effects. The derived masses are only typically ≃ 10^{10} M⊙ within the central ≃5 - 10kpc and could be even lower if outflows also contribute to the observed gas motions.

1 Introduction

Infrared spectroscopy now provides new opportunities for measuring velocity dispersions and possibly rotation curves and hence for estimating the dynamical masses of emission line galaxies to much higher redshifts than the z ≃ 1 achieved so far in the visible (Vogt et al. 1996). It has already been demonstrated on small samples that the new generation of infrared spectrometers on large telescopes, ISAAC at the VLT and NIRSPEC at Keck, can both achieve adequate s/n and resolve the brightest redshifted optical lines such as [OIII]5007/4959Å and Hα to z ≃ 3. Typical velocity dispersions are found to be ≃ 80 km s^{-1} implying dynamical masses of ∼10^{10}M⊙ within the central ∼ 5 - 10 kpc (Moorwood et al. 2000, Pettini et al. 2001). Evidence that the gas motions are dominated by ordered rotation has been inferred from position - velocity line tilts observed in a few cases but this is not yet totally convincing and the possibility that the linewidths may contain contributions from or even be dominated by outflows cannot be excluded. In this case, the fact that ≃ 10^{10} M⊙ then becomes an upper limit to the mass of these systems is perhaps of even greater interest in relation to galaxy evolution models. We have therefore followed up this issue by obtaining improved infrared spectra with ISAAC at the VLT of several galaxies originally found by narrow band infrared filter imaging with SOFI at the NTT (Moorwood et al. 2000). Our primary aim was to observe rotation curves both to directly estimate dynamical masses and to test whether or not the velocity dispersions measurable on much larger samples are likely to be reliable mass tracers. The exposures were thus relatively long and the observations were conducted in service mode in order to achieve the seeing conditions ($\leq 0.6''$) necessary to spatially resolve the galaxies. Here we present a first and still incomplete analysis of the data and discuss its significance for determining the dynamical masses of high z galaxies. The cosmology adopted in this paper assumes $H_0 = 65$ km s^{-1} Mpc^{-1}, $\Omega_m = 0.3$ and $\Omega_\lambda = 0.7$.

2 Observations and Results

Fig. 1 shows spectra obtained at R $\simeq$ 3000 around 2.1 μm with ISAAC at the VLT of five of the galaxies originally found in our narrow band infrared filter imaging survey (Moorwood et al. 2000) plus a sixth found serendipitously in the slit during the observation of 328.288-60.577. The observations were made in service mode during the period July - Sept. 2001 with clear sky and a seeing constraint of $\leq 0.6''$. Total integration times ranged from 3-6 hours/galaxy and the best seeing values were around 0.35$''$ in the visible. The slit width was 1$''$ in all cases and acquisition of the target galaxies was made by setting the rotation angle relative to a brighter reference galaxy (typically K$\sim$19) positioned accurately in the slit. As the prime aim was to detect rotation curves, the reference galaxies were chosen such that the slit was aligned as closely as possible with the major axes of the target galaxies as deduced from our original narrow band infrared discovery line images.

One consequence of the better seeing and higher s/n ratios is that one of the sample galaxies, 338.165-60.518, is now clearly seen to be double. Another is that the detected 'line' in most cases is clearly the [OIII]5007/4959Å doublet at a redshift of z $\simeq$ 3.2. Because of noise associated with the OH sky lines it is not possible to exclude Hα at z $\simeq$ 2.2 in all cases. The fact that an unbiased narrow band 2μm filter survey appears to yield more galaxies emitting [OIII]5007/4959Å at z $\simeq$ 3.2 rather than Hα at z $\simeq$ 2.2 is interesting, however, because it implies an intrinsic [OIII]/Hα intensity ratio typically $\geq$ 2 or about the largest value measured or inferred for Lyman break selected galaxies (Pettini et al., 2001, Kolbunicky et al. 2001).

The 2D spectra required to see rotation curves are shown in Fig. 2 together with the original 2.09 μm discovery images and WFPC2 F814W images available for 3 of the galaxies in the HDFS flanking fields. These images illustrate quite well why establishing the true major axis remains one of the major uncertainties for this type of work at high redshift!

Inspection of Fig. 2 shows that the clearest example of possible rotation is the larger component of 338.165-60.518 whose enlarged 2D spectrum and extracted 'rotation curve' are shown in Fig. 3. The line emission is traced over 1.6$''$ or

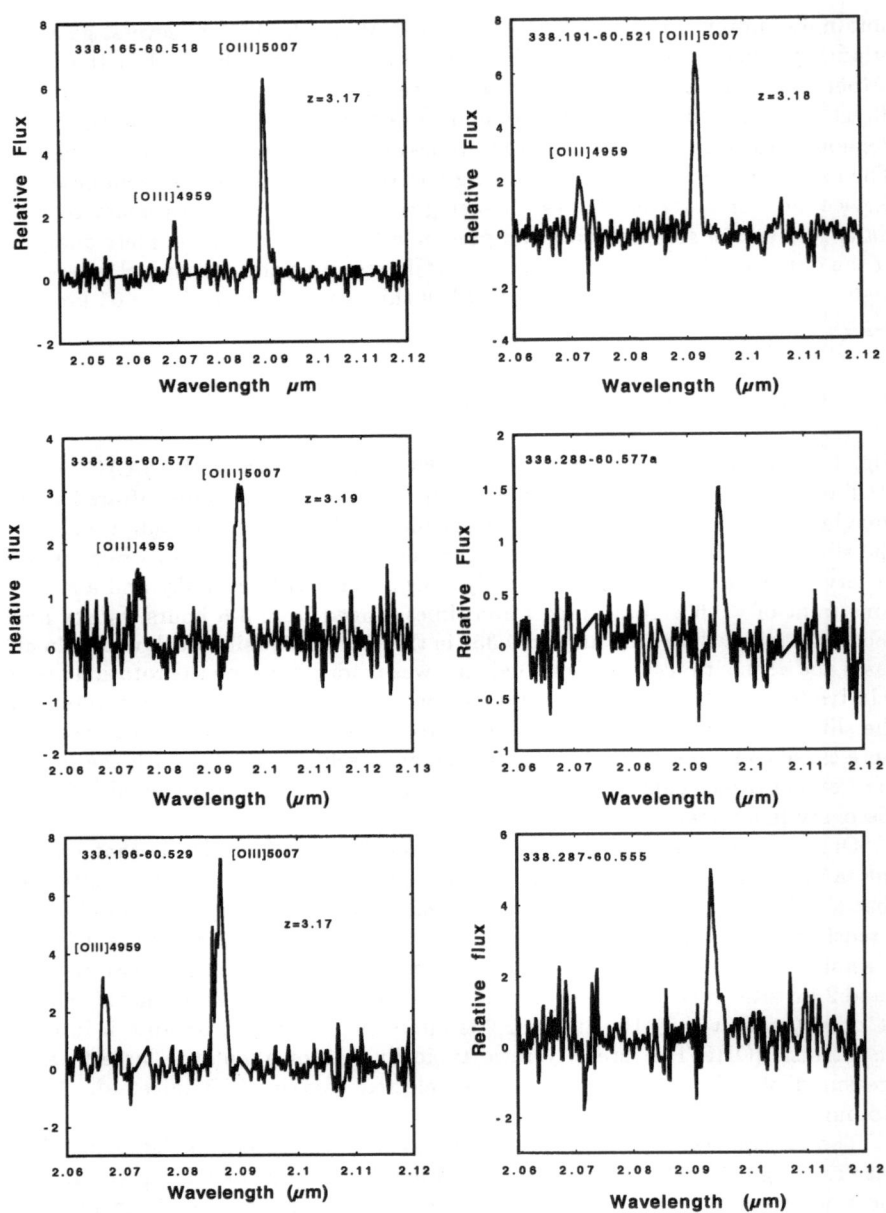

Fig. 1. Spectra at R ≃ 3000 around 2.1μm of high redshift galaxies obtained with ISAAC at the VLT. Where redshifts are given the emission lines are the [OIII]5007/4959Å doublet. Where redshifts are not given the line identification is uncertain but most likely to be [OIII] at $z \simeq 3.2$ or Hα at $z \simeq 2.2$

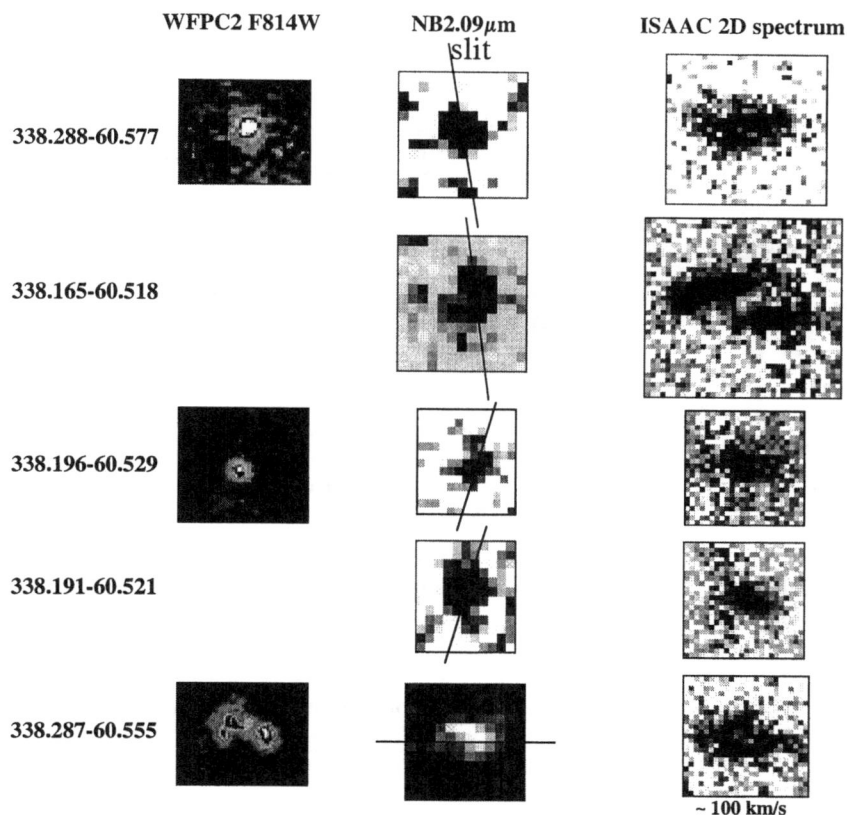

WFPC2 F814W   **NB2.09μm**   **ISAAC 2D spectrum**

338.288-60.577

338.165-60.518

338.196-60.529

338.191-60.521

338.287-60.555

~ 100 km/s

Fig. 2. The right hand column shows the 2d spectra with position along the slit in the vertical direction and wavelength/velocity horizontal. The middle column shows the original 2.09μm narrow band discovery images and the left hand column shows F814W images of 3 of the galaxies observed with WFPC2 in the HDFS flanking fields

$\simeq$12kpc and the terminal velocity exceeds 100km s^{-1}. The rotation curve has been obtained by first averaging pairs of rows in the 2D spectrum to obtain a resolution more consistent with the seeing and then determining the velocity centroids by gaussian fitting. Rotation is also confirmed in 338.288-60.577, albeit at a lower velocity than claimed on the basis of our earlier lower s/n spectra (Moorwood et al. 2000). The line in 338.191-60.521 also shows a similar tilt but the emission is either only present or can only be detected over a smaller region and the maximum rotational velocity is lower. 338.287-60.555 presents a complex picture due to the presence of several components within the central $\simeq$1$''$ and has not yet been analysed in detail. No tilt can be measured in either 338.196-60.529 or 338.288-60.577a - the latter being the serendipitously discovered object which has the lowest s/n and whose orientation is totally unknown.

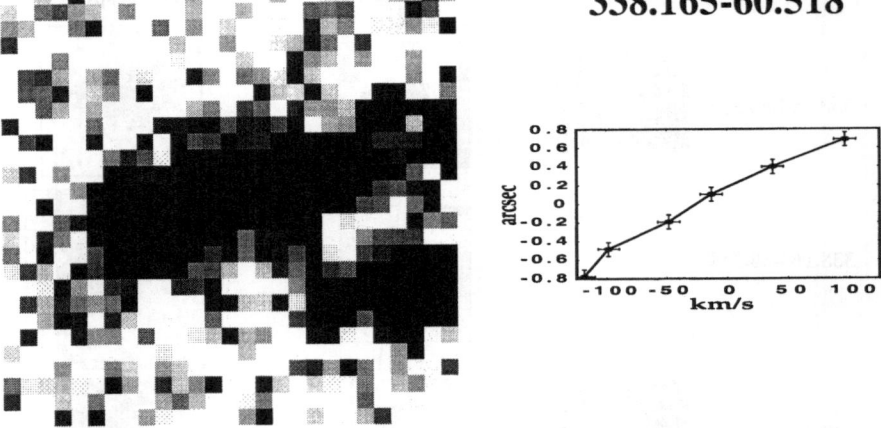

Fig. 3. 2D spectrum of 338.165-60.518 (left) together with the rotation curve obtained by fitting Gaussians to pairs of averaged rows

Table 1. summarizes the velocity dispersions (measured within the central $\simeq$ $1 \times 1''$), maximum rotational velocities and corresponding radii plus the estimated masses. The masses given first have been derived from the velocity dispersions using $M = 4.7 \times 10^5\ \sigma^2(km/s)r(kpc)\ M_\odot$ (Devereux et al. 1987) and the second mass, when given, has been computed from the estimated rotational velocities as $V_r^2 r/G$.

Table 1. Summary of measured velocities and mass estimates

Galaxy	σ(km/s)	$V_r \sin i$(km/s)	r(kpc)	M ($10^{10} M_\odot$)
338.165-60.518	73	108	6	1.5/1.7
338.288-60.577	123	65	5	3.5/0.5
338.288-60.577a	74	-	(3.75)	1
338.287-60.555	84	-	(3.75)	1.3
338.196-60.529	83	-	(3.75)	1.3
338.191-60.521	73	25	2.5	0.9/0.04

3   Discussion

The galaxy 338.165-60.518, which was observed at the highest s/n ratio and with the best seeing limited spatial resolution, does actually provide the most convincing case for equating velocity dispersion with mass. It is (relatively) well resolved spatially and the maximum rotational velocity of 108 km s^{-1} at $\simeq$ 6 kpc is larger than the velocity dispersion by about the amount expected for an edge-on disk. Deriving the actual mass is subject to many uncertainties and it is worth remembering here that deriving accurate masses is difficult even for nearby galaxies whose complete rotation curves, orientation and morphology are known. Here we are dealing with, at best, velocity dispersions and the inner part of the rotation curve of objects which are barely resolved spatially and whose orientations are basically unknown. Nevertheless, with these caveats, the simplest formulae relating the velocity dispersion and rotational velocities to mass both yield values $\simeq$ 1.5 10^{10}M$_\odot$ for 338.165-60.518. This is comforting as the velocity dispersion should yield an upper limit (if part of the velocity is due to outflows) and the rotational velocity should provide a lower limit (due to inclination and other corrections).

The situation is less convincing for the other members of the sample. Rotation already seen previously in our earlier observations of 338.288-60.577 is confirmed but at a lower level and with a value lower than the velocity dispersion. This latter is the highest in the sample and its value does agree perfectly with our previous data. Rotation is also seen in 338.191-60.521 but can only be traced out to a much smaller radius and again the velocity is smaller than the velocity dispersion. 333.196-60.529 shows little or no rotation, 338.287-60.555 is confused by the presence of several, probably interacting, components and 338.288-60.577a is the serendipitously detected galaxy observed at relatively low s/n and whose orientation is totally unknown. It is still possible therefore to invoke contributions to the velocity dispersion from outflows or turbulence in most of these objects. The difficulties of seeing rotation curves at this redshift are so large and various, however, that we believe that the fact that any are seen at all provides a strong argument in favour of the velocity dispersions reliably reflecting the gravitational mass of these systems. If that is the case it should be noted that the masses estimated within the central 5-10 kpc regions are still only $\simeq$10^{10}M$_\odot$ i.e $\sim$10% of that within a similar region of the Milky Way. If outflows are important then these galaxies are even less massive than that.

References

1. N.A. Devereux, E.E. Becklin, N. Scoville.: Ap.J., 312 (1987),529
2. H.A. Kobulnicky, A. Henry, K. E. Johnson: ApJ, 539 (2000), 712
3. M. Pettini, A. E. Shapley, C.C. Steidel et al.:Ap.J.,554 (2001), 981,
4. A.F.M. Moorwood, P.P. van der Werf, J.G., Cuby, E. Oliva: A&A, 362 (2000), 9
5. N.P. Vogt, D.A. Forbes, A.C. Phillips et al.:Ap.J (1996), 465, L15

Emission Line Measurements in the Magnified $z = 3.357$ HII Region Behind a Cluster at $z = 0.54$

R.A.E. Fosbury[1], A. Humphrey[2], M. Villar-Martín[2], P. Rosati[3], G. Squires[4], S.A. Stanford[5], B.P. Holden[5], and M. Rauch[6]

[1] Space Telescope - European Coordinating Facility, D
[2] University of Hertfordshire, UK
[3] European Southern Observatory, D
[4] Caltech, USA
[5] University of California, Davis, USA
[6] Carnegie Observatories, USA

Emission line spectroscopy of stellar-photoionized nebulae (HII regions/galaxies) at high redshift is crucial for tracking the chemical evolution of galaxies at early epochs. The recognition of 'primordial' HII regions will be based on both the gas-phase element abundances and on the effective temperature of the ionizing stellar radiation field which is dependent on the upper stellar mass function. The desire to find and analyse the emission lines from such objects is one of the primary scientific drivers for NGST NIR spectroscopy [1] which will allow measurements to be made of the restframe optical spectrum. In addition to their obvious importance for element abundance studies, such objects offer the intriguing possibility of studying the size and mass of objects seeing perhaps their first burst of star formation. While the calibration of the relationship between HII region/galaxy emission line velocity dispersions and $H\beta$ luminosity has been carried out in the local Universe ([2] hereafter MTT, and references therein), the extension of this technique to high redshifts (and hence, presumably, low metallicities) can be of real value as a cosmological tool. This poster reports measurements of restframe UV emission line strengths and widths in an HII region at $z = 3.357$, lensed by a cluster at $z = 0.543$.

Emission line objects with weak continua are difficult to find using broadband imaging techniques. This has stimulated spectroscopic searches from the ground, notably ones using clusters as magnifying lenses [3]. Holden et al. [4] have reported the serendipitous discovery of a bright emission line arc in southern, slightly lower redshift part of the cluster RXJ 0848+4456. This shows narrow emission lines of NIV] 1486; CIV 1550; HeII 1640 and OIII] 1664 at a redshift of 3.357. The spectrum suggests that this is a giant HII region-like object with ionizing stars with an effective temperature of at least 75,000K. Such stellar temperatures and low gas metallicities are approaching the extreme values expected for a primordial HII region ionized by first generation stars (see eg. [5]).

We have obtained a high ($R \sim 4000$) resolution optical spectrum with the Keck ESI spectrograph – a one hour exposure of the brighter of the two arc components (Fosbury et al. in prep.) – and are able to measure additional emission lines, notably Lyα and CIII] 1909.

This allows us to derive accurate line widths for all lines. Prominent absorption components in both Lyα and CIV indicate the presence of a highly ionized outflow.

The two components of the arc and the extended morphology imply that the source object lies along a fold caustic. A first approximation to a strong lensing model (M. Lombardi, priv. comm.) with $z_{\text{lens}} = 0.543$, $z_{\text{source}} = 3.357$ suggests that the magnification could be as high as 20–40.

From these data, we deduce the following:

- From photoionization modelling, the nebula is ionized by stars with an effective temperature of at least 75,000K. The gas metallicity is a few percent of Solar. The ionization parameter is quite high, $U \approx 0.1$.
- From the CIII] doublet ratio, the electron density is near the low density limit, ie. $n_e \leq 10^{3.5} \text{cm}^{-3}$.
- The velocity dispersions measured for the doubly ionized lines OIII] and CIII] are around 37 km/s while the triply ionized values for NIV] and CIV are larger, ~ 90 km/s.
- The absorptions in Lyα and CIV indicate an outflowing wind in a high state of ionization.
- Using the width of the OIII] lines in the MTT calibration $\log L(H_\beta) = 5 \log \sigma - \log O/H + 29.5$ allows us to estimate the intrinsic H_β luminosity assuming $[O/H] = -1.3$. The resulting value, $\log L(H_\beta) = 41.7$, can be used as a 'standard candle' to investigate the combination of the cosmology and the lensing model.
- Inferring the H_β flux from the UV line spectrum and our photoionization model, using a cosmology with $H_0 = 65 \text{ km s}^{-1} \text{ Mpc}^{-1}$, $\Omega_m = 0.3$ and $\Omega_\Lambda = 0.7$ and using our initial magnification estimate of ~ 30, this translates to a measured value of $\log L(H_\beta) = 41.9$. This good agreement suggests that the calibration of the MTT $\sigma - L$ relationship at low metallicity is not too inaccurate.

The refining of this technique should offer access to a powerful probe of both astrophysics at high redshift and of cosmological parameters.

Acknowledgments We thank Marco Lombardi for access to his preliminary work on the lensing model for the cluster. Some of the data presented herein were obtained at the W.M. Keck Observatory. Based partly on observations made with the NASA/ESA Hubble Space Telescope, obtained from the Data Archive at the ESA/ESO Space Telescope - European Coordinating Facility.

References

1. Kennicutt, R. 1998, ESA SP-429, p81
2. Melnick, J. Terlevich, R. & Terlevich, E., 1999, MNRAS, 311, 629
3. Ellis, R., Santos, M.R., Kneib, J-P. & Kuijken, K., 2001, ApJ, 560L, 119
4. Holden, B. P. et al., 2001, AJ, 122, 629
5. Bromm, V., Kudritzki, R-P. & Loeb, A., 2001, ApJ, 552, 464

SCUBA Sources: Massive Galaxies at High Redshifts?*

Ian Smail

Department of Physics, University of Durham, Durham DH1 3LE

Abstract. I discuss the results of a multi-wavelength survey of the submillimetre-selected galaxy population. I describe the properties of this population, in particular the limited information we have on their masses. Finally, I briefly comment on the interpretation of this population within the framework of our understanding of galaxy formation and evolution.

1 The Submillimetre Galaxy Population

Submillimetre-wave surveys undertaken with SCUBA on the JCMT have produced a revolution in our understanding of the sources which make up the extragalactic background in this waveband [6,10,14,15]. The cumulative flux in sources detected at 850μm by SCUBA, down to blank-field confusion limit of SCUBA of 2 mJy [10], corresponds to roughly half of the diffuse background seen by *COBE* [7]. Moreover, by exploiting the gravitational amplification of massive cluster lenses, we can extend these limits to reach a cumulative surface density of ~ 10 arcmin^{-2} at a flux limit of 0.5 mJy, and hence resolve 80% of the 850-μm background at this depth [2].

SCUBA's coarse angular resolution (15″) hampers the identification of counterparts to submm sources [16]. However, by exploiting the tight far-infrared/radio correlation for star-forming galaxies we can employ sensitive, higher resolution VLA maps to identify the source of the submm emission in these regions [17]. Using this approach we uncover a wide variety of infrared luminosities and optical-infrared colours for submm galaxies (Fig. 1). The classification of counterparts of a small sample of submm sources [18] brighter than ~ 1 mJy (after correcting for lens amplification), breaks down as: 20% optically-bright, 20% near-infrared bright and extremely red (I–K>5), with 60% faint/invisible in both the optical and infrared (I>26, K>21).

Importantly, the submm fluxes of the sources in Fig. 1 span only a factor of four, while their optical (restframe UV) fluxes span a range of 10^3, only part of which can be ascribed to differential K-corrections. This underlines the difficulties of using the UV properties of obscured galaxies to estimate their submm

---

* This paper is based on a collaborative project with Rob Ivison, Dave Frayer, Andrew Blain and Jean-Paul Kneib.

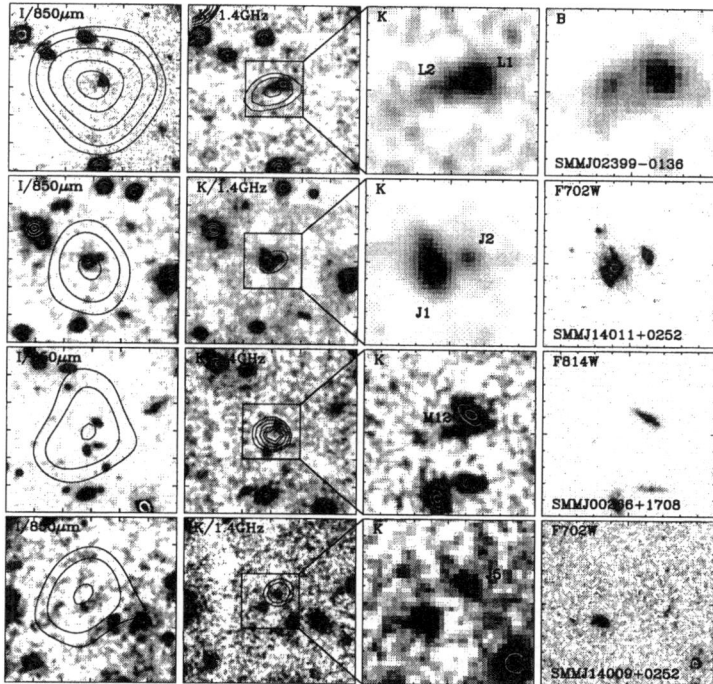

Fig. 1. A selection of the more luminous submm galaxies from SCUBA Lens Survey[18]. The top two rows show optically-bright submm sources: SMM J02399−0136 [11] and SMM J14011+0252 [12]. These bright optical counterparts have facilitated the detailed study of the galaxies, including the measurement of their dynamical masses from CO line mapping. However, these galaxies are not typical of the majority of the submm population and we illustrate more representative examples in the two lower panels. Each panel shows (from left to right): a 30″-wide I-band image with the 850-μm map overlayed; K-band image with a deep 1.4-GHz VLA map; an enlarged view of the K-band image (6″×6″) with the submm counterpart marked; and the equivalent area as seen in deep *HST* imaging (except for SMM J02399−0136 which is a high-resolution ground-based image). More details of these data are given in [4,18]

fluxes or total star formation rates. The high fraction of optically faint counterparts also hampers the measurement of a complete redshift distributions. Nevertheless, we can obtain crudely estimate redshifts from the radio/submm colours of the sources [5], assuming their dust temperatures are similar to those seen in luminous ultraluminous infrared galaxies (ULIRGs) at $z = 0$. Applying this approach to our sample we estimate that the median redshift for $\gtrsim 1$ mJy submm population is $< z > \sim 2.5$–3. This confirms that SCUBA is uncovering a population of highly luminous ($> 10^{12}$–$10^{13} L_{\odot}$), high-z galaxies with a space density which is several orders of magnitude above the equivalent local population.

To understand where this population goes in our picture of galaxy formation we need masses for these systems. Do these luminous sources represent violent, but short-lived bursts in low mass halos, or more extended and more significant star formation events in massive galaxies?

2   The Masses of Submm Galaxies

Of the ∼ 100 submm-/mm-selected galaxies known at the current time, a mere half-dozen have reliable counterparts with accurate redshifts. The most intensively studied examples from this small sample are the optically-luminous submm galaxies: SMM J02399−0136 at $z = 2.80$ [11] and SMM J14011+0252 at $z = 2.56$ [12] (Fig. 1).

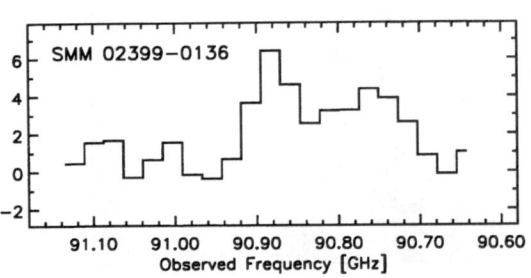 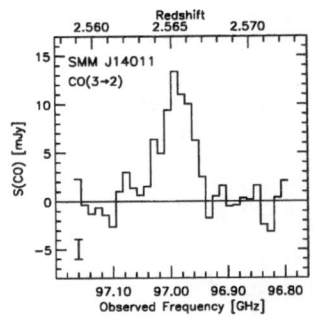

Fig. 2. The CO(3→2) spectra of SMM J02399−0136 (left, [8]) and SMM J14011+0252 (right, [9]) from the OVRO Millimeter Array. These spectra represent a total of 55 and 41 hrs of integration on the respective sources. In each case the CO emission is positionally coincident with the SCUBA source (Fig. 1) but is redshifted relative to the UV absorption lines in the spectra of the optical counterparts (Table 1). Systematic blueshifts of UV lines are not unusual for starbursts and have been interpreted as a signature of outflows in the optically emitting gas, where the systemic and redshifted emission is obscured by dust

The star formation rates in these galaxies can be estimated from their far-infrared luminosities. There is evidence of some non-thermal contribution to L_{FIR} in SMM J02399−0136, however, all estimates suggest that this contributes no more than half of the bolometric luminosity [1,8]. Assuming the remainder is powered by star formation, the implied SFR of massive stars (M> 5M⊙) in both systems are very high (Table 1) [11,12]. Including low mass stars, we expect a total SFR in these galaxies of ∼ 1–2 × 10³M⊙ yr⁻¹, depending on the exact IMF. If their gas reservoir are sufficient, then these rates would add a further ≫ L^* of stars to these already luminous systems on a characteristic starburst timescale (100 Myrs).

To assess the masses of these galaxies and their gas reservoirs we need dynamical information. Unfortunately, the optical spectra of such morphologically-complex galaxies (Fig. 1) provide little useful dynamical information and moreover, as was stressed at this conference, such restframe UV observations of strongly star forming systems are notoriously difficult to correctly interpret. For this reason, we instead use observations of molecular line emission from the OVRO Millimeter Array to directly estimate both the dynamical *and* gas masses of these galaxies (Fig. 2).

Table 1. Properties of SMM J02399−0135 and SMM J14011+0252

Property	SMM J02399−0135 L1/L2	SMM J14011+0252 J1/J2	
z	2.808	2.556	
Amp.	2.5	2.8	Lens amplification
L_{opt}	4	7	$(10^{11} L_\odot)$
L_{FIR}	10	6	$(10^{12} L_\odot)$
SFR $(M>5M_\odot)$	1.2	0.7	$(10^3 M_\odot)$
$M(H_2)$	1.8	1.1	$(10^{11} M_\odot)$
θ_{CO}	< 15	< 20	(kpc)
$FWHM_{CO}$	710	200	$(km\,s^{-1})$
M_{dyn}	< 2.3	< 0.3	$(10^{11} \sin^{-2}(i) M_\odot)$
f_{gas}	$> 0.8 \sin^2(i)$	$> 3.0 \sin^2(i)$	Gas fraction
$M(H_2)/M(dust)$	140–700	200–900	$T_d = 30$–70 K
$\Delta v(UV - Submm)$	-400	-1100	$(km\,s^{-1})$

The observed CO(3→2) fluxes indicate intrinsic line luminosities, $L'(CO) =$ 3–4 × $10^7 L_\odot$ K km s^{-1} pc^{-2}. We estimate the mass of molecular gas (including He) from the CO line through $M(H_2)/L'(CO) = \alpha$. The value for α should lie between $\alpha \sim 1$ and the Galactic value of $\alpha \sim 5$: we adopt $\alpha = 4$, which is consistent with that estimated for Arp 220. Based on this we infer molecular gas masses of $M(H_2) \sim 1$–$2 \times 10^{11} M_\odot$ for these galaxies ([8,9], Table 1), giving them higher gas masses than typical ULIRGs at $z = 0$.

The CO emission in the OVRO maps of both submm sources is un- or marginally-resolved on scales of $\theta < 5$–$7''$, corresponding to a size of $D < 15$–20 kpc after correcting for lensing. We can use this limit to constrain the dynamical masses of the galaxies. The dynamical mass within the CO emission regions is $M_{dyn} = R(\Delta V/[2\sin(i)])^2/G$, where ΔV is the observed line width in Table 1. The inclinations are unknown. The estimated dynamical masses, within the central 20 kpc, are listed in Table 1. These results indicate very high gas fraction, consistent with young, gas-rich systems, indeed for SMM J14011+0252 we require that this system must be close to face-on, $i \leq 35°$, or else the gas mass will exceed the dynamical mass.

We have recently undertaken a higher-resolution study of SMM J14011+0252 [13] using *HST*, OVRO/BIMA and the VLA in A-array (Fig. 3). After deconvolving, the CO emission subtends $6.6'' \pm 1.4''$ (Fig. 3a), conservatively equivalent

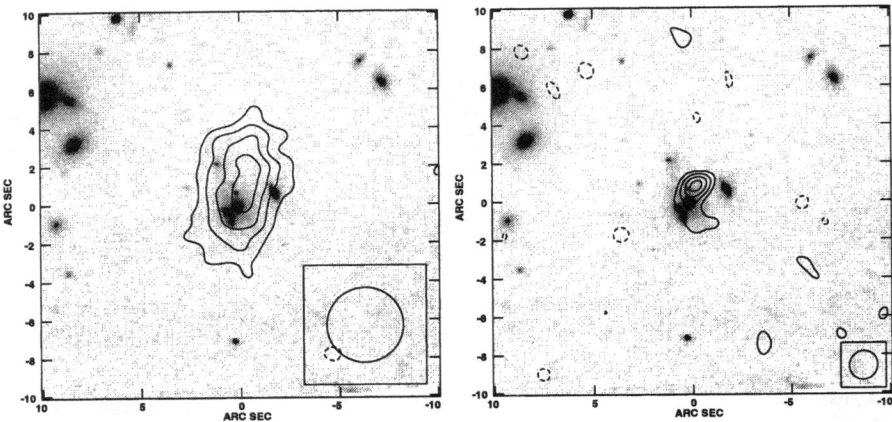

Fig. 3. High resolution maps of SMM J14011+0252. Each panel shows an *HST* WFPC2 F702W image of the galaxy, overlayed on this are, in the left-hand panel, a high-resolution OVRO/BIMA millimetre CO map and in the right-hand panel a high-resolution 1.4-GHz VLA radio map. The deep radio map shows the location of the obscured starburst in this system: outside the optical extent of the galaxy. Clearly the complex morphology we see in the restframe UV view of this galaxy results in large part from dust obscuration. The UV emission arises from relatively lightly reddened regions, whereas the bulk of the far-infrared emission comes from highly obscured regions which are effectively invisible even in the restframe optical (observed K-band). The insets in both panels show the effective beam size for the millimetre/radio maps. This figure is adapted from [13]

to 20 kpc. In addition, the 1.4-GHz emission has a FWHM of $\sim 2''$, indicating that the starburst extends across a region of ~ 5 kpc. Both these scales are significantly larger than is typically seen in comparable ULIRGs at low redshifts, $\lesssim 1$ kpc. This suggests that there may be real, physical differences in the macrophysics of merger-induced activity at low- and high-z. The most likely cause of this is the increased instability of the high-z systems [3] arising from the absence of massive bulges (which stabilise the gas disks against bar-formation) in their progenitors and their higher gas fractions. The latest generation of mm interferometers such as PdB, SMA or CARMA will address this issue by mapping the internal structure of these systems in CO at arcsecond resolution.

Another striking result from the high resolution study, is the disagreement between the location of the starburst, identified at long wavelengths, and the restframe UV morphology of SMM J14011+0252: the starburst peaks outside the optical extent of the galaxy – close to the position of an extremely red feature. This illustrates the radical differences in the spatial distribution of highly-obscured star formation and the lightly reddened regions visible in the restframe UV. It is not clear whether the dynamics or star formation histories of the regions visible in the UV can be simply related to the properties of the underlying galaxy. This stresses the importance of obtaining full 2-d dynamical information (i.e. requiring IFUs) and employing dynamical tracers such as $H\alpha$ emission, or better-yet CO mapping, which minimise the effects of dust.

3 Discussion and Conclusions

We conclude that deep submm/mm surveys have uncovered a population of highly obscured, ULIRG-like galaxies lying at high redshifts. This population spans a redshift range which overlaps the Lyman-break sources, has a somewhat lower surface density, but are individually more strongly star forming, $\gtrsim 10^3 L_\odot \, yr^{-1}$. As a whole the submm-selected galaxies probably dominate the massive star formation at $z \gtrsim 2$.

The mass constraints available for a very small number of these galaxies indicate that their masses exceed $> 10^{11} M_\odot$ within the central 20 kpc, implying that they are more massive than typical Lyman-break sources. Moreover, a large fraction of this mass is in cold gas – providing a reservoir sufficient to fuel their very high SFR for 100 Myrs and in doing so increase their already considerable stellar luminosities, as well as chemically-enriching their environments. The present-day descendents of these systems will be amongst the most luminous, and presumably massive, galaxies in the local universe: giant ellipticals.

Acknowledgements

I acknowledge support through a University Research Fellowship from the Royal Society and a Philip Leverhulme Prize Fellowship from the Leverhulme Trust.

References

1. Bautz, M. W., et al., *ApJL*, **543**, L119, (2000)
2. Blain, A. W., et al., *ApJL*, **512**, L87, (1999)
3. Blain, A. W., et al., *MNRAS*, **309**, 715 (1999)
4. Blain, A. W., Smail, I., Ivison, R. J., Kneib, J.-P., Frayer, D. T., *Phys. Rep.*, submitted, (2002)
5. Carilli, C. L., Yun, M. S., *ApJ*, **530**, 618, (2000)
6. Eales, S. A., et al., *ApJ*, **515**, 518 (1999)
7. Fixsen, D. J., et al., *ApJ*, **508**, 123 (1998)
8. Frayer, D. T., et al., *ApJL*, **506**, L7, (1998)
9. Frayer, D. T., et al., *ApJL*, **514**, L13, (1999)
10. Hughes, D. H., et al., *Nature*, **394**, 241 (1998)
11. Ivison, R. J., et al., *MNRAS*, **298**, 583, (1998)
12. Ivison, R. J., et al., *MNRAS*, **315**, 209, (2000)
13. Ivison, R. J., et al., *ApJ*, **561**, L45, (2001)
14. Scott, S., et al., *MNRAS*, submitted, (2001)
15. Smail, I., Ivison, R. J., Blain, A. W., *ApJ*, **490**, L5, (1997)
16. Smail, I., et al., *MNRAS*, **308**, 1061, (1999)
17. Smail, I., et al., *ApJ*, **528**, 612, (2000)
18. Smail, I., Ivison, R. J., Blain, A. W., Kneib, J.-P., *MNRAS*, submitted, (2002)

The Assembly of the First Galaxies

Zoltán Haiman[1,2]

[1] Princeton University Observatory, Peyton Hall, 08544 NJ, USA
[2] Hubble Fellow

Abstract. The first galaxies formed at high redshifts, and were likely substantially less massive than typical galaxies in the local universe. We argue that (1) the reionization of a clumpy intergalactic medium (IGM) by redshift $z \approx 6$, (2) its enrichment by metals by $z \approx 3$ without disturbing the Lyα forest, and (3) the presence of supermassive black holes powering the recently discovered bright quasars at $z \sim 6$, strongly suggest that a population of low–mass galaxies exists beyond redshifts $z \gtrsim 6$. Although the first stars could have been born in dark matter halos with virial temperatures as low as $T_{\mathrm{vir}} \approx 200\mathrm{K}$, collapsing as early as $z \sim 25$, the first galaxies likely appeared in significant numbers only in halos with $T_{\mathrm{vir}} > 10^4\mathrm{K}$ that collapsed later ($z \sim 15$). The gas in these more massive halos initially contracts isothermally to high densities by atomic Lyα cooling. H_2 molecules can then form efficiently via non–equilibrium gas–phase chemistry, allowing the gas to cool further to $T \sim 100\mathrm{K}$, and fragment on stellar mass scales. These halos can harbor the first generation of "mini-galaxies" that reionized the universe. The continuum and line emission from these sources, as well as their Lyα cooling radiation, can be detected in the future by *NGST* and other instruments.

1 Introduction

Recent measurements of the cosmic microwave background (CMB) temperature anisotropies, determinations of the luminosity distance to distant type Ia Supernovae (SNe), and other observations have led to the emergence of a robust "best–fit" cosmological model with energy densities in cold dark matter (CDM) and "dark energy" of $(\Omega_{\mathrm{M}}, \Omega_{\Lambda}) \approx (0.3, 0.7)$. The growth of density fluctuations, and their evolution into non–linear dark matter structures can be followed in detail from first principles by semi–analytic methods [1,2] and N–body simulations [3]. Structure formation is "bottom–up", with low–mass halos condensing first. Halos with the masses of globular clusters, $10^{5-6}M_{\odot}$, are predicted to have condensed as early as ~1% of the current age of the universe, or redshift $z \sim 25$. It is natural to identify these condensations as the sites where the first astrophysical objects, such as stars, or quasars, were born.

2 Current Evidence for High Redshift Galaxies

Current observations directly probe the universe out to redshift $z \sim 6$, with the record–holder quasar at $z = 6.28$ [4], and the history of star–formation and of quasar activity mapped out to $z \sim 5$ [5]. However, there is convincing

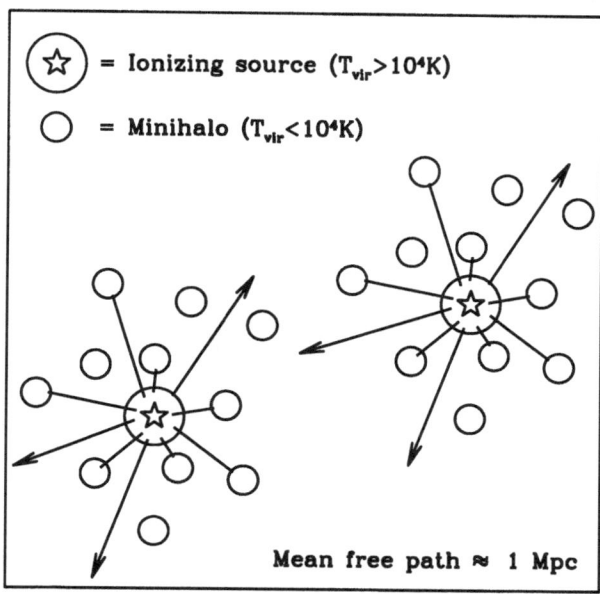

Fig. 1. Dense minihalos with virial temperatures $10^3\,\mathrm{K} \lesssim T_{\mathrm{vir}} \lesssim 10^4\,\mathrm{K}$ are an important sink of ionizing radiation as they surround UV sources at high redshifts ($z \gtrsim 6$). As a population, such minihalos raise by about an order of magnitude the budget of ionizing photons necessary to fully reionize the universe by redshift $z \approx 6$ [6]

observational evidence that an additional, yet undiscovered population of low-mass galaxies exists at these redshifts and beyond.

2.1 Reionization of a Clumpy Universe

The lack of a Gunn–Peterson trough in the spectra of all quasars to date, except for the record holder quasar at $z = 6.28$, imply that the universe is highly ionized prior to redshift $z \sim 6$ (perhaps near $z \sim 6.3$). An early population of stars or quasars could have photoionized the IGM, if they produced at least one ionizing photon per hydrogen atom in the universe. A naive extrapolation of the luminosity density of bright quasars towards $z = 6$ reveals that these sources fall short of this requirement. Extrapolating the known population of Lyman Break Galaxies (LBGs) towards $z = 6$ comes closer: assuming that 15% of the ionizing radiation from LBGs escapes into the IGM (on average, relative to the escape fraction at 1500Å), a naive extrapolation shows that LBGs emitted $\sim$ one ionizing photon per hydrogen atom prior to $z = 6$ [6]. Although this would be sufficient to ionize every H atom once, the required photon budget exceeds this value. The earliest ionizing sources are likely surrounded by numerous "minihalos" that had collapsed earlier, but had failed to cool and form any stars or quasars[1]. This is illustrated in Figure 1. The minihalos have typical masses below $10^7 M_\odot$, and represent a population of dense clumps that is currently unresolved in large three–dimensional cosmological simulations. The UV radiation incident on the minihalos heats their gas to $T \approx 10^4$K, causing it to photo–evaporate [7,8]. The mean free path of ionizing photons, before they are absorbed by an evaporating minihalo, is about ~ 1 (comoving) Mpc. As a result, the typical fate of an ionizing photon, emitted at $z \gtrsim 6$, is to be absorbed by a minihalo within a small fraction of the Hubble distance – before it could contribute to the reionization of the bulk of the IGM. A simple model of the photoevaporation process [6], summed over the expected population of minihalos, reveals that on average, an H atom in the universe recombines $\gtrsim 10$ times before redshift $z = 6$. The implication is that the ionizing emissivity at $z > 6$ was ~ 10 times higher than provided by a straightforward extrapolation back in time of known quasar and galaxy populations.

2.2   Metal Enrichment of the Intergalactic Medium

Recent detections of CIV and SIV absorption associated with low column density Lyα absorption lines in the spectra of distant quasars imply that the universe was enriched by metals to a mean level of $\sim 10^{-3} Z_\odot$ prior to redshift $z \sim 3$. These absorbers, whose HI column densities are as low as $10^{14.2}$ cm^{-2} [9], are identified in cosmological hydrodynamical simulations as regions with typical densities only a few times above that of the mean IGM. This low density rules out "in–situ" metal enrichment, and raises the question: where did these metals come from? LBGs discovered at redshifts $z = 3 - 4$ would be natural candidates for producing and dispersing heavy elements in galactic winds [10]. In order for a significant fraction of the volume to be enriched within a Hubble time, the winds need to move heavy elements at mean speeds close to 1000 km/s (the typical separation of LBGs is a few Mpc). While outflows at such high speeds are known to occur, it is unclear whether this scenario can be reconciled with the observed line–widths of the absorbers, which are as narrow as ~ 20 km/s, when the cooling time in the low–density IGM is exceedingly long at $z \sim 3$. An alternative possibility is early enrichment by much more numerous low–mass systems, with typical separations of < 0.1Mpc, which can drive outflows at their escape velocities of a few$\times 10$ km/s (see also [11]). In Figure 2, we show the maximum masses of halos (using the halo mass function from [3]) whose mean separation is small enough so that their distance can be crossed in the age of the universe (at each redshift) at speeds of 10, 100, or 1000 km/s. In all three cases, the shaded regions correspond to requiring that the dispersed metals fill 3-100% of the volume. The figure reveals that at $z \approx 3 - 4$, systems with $M \approx 10^{12} M_\odot$ are sparse, and require speeds of ~ 1000km/s. Only systems with masses as low as $M \approx 10^{6-7} M_\odot$ (lowest shaded curve) are sufficiently abundant to widely disperse metals with outflow speeds of a few $\times 10$ km/s. However, such

---

[1] If minihalos are themselves the sources of ionizing radiation, and their gas is photo–evaporated "inside–out", then the required photon budget is likely to be even higher.

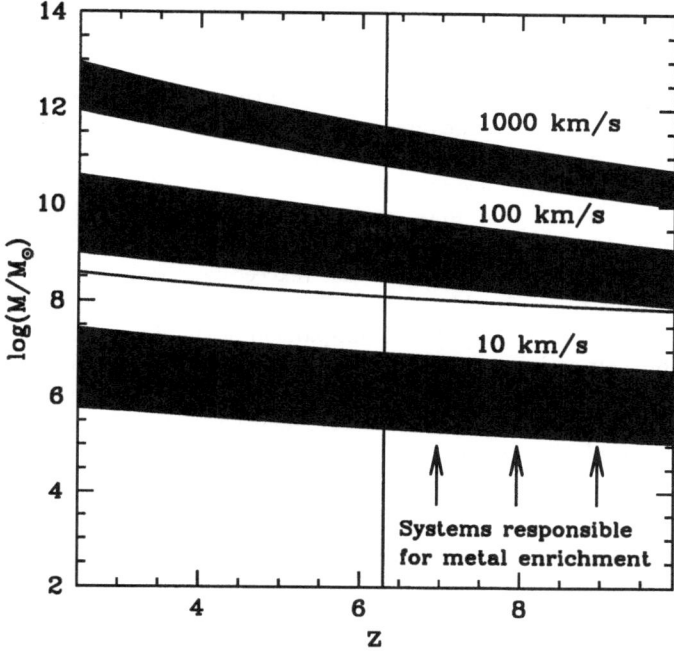

Fig. 2. Typical masses of halos responsible for dispersing metals into the IGM if metals travel at mean speeds of 10, 100, or 1000 km/s. The upper (lower) envelope for each shaded curve corresponds to the requirement that 3 (100)% of the volume is enriched within the age of the universe at redshift z. The solid horizontal curve shows the halo that has a velocity dispersion of $\sigma = 10$km/s; the vertical line shows the minimum reionization redshift at $z = 6.3$. Widely dispersing the metals at low ($\sim$ 10km/s) speeds requires enrichment by low–mass systems, $M \approx 10^6 M_\odot$, which produce metals only prior to reionization

low–mass systems have velocity dispersions below $\sigma = 10$km/s, and are unlikely to have formed metals in the presence of the post–reionization UV background (the horizontal solid curve shows the halo mass corresponding to $\sigma = 10$km/s). The natural conclusion is that the metals had to be produced and dispersed by low–mass systems, $M \approx 10^6 M_\odot$, prior to reionization at $z \gtrsim 6.3$.

2.3   The Growth of Supermassive Black Holes

A third line of evidence for high–redshift activity comes from the sheer size of supermassive black holes (BHs) required to power the recently discovered bright quasars near $z \approx 6$. Assuming that these quasars are shining at their

Eddington limit, and are not beamed or lensed[2], their BH masses are inferred to be $M_{bh} \sim 4 \times 10^9 M_\odot$. The Eddington–limited growth of these supermassive BHs by gas accretion onto stellar–mass seed holes, with a radiative efficiency of $\epsilon \equiv L/\dot{m}c^2 \approx 10\%$, requires ~ 20 e–foldings on a timescale of $t_E \sim 4 \times 10^7 (\epsilon/0.1)$ years. While the age of the universe leaves just enough time ($\lesssim 10^9$ years) to accomplish this growth by redshift $z = 6$, it does mean that accretion has to start early, and the seeds for the accretion have to be present at ultra–high redshifts: $z \gtrsim 15(20)$ for an initial seed mass of $100(10) M_\odot$. Furthermore, the radiative efficiency cannot be much higher than $\epsilon \approx 10\%$ [13,14]. Since an individual quasar BH could have accreted exceptionally fast (exceeding the Eddington limit), it will be important to apply this argument to a larger sample of high–redshift quasars. Nevertheless, we note that a comparison of the light output of quasars at the peak of their activity ($z \sim 2.5$) and the total masses of their remnant BHs at $z = 0$ shows that during the growth of most of the BH mass the radiative efficiency cannot be much smaller than 10%, and hence any "super-Eddington" phase must be typically restricted to building only a small fraction of the final BH mass.

3   The Assembly of the First Galaxies

While the formation of non–linear dark matter halos can be followed from first-principles, the formation of stars or BHs in these halos is much more difficult to model ab–initio. Nevertheless, we may identify three important mass–scales, which collapse at successively smaller redshifts,

1. Gas can only condense in dark halos above the cosmological Jeans mass, $M_J \approx 10^4 M_\odot [(1+z)/11]^{3/2}$, so that the gravity of dark matter can overwhelm gas pressure in the IGM.
2. Gas that condensed into Jeans–unstable halos can cool and contract further in halos with masses above $M_{H2} \gtrsim 10^5 M_\odot [(1+z)/11]^{-3/2}$ (virial temperatures of $T_{vir} \gtrsim 10^2 K$), provided there is a sufficient abundance of H_2 molecules (at a level of at least $n_{H2}/n_H \sim 10^{-3}$).
3. In halos with masses above $M_H \gtrsim 10^8 M_\odot [(1+z)/11]^{-3/2}$ (virial temperatures of $T_{vir} \gtrsim 10^4 K$), gas can cool and contract via excitation of atomic Lyα, even in the absence of any H_2.

The first stars or BHs likely formed in gas that cooled and condensed via excitations of roto-vibrational levels of H_2 molecules in dark matter condensations with $M \sim 10^5 M_\odot$ at redshift $z \sim 25$ (corresponding to $\sim 3\sigma$ peaks in the primordial density field) [15,16]. Unless the first rare objects were significant sources of X–ray photons with energies $E \gtrsim 1 keV$, their soft UV radiation, permeating the distant universe at photon energies $E < 13.6 eV$, caused a negative feedback, strongly suppressing H_2 cooling and star-formation in clumps

---

[2] Strong lensing or beaming would contradict the large proximity effect around these quasars [12].

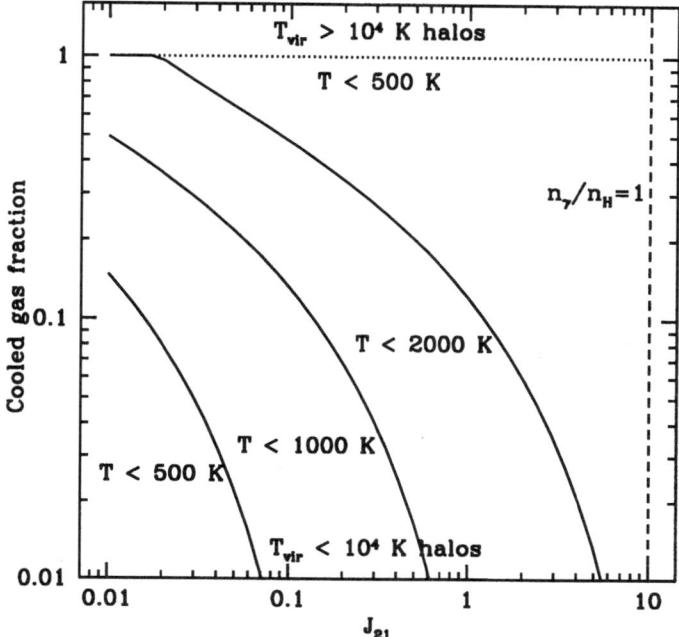

Fig. 3. The fraction of gas in a halo that can cool below a given temperature T, as a function of the external UV radiation field J_{21}, for halos at z=15. In halos with $T_{\rm vir} < 10^4\,{\rm K}$ (lower three solid curves), only a small fraction of the gas is at sufficiently high density to be unaffected by external radiation fields and cool. By contrast, in halos with $T_{\rm vir} > 10^4\,{\rm K}$, where atomic cooling operates, the gas contracts to high densities, and virtually all of the disk gas can form H_2 and cool to low temperatures (dotted horizontal line). The UV intensity corresponding to one ionizing photon per baryon in the universe is marked by the dashed vertical line on the right. Adopted from [19]

that collapsed subsequently [17,18]. Efficient and widespread star (and/or BH) formation, capable of reionizing the universe, had to then await the collapse of halos with $T_{\rm vir} > 10^4{\rm K}$, or $M_{\rm halo} > 10^8[(1+z)/11]^{-3/2}\,{\rm M}_\odot$.

The evolution of halos with $T_{\rm vir} > 10^4{\rm K}$ differs from their less massive counterparts [19]. Efficient atomic line radiation allows rapid cooling to ~ 8000 K; subsequently the gas can contract to high densities nearly isothermally at this temperature. In the absence of H_2 molecules, the gas would likely settle into a locally stable disk and only disks with unusually low spin would be unstable. However, the initial atomic line cooling leaves a large, out–of–equilibrium residual free electron fraction. This allows the molecular fraction to build up to a universal value of $x_{H_2} \approx 10^{-3}$, almost independently of initial density and temperature (this is a non–equilibrium freeze–out value that can be understood in terms of timescale arguments [19]).

Unlike in less massive halos, H_2 formation and cooling is much less susceptible to feedback from external UV fields. This is because the high densities n that can be reached via atomic cooling. The H_2 abundance that can build up in the presence of a UV radiation field J_{21}, and hence the temperature to which the gas will cool, is controlled by the ratio J_{21}/n. For example, in order for a parcel of gas to cool down to a temperature of 500K, this ratio has to be less than $\sim 10^{-3}$ (where J_{21} has units of $10^{-21}\mathrm{erg\,s^{-1}\,cm^{-2}Hz^{-1}\,sr^{-1}}$, and n has units of $\mathrm{cm^{-3}}$). In Figure 3, we show, as a function of the external UV radiation field J_{21}, the mass fraction of gas which is able to cool to a temperature T in $T_{\mathrm{vir}} < 10^4\mathrm{K}$ halos (the gas is assumed to be in hydrostatic equilibrium within an NFW halo [20]), and $T_{\mathrm{vir}} > 10^4\mathrm{K}$ halos (the gas is assumed to have cooled via atomic $Ly\alpha$ to a $10^4\mathrm{K}$, rotationally supported disk [21]).

The figure reveals that flux levels well below that required to fully reionize the universe strongly suppresses the cold gas fraction in $T_{\mathrm{vir}} < 10^4\mathrm{K}$ halos. By comparison, the UV flux has nearly negligible impact on H_2 formation and cooling in $T_{\mathrm{vir}} > 10^4\mathrm{K}$ halos, where all of the gas is able to cool to $T = 500\mathrm{K}$. Indeed, under realistic assumptions, the newly formed molecules in the dense disk can cool the gas to ~ 100 K, and allow the gas to fragment on scales of a few $\times 100$ $M_\odot$. Various feedback effects, such as H_2 photodissociation from internal UV fields, and radiation pressure due to $Ly\alpha$ photon trapping, are then likely to regulate the eventual efficiency of star formation in these systems.

4   Future Observational Signatures

Although the first generation of galaxies are distant, and intrinsically faint objects, they should be within reach of *NGST*. If $\sim 10\%$ of the gas turns into stars in a $10^8 M_\odot$ halo, it should be detectable in the $1 - 5\mu m$ band with *NGST* out to redshifts beyond $z = 10$ in a 10^4 second integration. A $\sim 10^5 M_\odot$ black hole, forming out of $\sim 1\%$ of the gas in such a halo, and shining at its Eddington limit, would be detectable to a similar redshift. Simple "semi–analytical" models [22,23] predict that *NGST* will either detect a significant number of ultra–high redshift sources, or else it will severely constraint any early model of early structure formation. Similarly optimistic conclusions can be drawn about detecting recombinant line emission from high–redshift sources: star formation rates as small as $\sim 1 M_\odot/\mathrm{yr}$, or BHs as small as $\sim 10^5 M_\odot$ translate into detectable $H\alpha$ and He line fluxes beyond $z = 10$, as long as the escape fraction of H– and He–ionizing radiation from these sources is low [24]. Finally, as the baryons cool and contract inside high–redshift halos with virial temperatures $T \gtrsim 10^4\mathrm{K}$, they likely channel a significant fraction of their gravitational binding energy into the $Ly\alpha$ line. At the limiting line flux $\approx 10^{-19}$ erg s^{-1} cm^{-2} asec^{-2} of the *NGST*, several sufficiently massive halos, with velocity dispersions $\sigma \gtrsim 120$ km s^{-1}, would be visible per $4' \times 4'$ field. The halos would have characteristic angular sizes of $\sim 10''$, and could be detectable in a broad–band survey out to $z \approx 6 - 8$ (but not beyond the reionization redshift, where $Ly\alpha$ photons are be resonantly scattered). Their detection would provide a novel and direct probe of galaxies caught in the process of their formation [25,26] – possibly before the first stars or quasars even lit up.

I thank the organizers of this workshop for their kind invitation, and Peng Oh for many recent fruitful discussions, and for permission to draw on our joint work. I acknowledge support from NASA through a Hubble Fellowship.

References

1. W.H. Press, P.L. Schechter: ApJ, **181**, 425 (1974)
2. R.K. Sheth, H.J. Mo, G. Tormen: MNRAS, **323**, 1 (2001)
3. A. Jenkins, et al.: MNRAS, **321**, 372 (2001)
4. R.H. Becker et al.: AJ, in press, preprint astro-ph/0108097 (2001)
5. P. Madau: Physica Scripta, volume T, **85**, 156 (2000)
6. Z. Haiman, T. Abel, P. Madau: ApJ, **551**, 599 (2001)
7. P.R. Shapiro, A.C. Raga, G. Mellema: in *Molecular Hydrogen in the Early Universe*, Memorie Della Societa Astronomica Italiana, Vol. 69, ed. E. Corbelli, D. Galli, and F. Palla (Florence: Soc. Ast. Italiana), p. 463
8. R. Barkana, A. Loeb: ApJ, **523**, 54 (1999)
9. S.L. Ellison, A. Songaila, J. Schaye, M. Pettini: AJ, **120**, 1167 (2000)
10. A. Aguirre, L. Hernquist, J. Schaye, D.H. Weinberg, N. Katz, J. Gardner: ApJ, **560**, 599 (2001)
11. P. Madau, A. Ferrara, M.J. Rees: ApJ, **555**, 92 (2001)
12. Z. Haiman, R. Cen: in preparation, to be submitted to ApJ
13. Z. Haiman, A. Loeb: ApJ, **552**, 459 (2001)
14. R. Barkana, Z. Haiman, J.P. Ostriker: ApJ, **558**, 482 (2001)
15. T. Abel, G.L. Bryan, M.L. Norman: ApJ, **540**, 39 (2000)
16. V. Bromm, P. Coppi, R.B. Larson: ApJ, **527**, 5 (1999)
17. Z. Haiman, T. Abel, M.J. Rees: ApJ, **534**, 11 (2000)
18. M.E. Machacek, G.L. Bryan, T. Abel: ApJ, **548**, 509 (2001)
19. S.P. Oh, Z. Haiman: ApJ, submitted, preprint astro-ph/0108071 (2001)
20. J.F. Navarro, C.S. Frenk, S.D.M. White: ApJ, **490**, 493 (1997)
21. H.J. Mo, S. Mao, S.D.M. White: MNRAS, **295**, 319 (1998)
22. Z. Haiman, A. Loeb: ApJ, **503**, 505 (1998)
23. Z. Haiman, D.N. Spergel, E.L. Turner: ApJ, submitted, preprint astro-ph/0110226 (2001)
24. S. P. Oh, Z. Haiman, M. Rees: ApJ, **553**, 73 (2001)
25. Z. Haiman, M. Spaans, E. Quataert: ApJ, **537**, L9 (2000)
26. Z. Haiman, M.J. Rees: ApJ, **556**, 87 (2001)

The Great Observatories Origins Deep Survey

Mark Dickinson[1], Mauro Giavalisco[1], and the GOODS team[2]

[1] Space Telescope Science Institute, Baltimore MD 21218, USA
[2] STScI/ESO/ST-ECF/JPL/SSC/Gemini/U. Fla./Yale/Boston U./U. Ariz./IAP/
    Saclay/UCLA/UC Berkeley/UCSC/U. Hawaii/U. Wisc./PSU/Caltech/GSFC/AUI

Abstract. The Great Observatories Origins Deep Survey (GOODS) is designed to gather the best and deepest multiwavelength data for studying the formation and evolution of galaxies and active galactic nuclei, the distribution of dark and luminous matter at high redshift, the cosmological parameters from distant supernovae, and the extragalactic background light. The program uses the most powerful space- and ground–based telescopes to cover two fields, each $10' \times 16'$, centered on the Hubble Deep Field North and the Chandra Deep Field South, already the sites of extensive observations from X–ray through radio wavelengths. GOODS incorporates 3.6–24μm observations from a SIRTF Legacy Program, four–band ACS imaging from an HST Treasury Program, and extensive new ground–based imaging and spectroscopy. GOODS data products will be made available on a rapid time–scale, enabling community research on a wide variety of topics. Here we describe the project, emphasizing its application for studying the mass assembly history of galaxies.

1 Introduction

This conference, and this volume of contributions, demonstrate the vital interest in understanding the mass assembly history of galaxies. Theory provides guidance about how dark matter halos are built up in a hierarchical process largely controlled by the power spectrum of density fluctuations and the parameters of the cosmological world model. The assembly of the stellar content of galaxies is governed by more complex physics, including gaseous dissipation, the mechanics of star formation itself, and feedback due to the energetic output from stars and AGN on the baryonic material within galaxies.

Deep imaging and spectroscopic surveys now routinely find and study galaxies throughout most of cosmic history, back to redshifts $z = 6$ and earlier. However, observations are only now beginning to provide constraints on galaxy mass assembly, particularly at $z > 1$. For example, the stellar mass assembly history is characterized by the evolving distribution of masses ($\mathcal{M}$) and star formation rates (SFR, or $\dot{\mathcal{M}}$) with time or redshift, $f(\mathcal{M}, \dot{\mathcal{M}}, t)$. Most investigations to date have considered only moments over this distribution, such as luminosity functions (an imperfect surrogate for the mass distribution) or the global star formation rate SFR(z). Moreover, at high redshift, $\mathcal{M}$ and $\dot{\mathcal{M}}$ are at best only imperfectly measured using currently available observables. Starlight traces stellar mass only in an indirect manner: the mass–to–light ratio ($\mathcal{M}/L$) of a mixed stellar population depends on many parameters, including its age, past star formation history, initial mass function (IMF), dust extinction, and metallicity. Locally, the best constraints come from measurements at near–infrared wavelengths

[1,2], where the longer–lived stars which dominate the mass contribute most to the galaxy luminosity. Moreover, the effect of dust extinction is smaller at redder wavelengths. For $\dot{\mathcal{M}}$, no one observable provides a direct and "universal" tracer of star formation in all circumstances. Ultraviolet, mid– and far–infrared, radio, and nebular line emission are all valuable tools for measuring star formation, with different dependences on extinction, IMF, etc., and thorough surveys of high redshift star formation require the use and cross–calibration of multiple indicators.

2   The Great Observatories Origins Deep Survey

The Hubble Deep Fields (HDF–N and HDF–S [3–5]) provided an invaluable resource of public data for studying faint, distant galaxies. Moreover, they served as a catalyst for follow–up observations at many wavelengths using the most powerful telescope facilities in space and on the ground. However, the HDFs have their limitations. First, they are very small fields, 5 arcmin^2 each, probing very small co–moving volumes. Second, the HDF–S followed the HDF–N by several years. This diluted its impact somewhat, and reduced motivation for the vital follow–up studies needed to verify HDF–N results and to test their robustness against line–of–sight variations due to galaxy clustering. Third, the wavelength range $\lambda\lambda 3$–$1000\mu m$ is the "weak link" in HDF coverage. ISO data at 7 and $15\mu m$ [6,7] and SCUBA measurements at $850\mu m$ [8] probe mid– and far–IR emission, but can detect only the most luminous dust–obscured objects at high redshift. HDF studies of galaxies at $z > 1$ are therefore missing important information at mid– and far–infrared wavelengths where most of the bolometric luminosity from star formation is believed to emerge, as well as the redshifted near–infrared rest–frame light ($\lambda\lambda 3$–$10\mu m$) which most nearly traces total stellar mass.

The Great Observatories Origins Deep Survey (GOODS) follows in the footsteps of the HDF projects, and is a campaign to unite the best, deepest data across the electromagnetic spectrum to create a community resource for exploring the distant universe. GOODS data will be used to study the formation and evolution of galaxies, the radiative output from active galactic nuclei and star formation at high redshift, the characteristics of the extragalactic background light, large scale structure and the distribution of dark matter, the values of the cosmological parameters, and many other projects outside the scope of its core design.

GOODS builds upon existing or ongoing surveys from space– and ground–based facilities, including NASA's Great Observatories, HST, Chandra and SIRTF. The program targets two fields, each $10' \times 16'$, around the Hubble Deep Field North (HDF–N) and the Chandra Deep Field South (CDF–S). These are the most data–rich and well–studied deep survey areas on the sky, with extensive near–infrared and optical imaging and spectroscopy, highly sensitive radio and sub–mm measurements, and the deepest X–ray observations from Chandra [9,10] and XMM–Newton (in progress; PIs: Bergeron (CDF–S); Jansen and Griffiths (HDF–N)). Two fields, one in each celestial hemisphere, provide insurance

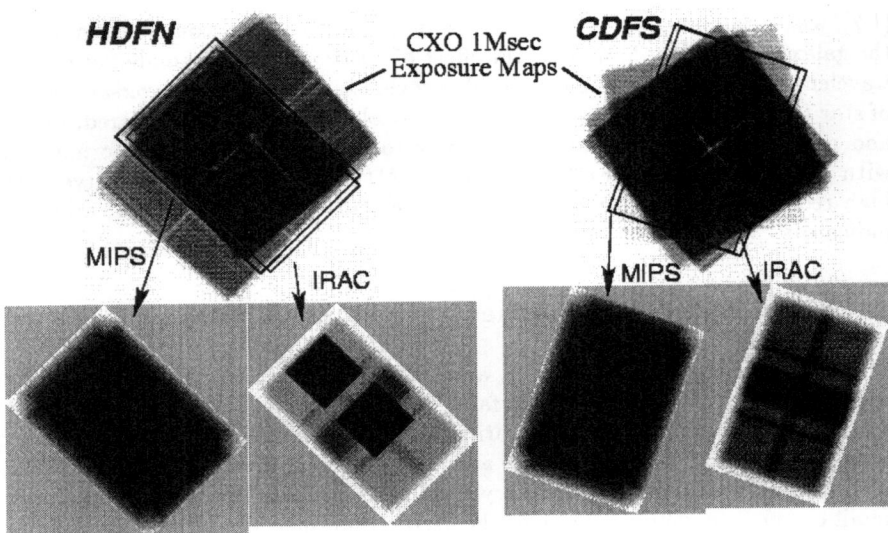

Fig. 1. Layout of the GOODS/SIRTF fields. The upper panels show the exposure maps for the 1 Msec Chandra X-ray observations of the HDF–N and CDF–S. Boxes show the approximate boundaries of the SIRTF IRAC and MIPS survey areas (roughly 10′ × 16′). Insets below show the nominal SIRTF exposure time maps. The HDF–N includes "ultradeep" IRAC observations, which are contingent upon on–orbit tests to establish the practical sensitivity limit of the instrument

against variance due to line–of–sight clustering effects, and enable follow–up programs by astronomers and observatories worldwide.

2.1   The SIRTF Legacy Program

The GOODS SIRTF Legacy Program (PI: Dickinson) will make the deepest observations with that facility at 3.6 to 24μm. Observations will be carried out in the first year of SIRTF operations, in 2003–2004. The bulk of the GOODS SIRTF program will use the Infrared Array Camera (IRAC), observing at 3.6, 4.5, 5.8 and 8.0μm with exposure times of 23.6 hours per band. A small overlap strip in each field will receive twice this integration time, and in the HDF–N only, a pair of 5′ × 5′ "ultradeep" IRAC fields are planned with exposure times of 70 hours (reaching 94 hours in the maximum overlap region atop the WFPC2 HDF–N).

These long exposure times are essential in order to reach sensitivities $\lesssim 1\mu$Jy with reasonably high S/N ratios. For example, the IRAC 8μm observations will sample the rest–frame K–band light from Lyman break galaxies (LBGs) at $z \approx 3$, and will thus provide an important handle on their total stellar content. The expected 8.0μm flux for an "L^*" LBG (with $\mathcal{M}_* \approx 10^{10}\mathcal{M}_\odot$ [11,12]) is 1.5μJy. At

3.6 and 4.5μm, the flux sensitivity achieved by the GOODS/IRAC observations will depend strongly on the achieved image quality (expected to be in the range 1.5–2.3 arcsec FWHM), since source confusion will be important – less so at 5.8 and 8.0μm, where the zodiacal background should set the flux limits. The ultradeep IRAC fields will probe farther down the luminosity and mass function at $z \sim 3$, and should detect typical objects at $z \approx 5$.

The GOODS fields will also be observed at longer wavelengths with the SIRTF/MIPS instrument. Deep ISOCAM 15μm imaging surveys were sensitive to redshifted 7.7μm PAH emission from star–forming galaxies at $z \approx 1$, and the GOODS 24μm observations are designed to detect objects with similar rest–frame luminosities at $z = 2$ to 2.5. In principle, they should be able to detect the mid–infrared emission from obscured star formation in typical Lyman break galaxies at these redshifts. The actual sensitivity achieved at 24μm will depend on the (presently uncertain) level of source confusion and on instrument performance. The GOODS program plans 10.4 hour exposures at 24μm, contingent upon on–orbit demonstration that they will reach substantially fainter flux limits than planned 20 min exposures from the MIPS GTO wide–field survey (PI: Rieke) which covers the GOODS fields. Test observations made early in the mission will be used to determine the longest exposure times practical for making confusion–limited observations of the GOODS fields. The SIRTF GTO program will also cover these fields at 70μm and 160μm, sensitive to the far–infrared thermal emission from high redshift, dust–obscured star formation.

2.2   The HST/ACS Treasury Program

The GOODS HST Treasury Program (PI: Giavalisco) will use the Advanced Camera for Surveys (ACS) to image the fields with four broad, non–overlapping filters, F435W (B), F606W (V), F775W (i), and F850LP (z), with exposure times of 3, 2.5, 2.5 and 5 orbits, respectively, reaching extended–source sensitivities within 0.5-0.8 mags of the WFPC2 HDF observations. The observations will be carried out during HST Cycle 11, in 2002–2003. GOODS is a deep survey, not a wide one, but is nevertheless much larger than most previous HST/WFPC2 programs. The GOODS fields cover 32× the solid angle of the combined HDF-N and S, and are 4× larger than the combined HDF Flanking Fields, and 2.5× larger than the WFPC2 Groth Strip Survey. The Viz observations will be taken in five repeat visits separated by approximately 45 days, enabling a search for SNe Ia at $1.2 < z < 1.8$ to test the apparent transition from cosmic deceleration to acceleration that is predicted in world models dominated by a cosmological constant (and suggested by current data [13]).

The z–band observations will image the optical rest–frame light from galaxies out to $z = 1.2$, with angular resolution superior to that from WFPC2. The $BViz$ imaging will also enable a systematic survey of Lyman break galaxies at $4 < z < 6.5$, reaching back to the suggested epoch of reionization [14,15]. The photometric depth and co–moving volume coverage will make it possible to quantify the LBG population in this redshift range with statistical accuracy comparable to that now available from large, ground–based LBG surveys at $z \approx 3$ [16]. The ACS

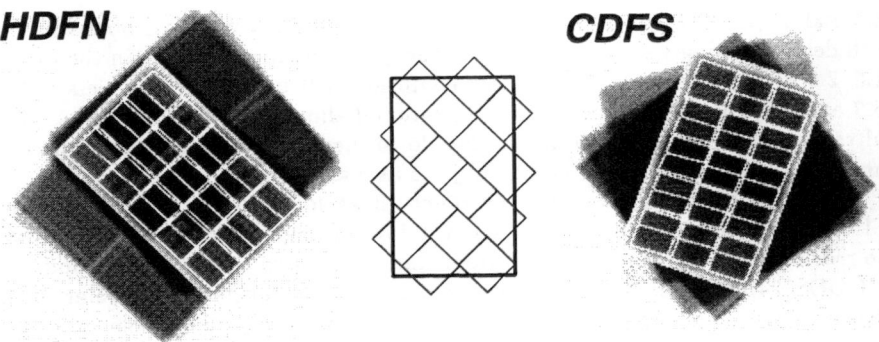

Fig. 2. Layout of the GOODS/HST observations. The grid of white boxes shows the tiling of HST/ACS fields at one telescope orientation, superimposed on the Chandra (outer greyscale) and SIRTF IRAC (inner greyscale) exposure maps. The fields will be revisited approximately every 45 days to enable a search for high redshift supernovae. The center inset schematically shows how the rotated ACS pointings from alternate visits will be tiled over the GOODS area

data will also provide a powerful tool for studies of gravitational lensing, low–mass stars in our galaxy, and perhaps objects in the outer solar system.

2.3   Ground-Based Observations

As noted above, the HDF–N and CDF–S are already among the most data–rich deep survey regions on the sky, and the GOODS program includes a large component of ground–based supporting observations to enable research on distant objects. In large, coordinated NOAO and ESO programs (PIs: Cesarsky, Dickinson), we are obtaining new optical and near–infrared imaging, including U–band imaging from the KPNO and CTIO 4m MOSAIC cameras, and JHK_s imaging with the KPNO 4m/FLAMINGOS and VLT/ISAAC instruments. We are also planning a spectroscopic campaign in the CDF–S using the new VIMOS and red–upgraded FORS–2 spectrographs on the VLT (see the contribution by Renzini et al. to these proceedings). This program will provide a public data resource of several thousand spectra and redshifts for galaxies in the southern GOODS field. We will also supplement the already–extensive HDF–N redshift data [17] and extend it over the whole GOODS/HDF–N using spectroscopy with Gemini–N/GMOS and Keck/DEIMOS+LRIS. We are collaborating on new ATCA radio observations of the GOODS/CDF–S (PI: Koekemoer), and JCMT/SCUBA observations of the GOODS/HDF–N (PIs: Barger, Scott). Table 1 provides a summary of observations being taken as part of the GOODS project, along with some other key data sets for the GOODS fields at other wavelengths.

Table 1. GOODS observations and complementary data sets. The top portion of the table lists GOODS space– and ground–based imaging observations at 0.36–24μm and their nominal sensitivities. The bottom portion is an incomplete list of additional observations available, in progress, or in preparation, which cover the GOODS areas.

Wavelength	Facility	Sensitivity (S/N=5)
0.36μm	KPNO+CTIO 4m	AB = 27.3 (U)
0.4-0.9μm	HST/ACS	AB = 27.9, 28.2, 27.5, 27.4[a] ($BViz$)
1.2-2.2μm	VLT, KPNO 4m	AB = 25.2, 24.7, 24.4 (JHK_s)
3.6-8.0μm	SIRTF/IRAC	AB = 24.5, 24.5, 23.8, 23.7 (0.6–1.2 μJy)[b]
24μm	SIRTF/MIPS	20-80 μJy[c]

Type	Facility	Notes
Spectroscopy	VLT, Gemini, Keck	Various PIs; GOODS programs & collabs.
X-ray	Chandra, XMM	Public Chandra data and XMM GTO progs.
70, 160μm	SIRTF/MIPS	SIRTF GTO program
Sub-mm	SCUBA, SEST	Various PI programs
Radio	VLA, ATCA	Various PIs; CDF–S observs. in progress

[a] For 0.5 arcsec diameter aperture
[b] IRAC deep survey, for "handbook" PSF; 3.6 and 4.5μm performance may be better
[c] Uncertain sensitivity; depends on instrument performance and source confusion

2.4 Data Products

In the spirit of the HDF projects, the GOODS team will make data products available to the community on a rapid time–scale. The raw SIRTF and HST data will be available upon ingestion into the SSC and STScI archives. The reduced data products from both facilities will be provided in a series of incremental releases: "best effort" (version 0.5) reduced images two to three months after each observing epoch; improved (version 1) image mosaics three months after the final observations; and reprocessed (version 2) data products and multiwavelength catalogs six to twelve months after the final observations. Similar release schedules apply to the ancillary data from ESO and NOAO, and we will generally follow similar procedures for GOODS–related data from other facilities.

3 Science Enabled by GOODS

A primary goal of the GOODS program is to provide observational data for tracing the mass assembly history of galaxies throughout most of cosmic history. First, this requires redshift information to sort galaxies by distance and cosmic time. Much of this will come from the existing and planned spectroscopic surveys of these fields. At fainter fluxes, the 13–band GOODS imaging data, covering 4.5 wavelength octaves from 0.36–8μm, will provide an exceptional resource for estimating photometric redshifts for galaxies of all types, calibrated by the extensive spectroscopy.

The SIRTF IRAC data is designed to measure the rest–frame K–band starlight from "ordinary" galaxies (e.g., the progenitor fragments of the Milky Way) at $z \approx 3$, and can detect rest–frame near–infrared light ($\lambda > 1\mu$m) from objects out to $z = 7$. Photometry will trace the spectral energy distributions of galaxies from UV through IR rest–frame wavelengths, and thus constrain their stellar populations and $\mathcal{M}/L$, providing the best estimates (modulo assumptions about the IMF) of their total stellar masses. GOODS data, as well as other observing programs covering these fields, will offer a wide array of star formation indicators, including rest–frame UV and mid–infrared photometry, far–infrared measurements from the SIRTF GTO MIPS program, very deep radio and sub–mm surveys, and nebular line spectroscopy from the redshift surveys and targeted follow–up programs. These different indicators can be applied and cross–calibrated for the *same* high redshift galaxies, guided by detailed knowledge from galaxy surveys in the local universe, such as the SINGS SIRTF Legacy Program (PI: Kennicutt).

The HST/ACS program will provide high resolution imaging needed to relate the morphological properties of galaxies (size, surface brightness, Hubble type, etc.) to their stellar populations, masses, star formation rates, and AGN activity, tracing the emergence of the Hubble sequence and its relation to the physical characteristics of galaxy assembly. Future programs of high–dispersion spectroscopy can be used to measure galaxy kinematics which trace gravitational mass, connecting the stellar population properties traced by SIRTF and the dark matter potential wells. Weak and strong lensing measurements, as well as galaxy clustering, will also provide statistical constraints on dark halo mass on larger physical scales.

GOODS data will also provide an important resource for studying the evolution of active galactic nuclei, in particular to identify both obscured and unobscured AGN with "typical" luminosities (i.e., not just the most powerful QSOs and radio galaxies) out to high redshifts, back to the "QSO era" at $z > 2$. Deep X–ray data will sort AGN from starbursts as the engines powering mid– and far–IR emission in distant objects, enabling a census of energetic output from these mechanisms over a broad range of redshift. The SIRTF data will also fill an important gap in our measurements of the discrete source component of the extragalactic background light (EBL), the integral record of emission and absorption of radiation throughout cosmic history. IRAC data at 3.6–8μm will trace the "downside" of the peak of the direct stellar contribution to the EBL, while 24μm measurements (and GTO SIRTF data at 70 and 160μm) will follow the "upside" of the far–infrared peak due to dust–absorbed starlight and AGN emission.

4   Conclusion

The installation of ACS on board HST, the launch of SIRTF, and the implementation of a new generation of massively multiplexed spectrographs on large
ground–based telescopes will open rich new opportunities for observing galaxy
formation and evolution, out to very high redshift and early cosmic epochs. The
GOODS project is designed to bring all of these tools to bear on common deep
survey fields, uniting the best observations at all accessible wavelengths. These
data will be gathered into a coherent archive for public release, enabling community research on a wide variety of topics. We hope that at future meetings
such as this one, GOODS data will have helped to advance our understanding
of galaxy masses at high redshift, and of the history of galaxy assembly.

More information about the project and observing program can be found on
the GOODS web sites at STScI (http://www.stsci.edu/science/goods) and
ESO (http://www.eso.org/science/goods).

The authors wish to thank the organizers for hosting a wonderful meeting
in a magical location (a former insane asylum, perhaps not coincidentally), and
for generous travel support. Additional support for the GOODS SIRTF Legacy
Program is provided by JPL contract 1224666.

References

1. G. Gavazzi, D. Pierini, & A. Boselli: A&A **312**, 397 (1996)
2. S. Cole et al.: MNRAS **326**, 255 (2001)
3. R. E. Williams et al.: AJ **112**, 1335 (1996)
4. R. E. Williams et al.: AJ **120**, 2735 (2000)
5. H. C. Ferguson, M. Dickinson, & R. E. Williams: ARAA **38**, 667 (2000)
6. S. B. G. Serjeant et al.: MNRAS **289**, 457 (1997)
7. S. Oliver et al.: MNRAS in press (astro-ph/0201506) (2002)
8. D. H. Hughes et al.: Nature **394**, 241 (1998)
9. R. Giacconi et al.: ApJS in press (astro-ph/0112184) (2002)
10. W. N. Brandt et al.: AJ **122**, 2810 (2001)
11. C. Papovich, M. Dickinson, & H. C. Ferguson: ApJ **559**, 620 (2001)
12. A. E. Shapley et al.: ApJ **562**, 95 (2001)
13. A. G. Riess et al.: ApJ **560**, 49 (2001)
14. R. H. Becker et al.: AJ **122**, 2850 (2001)
15. S. G. Djorgovski, S. Castro, D. Stern, & A. A. Mahabal: ApJ **560**, L5 (2001)
16. C. C. Steidel et al.: ApJ **519**, 1 (1999)
17. J. G. Cohen et al.: ApJ **538**, 29 (2000)

ESO for GOODS' Sake

A. Renzini[1], C. Cesarsky[1], S. Cristiani[2], L. da Costa[1], R. Fosbury[2], R. Hook[2], B. Leibundgut[1], P. Rosati[1], and B. Vandame[1]

[1] ESO, 85748 Garching b. München, Germany
[2] ST-ECF, 85748 Garching b. München, Germany

Abstract. Currently public ESO data sets pertinent to the CDFS/GOODS field are briefly illustrated along with an indication on how to get access to them. Future ESO plans for complementing the GOODS database with optical/IR imaging and optical spectroscopy are also described.

1 Introduction

Recognizing the unique and long-lasting scientific value of the "SIRTF Legacy Programs", the Director General of ESO issued an "Open Letter to the ESO Community" to ensure "adequate coverage from the observatories at La Silla and Paranal for all the approved Legacy Programs which have a substantial participation from the ESO community. ESO will ensure that appropriate allocation of time on relevant instruments, in line with the scientific goals of approved SIRTF Legacy programmes, is made in a timely manner. In the spirit of all the Legacy Programs, the resulting data will be immediately made public worldwide." (see http://www.eso.org/observing/misc/20000824.sirtf.html). In this spirit, a team of ESO astronomers joined forces with the North American team led by Mark Dickinson at STScI, and the GOODS SIRTF Legacy Proposal was submitted in 2000, then complemented in 2001 by the HST/ACS Treasury Proposal led by Mauro Giavalisco. Both projects are now being implemented. The accompanying paper by Dickinson & Giavalisco describes the main scientific goals of the GOODS project, along with the planned observations with SIRTF and HST. This paper complements it with a description of the data being provided by ESO, and of the possibilities of spectroscopic follow up with the VLT. Involvement of the ESO community in the planning of the observations as well as in the reduction effort will be actively pursued.

2 Optical and Near-IR Imaging

There are several imaging data sets from observations taken at ESO telescopes that include the CDFS/GOODS field and are already publicly available. Some are part of the public ESO Imaging Survey (EIS), and some were obtained as part of "private" projects, but the data have become available after the one year proprietary period has expired. Other data have been obtained specifically for the GOODS project as a public survey (VLT Large Programme 68.A-0485, PI C. Cesarsky). Table 1 summarizes the situation as of February 2002. Fig. 1

shows the CDFS/GOODS field with superimposed the *brickwall* mosaic of the VLT/ISAAC observations currently under way. The raw and calibration JHK data having the planned integration time are already available for those bricks which identification number has the large format in Fig. 1. JK data for bricks 10, 11, 15 and 16 were secured by ESO programmes 64.O-0643, 66.A-0572 and 68.A-0544, whose PI (E. Giallongo) has kindly agreed to waive the residual proprietary time in such a way to have the reduced data being publicly released at the same time as the reduced GOODS data. For these bricks GOODS also provides the H-band data. The reduction of coadded, flux and astrometrically calibrated data is underway by the EIS Team, with the data release being planned for the end of March, 2002. The first public release of mosaiced images and source catalogs is planned for the end of July, 2002. ISAAC observations to complete the whole brickwall will continue in 2002 and 2003 as part of the GOODS VLT Large Programme. Real time information on the progress of the ESO/GOODS observations can be obtained from the URL http://www.eso.org/science/goods/, with links to the EIS and other databases.

While deep optical imaging was initially also envisaged with the VLT, this has become partly redundant after the approval of the GOODS/ACS Treasury Program, which will provide deep $BViz$ imaging. Exception refers to deep U-band imaging with the VIMOS instrument and/or with a UV-optimized FORS-1 (if and when it becomes available). Depending on the public availability of data, additional VIMOS imaging may or may not be also needed for the selections of spectroscopic targets in the field surrounding the GOODS field (see Section 3.1).

Table 1. ESO imaging data set on CDFS/GOODS (February 2002)

Bands	Tel./Instrument	Area	5-σ AB mag limits	Programme
$UBVRI$	2.2m/WFI	$30' \times 30'$	$U < 26.0,\ B < 26.4,\ V < 25.4;$ $R < 25.5;\ I < 24.7$	EIS/DPS
JK_s	NTT/SOFI	$20' \times 20'$	$J < 23.4;\ K_s < 22.6$	EIS/DPS
RI	VLT/FORS	$15' \times 15'$	$R < 27.0;\ I < 26.0$	64.O-0621
JHK_s	VLT/ISAAC*	$10' \times 16'$	$J < 25.3;\ H < 24.8;\ K_s < 24.4$	GOODS
BVR	2.2m/WFI	$30' \times 30'$	$B < 27.0;\ V < 26.5;\ R < 26.5$	GOODS

* See text.

3 VLT Spectroscopy

The most important ESO contribution to the GOODS project will certainly be the wide spectroscopic coverage of the field. No other Southern observatory has high-multiplex spectroscopic capabilities comparable to those of the VLT. As of February 2002, more than 30 nights of multi-object spectroscopy have already

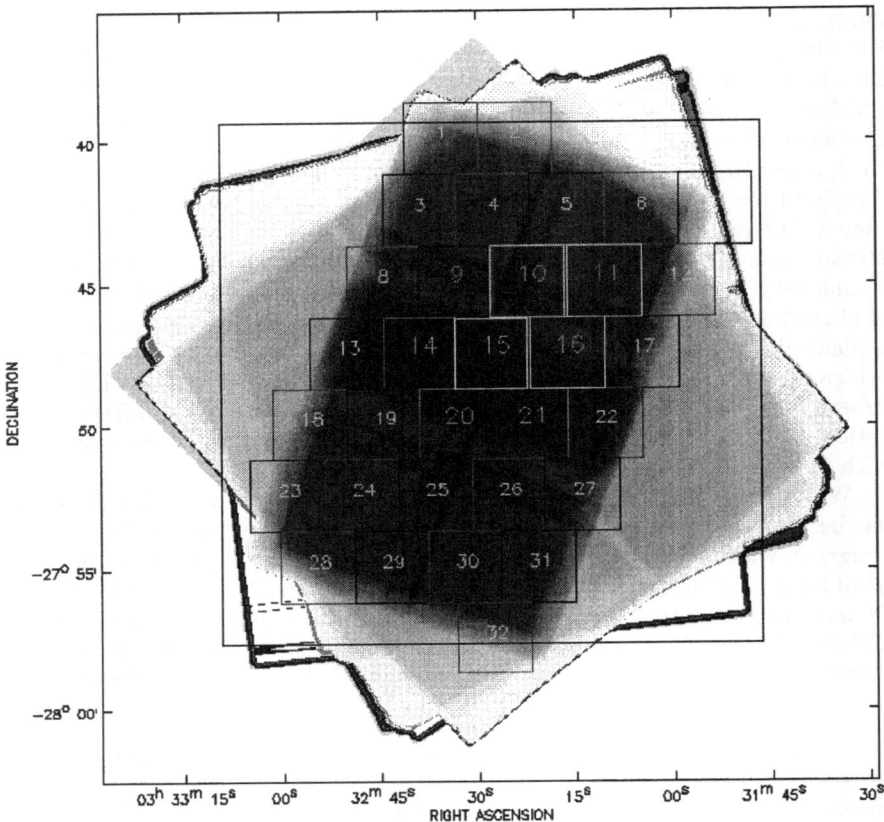

Fig. 1. The ISAAC mosaic coverage of the CDFS/GOODS field (the shaded IRAC exposure map) is shown by the *brickwall* with individual pointings numbered from 1 to 32. Bricks with a large-size identification number have already been completed. The Chandra coverage is shown by the set of shaded squares with various orientations. The centered, unshaded square shows the EIS/DPS Deep-2c field as covered by SOFI

been allocated at VLT/FORS1-2 for several scientific programs (e.g. the $K < 20$ redshift survey; identification of Chandra sources; study of high-z dropouts). To date, several hundred redshifts have being measured in the CDF-S area. This increasing spectroscopic data set amassed with the VLT is already public in the ESO Science Archive or will become public before the SIRTF data are taken.

However, these data either cover only a small fraction of the GOODS field, or only very specific types of targets, such as e.g. X-ray sources or a few Lyman-break galaxies. Table 2 gives the cumulative source number counts in the R and I bands (columns 2 and 3, respectively) as derived from the EIS/DPS data for

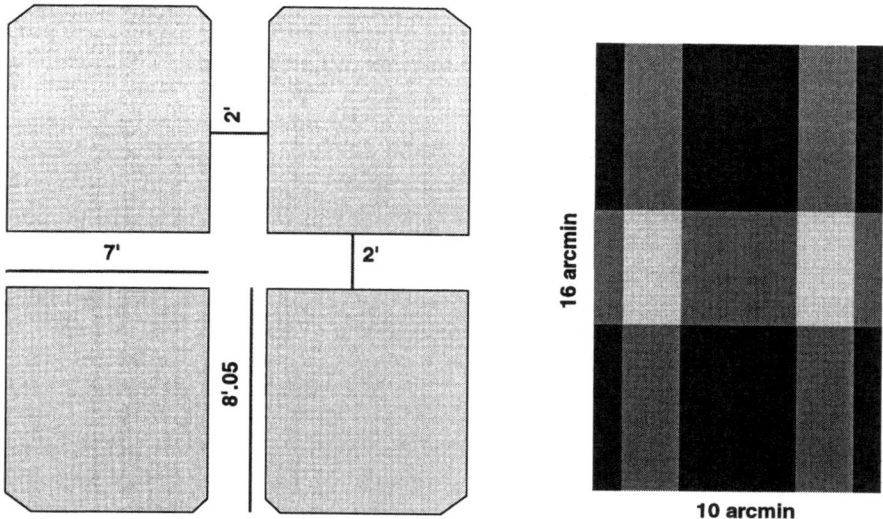

Fig. 2. *Left*: the geometry of the VIMOS field of view, for either imaging or multiobject spectroscopy.   *Right*: the CDFS/GOODS 10′ × 16′ field, in the same scale as the VIMOS FoV (left). The VIMOS imprinting on the GOODS field is rendered with different shadings. Dark areas: areas common to all 4 default VIMOS pointings (A set of subfields). Shaded areas: areas common to 2 default VIMOS pointings (B set of subfields. Lightly shaded areas: areas covered by only one VIMOS pointing (C set of subfields)

Table 2. Cumulative Source Counts in the CDFS Field

M_{AB}	N_R^{EIS}	N_I^{EIS}	N_R^{GOODS}	N_I^{GOODS}
22	3628	5794	600	1000
23	8009	11096	1300	2000
24	17962	20755	3000	3500
25	35949	31148*	6000	5500*
26	52840*		9400*	

* Incomplete counts in these magnitude bins.

the Deep-2c field that includes the CDFS/GOODS field [1]. Scaling by the areas, columns 4 and 5 give the number of sources within the GOODS field.

It appears from Table 2 that no more than ∼ 6000 objects in the GOODS field are bright enough (i.e. down to R_{AB} or $I_{AB} = 25$) for a low-resolution spectrum to be useful, e.g. to provide the redshift. Hence, the multiplex capabilities of the VLT instruments ensure the possibility to observe them all in a quite reasonable amount of telescope time. This is now explored in more quantitative terms.

3.1 VIMOS Spectroscopy

The layout of the VIMOS FoV is shown in Fig. 2 (left panel). A minimum of 4 VIMOS pointings is necessary to cover the whole GOODS field. A default pattern may consist of 4 VIMOS pointings in which in turn each of the 4 outer corners of the VIMOS FoV coincides with each of the 4 GOODS field corners, with the long (dispersion) axis of VIMOS parallel to the long side of GOODS. The missing triangles at the VIMOS corners are ignored.

For each pointing, the central cross gaps of VIMOS separate 4 sub-fields over GOODS, for a total coverage of 112 arcmin². Hence, for each pointing, a fraction 112/160 (70%) of the GOODS field is accessible to spectroscopy. The fraction of the VIMOS multiplex expendable on GOODS is 112/224 (50%,), i.e., with an *effective multiplex* 0.5 × 800 = 400 (in low resolution mode).

Moreover, the 4 default pointings of VIMOS imprint on the GOODS field a set of subfields, with some being common to all 4 pointings, some only to 2 pointings, and 2 subfields are covered by only one pointing. One defines these 3 sets of subfields as set A, B and C, respectively. The area of set A and B is 72 arcmin² each (45% of GOODS each) while set C has an area of 16 arcmin² (10% of GOODS). The default (A,B,C) pattern imprinted on the GOODS field is shown in Fig. 2 (right panel).

The surface density of targets (6000/160=37.5 arcmin^{-2}) is ∼ 10 times larger than the surface density of VIMOS slits (800/224= 3.6 arcmin^{-2}). Hence, if one insists on having completeness in set C one needs at least 10 pointings to cover this small (10%) part of the GOODS field, while automatically set B and A would be observed two times and four times more than strictly needed to ensure completeness, respectively.

In a more time-saving approach, one may ensure completeness in set B, which requires this set to be covered by at least 12 pointings. Automatically, set C will be covered by 6 pointings (60% completeness), and set A by 24 pointings (twice more than strictly needed to ensure completeness). Hence, after having achieved completeness in set A, the 12 additional pointings will allow the observation of a subset of the targets in this field for longer integration times. In the limit, some targets could be observed for up to ∼ 13 times the basic exposure time. In this scheme, 24 pointings are required and assuming 4 hours integration time per pointing this makes a total of 24 × 4=96 hours of integration, plus 20% overhead makes ∼ 120 hours, or ∼ 13 nights. These 13 nights will ensure that ∼ 96% of the targets are observed at least once with 4 hours integration, while 45% of the targets (those in set A) could be observed with an integration of 8 hours each, or a smaller fraction with even longer integrations.

This somewhat laborious exercise demonstrates that the geometries of the GOODS field and the VIMOS FoV combine in such a way that, when ensuring the complete spectroscopic coverage of the field, one has automatically the opportunity to integrate on some of the targets for longer times than others. For this to be achieved some targets will have to be kept in more than one VIMOS multislit mask.

Very simple color criteria can be adopted to assign targets to either observations with the red or the blue low-resolution grism of VIMOS, in such a way to maximize the chance to obtain a high S/N spectrum. Similarly, just a magnitude limit criterion can be adopted for the selection of targets to be observed for multiples of the basic exposure time. Finally, having only 50% of the VIMOS spectroscopic multiplex used on the GOODS field, the other 50% will be available to get spectra of objects in the accessible area around GOODS. This should allow to get spectra of nearly as many objects outside GOODS as inside it, i.e. 6000 objects.

3.2 FORS2 Spectroscopy

While having a smaller FoV and lower multiplex compared to VIMOS, FORS2 offers an attractive complementary capability. Its throughput is about twice that of VIMOS longward of $\sim$ 800 nm, with the additional advantage of providing virtually fringe-free CCD frames. On red objects, FORS2 may go $\sim$ 1 mag or more deeper than VIMOS, or could reach much better S/N, given also the higher resolution ($\times 3$) that should help resolve the OH lines. For multiobject spectroscopy, the effective FoV of FORS2 is $\sim$ 6.8 $\times$ 3.2 $\simeq$ 22 arcmin^2 and 9 pointing are sufficient to cover the whole GOODS field, with minimal overlap between pointings. With a multiplex $\sim$ 40, such 9 pointings enables $\sim$ 360 objects to be observed, selected according to a simple color criterion. 18 masks exposed for 4 hours each would permit observations of > 500 faint, red objects, with galaxies in faintest magnitude bin observed on two masks for a total of 8 hours. The integration time for the faintest objects would be longer than for typical VIMOS exposures, ensuring that features will be recognizable in spectra with no emission lines. This FORS2 program would require $9 \times 2 \times 4 \times 1.2 = 86$ hours of telescope time including overhead, or $\sim$ 10 nights.

4 Conclusions

ESO is committed to provide its best possible contribution to the scientific effort started with the GOODS SIRTF Legacy and HST/ACS Treasury projects. This will include complementary optical and near-IR imaging, while a proposal will be submitted to the ESO Observational Programme Committee aimed to provide the whole astronomical community with the complete spectroscopic coverage of the CDFS/GOODS field. This proposal will follow the lines sketched in this paper, and should result in a complete, homogeneous database obtained with the minimum possible VLT time. Together with the complementary data from space, the ESO contribution is meant to establish a unique, long-lasting set of tools for the study of galaxy formation and evolution, hence preparing the way for further advances in the next decade with ALMA and NGST.

References

1. Arnouts, S., et al. 2001, A&A, 379, 740

The SWIRE SIRTF Legacy Program: Studying the Evolutionary Mass Function and Clustering of Galaxies

Alberto Franceschini[1], Carol Lonsdale[2], and the SWIRE Co-Investigator Team

[1] Astronomy Department, Padova University, Vicolo Osservatorio 2, I-35122 Padova, Italy
[2] IPAC, Pasadena, USA

Abstract. The SIRTF Wide-area Infrared Extragalactic (SWIRE) survey is a *Legacy Program* using 851 hours of SIRTF observing time to conduct a set of large-area ($\sim$ 67 square degrees split into 7 fields) high Galactic latitude imaging surveys, achieving 5-sigma sensitivities of 0.45/2.75/17.5 mJy at 24/70/160 μm with MIPS and of 7.3/9.7/27.5/32.5 μJy at 3.6/4.5/5.8/8.0 μm with IRAC. These data will yield highly uniform source catalogs and high-resolution calibrated images, providing an unprecedented view of the universe on co-moving scales up to several hundreds Mpc and to substantial cosmological depths ($z \simeq 2.5$ for luminous sources). SWIRE will, for the first time, study evolved stellar systems (from IRAC data) versus active star-forming systems and AGNs (from MIPS data) in the same volume, generating catalogues with of order of 2 million infrared-selected galaxies. These fields will have extensive data at other wavebands, particularly in the optical, near-IR and X-rays. SWIRE will provide a complement to smaller and deeper observations in the SIRTF Guaranteed Time and the Legacy Program GOODS, by allowing the investigation of the effect of environment on galaxy evolution. We expand here on capabilities of SWIRE to study with IRAC the evolution of the bright end of the galaxy mass function as a function of cosmic time.

1 Introduction

Measuring the mass function of distant galaxies and its evolution with cosmic time is an ultimate task when attempting to recover the origin of the presently observed cosmic structure. The study of this integral of the stellar mass content as a function of time makes a complementary approach to the usual determination of the instantaneous rate of star formation, and has the advantage over the latter of not being sensitive at all to the effects of dust extinction, otherwise a quite fundamental limitation. In particular, the evolution of the massive end of the galaxy population, tracing the assembly of present-day luminous galaxies, is a critical cosmogonic observable.

Current large optical/NIR telescopes allow in principle the measurement of dynamical masses through high spatial and spectral resolution spectroscopy of selected samples of high-redshift galaxies (e.g. Koo; Rigopoulou et al., these Proceedings). However, the long time integrations needed, and the difficulty to sample all but the inner galaxy cores due to the cosmological surface brightness dimming, hamper systematic studies of sizeable galaxy samples over large areas.

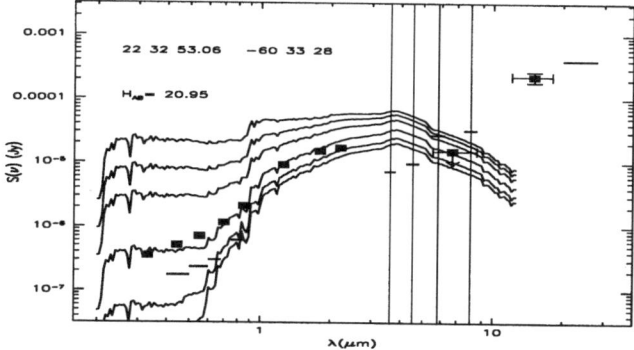

Fig. 1. Synthetic spectra of galaxies as a function of the age of the stellar population. Ages are 1, 2, 3, 5, 7 and 9 Gyr from top spectrum to bottom. The synthetic spectra were computed with the PEGASE code (Fioc & Rocca-Volmerange 1997), and assume a Salpeter IMF, the Padova stellar tracks, an exponentially decreasing star-formation rate with 1 Gyr time-scale, no extinction and no nebular emission. The 5 Gyr spectrum is fit to photometric data on a z=1.27 ISO-detected galaxy in the HDF South. The vertical lines with horizontal ticks indicate the λ_{eff} and 5σ sensitivities of SWIRE

Substantial sky areas and numerous samples would be required on consideration that galaxy evolution is evidently a function of the environment. Although the details may still be missing, there are indications that the amount of evolution of massive galaxies in rich clusters is different from what could have happened in the galaxy field (e.g. Franceschini et al. 1998; Stanford et al. 1998). In general, old red galaxies appear to cluster more strongly than the younger blue galaxy population. So any conclusions based on even deep investigations of small sky areas have to be considered as tentative until they are proven in representative cosmic volumes.

In principle, a powerful alternative to the time-expensive spectroscopic investigations for the study of the evolutionary mass function exploits observations of the spectral intensity of galaxies in the near-IR and its weak dependence on the age of the contributing stars (compared with that of the optical SED). An illustration of the robustness of such determination of the baryonic mass for an ISO-discovered galaxy at z=1.27 is given in Fig. 1. The same figure also shows, however, that sampling the emission longwards of the atmospheric limit at $\lambda = 2.5$ μm is needed for a proper mass determination, particularly at such high or higher redshifts.

SIRTF, to be launched end of 2002, will soon allow imaging with unprecedented sensitivity in this poorly known part of the e.m. spectrum. This will be achieved with the Infrared Array Camera (IRAC), one of the three SIRTF science instruments, providing simultaneous 5.12x5.12 arcmin images at 3.6, 4.5, 5.8, and 8 μm, using two InSb and two SiAs 256x256 detector arrays.

2   The SWIRE Project

The SWIRE survey is the largest SIRTF Legacy Program, devoted to image the evolution of dusty star-forming galaxies, evolved stellar populations, and AGNs as a function of environment, from redshifts $z \sim 2.5$ to the current epoch (Lonsdale et al. 2001). Building on ISO heritage, SWIRE complements smaller and deeper Guaranteed Time SIRTF surveys, and paves the way for Herschel-FIRST ($50 < \lambda < 670\mu$m) and later NGST ($0.5 < \lambda < 30\mu$m).

The key scientific goals of SWIRE are to view infrared emission from cosmic structures in the mid- and far-infrared, within the key redshift range $0.5 < z < 2.5$ where much of their formation has occurred. The main themes under study will be:

a) The evolution of active star-forming galaxies and passively evolving systems in volumes large enough to place it in the context of structure formation and to study the effect of galaxy environment.

b) The spatial distribution and clustering of starburst galaxies, evolved galaxies, and AGN, and the clustering evolution.

c) The identification and characterization of rare classes of sources, like the most luminous powerhouses at high-redshifts (the Ultra-luminous and Hyper-luminous IR sources) and the reddest (presumably the oldest, most massive and highest redshift) galaxies.

d) The evolutionary relationships between galaxies and AGN, and the contribution of accretion energy from obscured AGNs to the cosmic backgrounds.

Table 1. SWIRE Sensitivities and Confusion Limits

Band (μm)	3.6	4.5	5.6	8	24	70	160
Sensitivity [mJy], 5σ	0.0073	0.0097	0.0275	0.0325	0.45	2.75	17.5
Confusion[a] [mJy], 5σ	$4\ 10^{-5}$	$1.5\ 10^{-4}$	$4.3\ 10^{-4}$	0.001	0.085	3.7	36.

[a] Based on models by Franceschini et al. (2001), and paper in preparation.

2.1   SWIRE Survey Design: Depth and Area Coverage

Various important cosmological topics have remained untouched even after the ISO mission and the deep surveys with large millimetric telescopes. These include probing the IR emissivity properties of cosmic sources at $z > 1$, detection at $\lambda > 3\mu$m of the redshifted stellar continuum by high-z galaxies, the clustering properties of IR sources, the effects induced by the environment, and the characterization of populations of rare sources at high-z. While the GOODS survey (M. Dickinson, these Proceedings) will be very effective in addressing the former two items, the latter will require the large survey area capability provided by SWIRE.

Table 2. SWIRE Survey Areas

Field name	Field centers (J2000)	β [deg.]	I(100μm) [MJy/sr]	E(B-V)	Area [sq.deg.]
ELAIS-S1	00h38m30s -44d00m00s	−43	0.42	0.008	14.8
XMM-LSS	02h21m00s -05d00m00s	−18	1.30	0.027	9.3
CHANDRA-S	03h32m00s -28d16m00s	−48	0.46	0.001	7.2
Lockman Hole	10h45m00s +58d00m00s	+44	0.38	0.006	14.8
Lonsdale Field	14h41m00s +59d25m00s	+68	0.47	0.012	6.9
ELAIS-N1	16h11m00s +55d00m00s	+74	0.44	0.007	9.3
ELASI-N2	16h36m48s +41d01m45s	+62	0.42	0.007	4.5

The depth of the SWIRE survey (see Table 1) was thus designed to probe over the largest possible sky area the ∼ 0.5 < z < 2.5 redshift interval, where much evolution must have occurred, so that the highly transient (∼ 10^{7-8} yr) starburst and AGN phenomena can be studied in the context of the major galaxy populations at the same epoch. The area of each SWIRE field was defined to provide coverage of angular scales in the sky corresponding to linear scales of ∼ 50 to 100 Mpc at z > 1. Seven such separate target fields will be surveyed to take care of the noise from cosmic variance.

The SWIRE survey fields (see Table 2) were chosen using the COBE-normalized IRAS 100 μm maps by Schlegel et al. (1998) as areas in the sky with: $I(100)$ < 0.4 MJy/sr; ecliptic latitude $|b|$ > 40°; and contiguous area > 8°. Within these, additional constraints were imposed by minimum cirrus contamination, absence of bright stars and galaxies, absence of bright radio sources and nearby galaxy clusters, and guided by availability of multi-wavelength supporting data. SWIRE also includes 10 contiguous sq.deg. in the XMM-LSS survey area that is going to be covered with XMM in the GTO and GO program.

2.2 Multi-Band Optical/NIR Identification Program

Complementary data at other wavelengths are needed for source identification and physical characterization. The optical/near-IR follow-up consists of various programs. The entire 70 sq.deg. survey is being imaged to r' ∼ 25, to identify ∼ 75% of the SWIRE population. Multi-band imaging (g' ∼ 26, r' ∼ 25, ∼ i' 24, K_s ∼ 19.5) over partial fields (∼ 4 sq.deg.) in several of the 7 survey fields will provide photometric redshifts. Deeper imaging (g' ∼ 27, r' ∼ 26, ∼ i' 25) over 1 sq. deg. in a couple of fields, and in particular the ESO Large Program ESIS for ELAIS-S1 (B=26, V,R=25.5, I=25.5, Z=24 over 6 sq.deg., see http://www.eso.org/observing/visas/lpsummary/168.A-0322.html) will allow the characterization of the faintest and highest-z SWIRE source populations, and the identification of high-redshift clusters and groups.

Radio data at moderate depths from ATCA are available on 4 sq.deg. of ELAIS-S1, (see Gruppioni et al. 1999), while a very deep VLA survey has been approved for the Lockman Hole area. Shallow X-ray XMM data may eventually cover the whole XMM-LSS field, while deeper data may become available for the inner LSS part, and for other fields, should pending proposals be accepted.

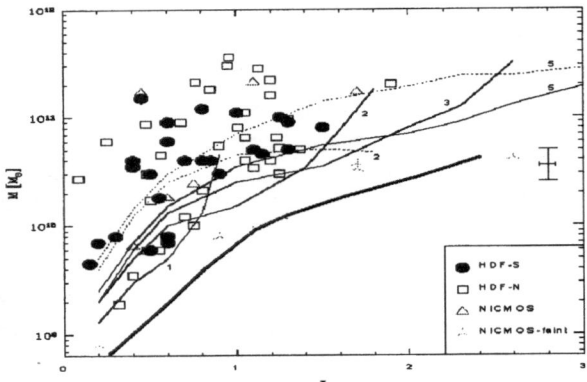

Fig. 2. Masses of morphologically-selected E/S0 galaxies with K< 20.15, based on fits to the optical/NIR SEDs, vs. redshifts (Rodighiero et al. 2001). The thin lines correspond to the K-band flux limit for different photometric models. The lower thick line corresponds to the equivalent "mass limit" for the SWIRE/IRAC surveys

2.3   SWIRE Data Products and Archiving Services

The SWIRE data products that will be released to the community on a 6-month timescale from observation will consist of preliminary – though highly reliable and complete – source lists, FITS images, cross-band identifications, sky coverage maps and documentation, as well as ancillary data that are available to a similar level of validation by the delivery date. Successive deliveries will take care of improving the area coverage, sensitivity limits, mosaicing, and cross-matching over more and more observing bands (including ancillary data).

The SWIRE dataset will be served through the IPAC InfraRed Science Archive (IRSA), which supports extensive data-analysis services, including fast catalogue and image sub-setting and correlation (see http://ipac.caltech.edu/IRSA/ for a preview).

3   Studying the Evolutionary Mass Function and Clustering of Galaxies with SWIRE

The two *Legacy Programs* GOODS and SWIRE will provide an overall complementary and very effective exploitation of SIRTF to study the evolutionary

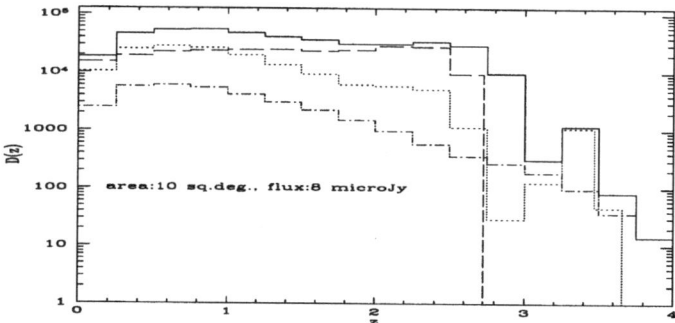

Fig. 3. Predicted redshift distribution for SWIRE-selected galaxies over 10 sq.deg. in the 4.5 μm channel. Dashed, dotted, and dot-dash lines correspond to E/S0, Sp/Ir and AGN sub-populations. The evolution model assumes $z_F = 2.5$ for E/S0 galaxies

mass function of galaxies through the analysis of the rest-frame near-infrared spectra. GOODS will be able to investigate deep into the galaxy mass function for galaxies up to quite high redshifts within two pencil-beam volumes centered on HDFN and CDFS. SWIRE will add to this the spatial dimension, which will be essential to disentangle the effect of environment on evolution, and to investigate rare classes of sources, like the most luminous & massive galaxies at high-z, and populations of strongly clustered galaxies, like ERO's (Daddi et al. 2001).

Although this capability for spatial sampling will be at the expense of sensitivity, still SWIRE will have enough of it to detect galaxies down to $M \simeq 10^{10} M_\odot$ in the critical redshift interval $1 < z < 2.5$, over an enormous sky area (67 sq.deg.). Fig. 2 compares this SWIRE sensitivity with results of current studies limited to areas of $\sim$ 10 sq.arcmin. Fig. 3 illustrates the potential quality of statistical investigations based on SWIRE-detected galaxies in any one of the 7 target fields.

References

1. Daddi, E., et al.: A&A **376**, 825 (2001)
2. Fioc M., Rocca-Volmerange B.: A&A **326**, 950 (1997)
3. Franceschini, A., et al.: ApJ **506**, 600 (1998)
4. Franceschini A., Aussel H., Cesarsky C., Elbaz D., Fadda D.: A&A **378**, 1 (2001)
5. Gruppioni, C., et al.: MNRAS **305**, 297 (1999)
6. Lonsdale, C., et al.: *The SWIRE Programme*,
   http://sirtf.caltech.edu/SSC/A_GenInfo/SSC_A1_legacy.html,
   see also http://www.ipac.caltech.edu/SWIRE/swire_v2.html (2001)
7. Rodighiero, G., Franceschini, A., Fasano, G.: MNRAS **324**, 491 (2001)
8. Schlegel, D.J., Finkbeiner, D.P., Davis, M. ApJ **500**, 525 (1998)
9. Stanford, S.A., Eisenhardt, P.R., Dickinson, M.: ApJ **492**, 461 (1998)

Lyman Break Galaxies in the NGST Era

Henry C. Ferguson[1], Mark Dickinson[1], and Casey Papovich[1,2]

[1] Space Telescope Science Institute, 3700 San Martin Drive, Baltimore, MD 21218
[2] Steward Observatory, University of Arizona, 933 North Cherry Avenue, Tucson, AZ 85721

Abstract. With SIRTF and NGST in the offing, it is interesting to examine what the stellar populations of $z \approx 3$ galaxies models imply for the existence and nature of Lyman-break galaxies at higher redshift. To this end, we "turn back the clock" on the stellar population models that have been fit to optical and infrared data of Lyman-break galaxies at $z \approx 3$. The generally young ages (typically $10^{8 \pm 0.5}$ yr) of these galaxies imply that their stars were not present much beyond $z = 4$. For smooth star-formation histories $SFR(t)$ and Salpeter IMFs, the ionizing radiation from early star-formation in these galaxies would be insufficient to reionize the intergalactic medium at $z \approx 6$, and the luminosity density at $z \approx 4$ would be significantly lower than observed. We examine possible ways to increase the global star-formation rate at higher redshift without violating the stellar-population constraints at $z \approx 3$.

1 Introduction

Near-infrared photometry provides access to the rest-frame optical portion of Lyman-break galaxy spectra; studies are beginning to provide some insight into the evolutionary status of LBGs. The stellar masses are reasonably well constrained at $M^* = 3 \times 10^{10} M_\odot \pm 0.5$dex [10,12,14]. These stellar masses are similar to estimates of the masses from kinematics [11,9], and together they suggest that Lyman break galaxies must grow substantially if they are to become L^* galaxies by redshift $z = 0$. Other stellar-population parameters such as ages, metallicities, extinction, star- formation rates and star-formation timescales are very poorly constrained by the fits to the spectral-energy distributions, leaving room for a variety of evolutionary paths both prior to and after $z \sim 3$.

At some point above redshift $z \sim 6$ the intergalactic medium was reionized. It is not known if the sources of ionization were stars or quasars, or whether the stars responsible for the reionization had a mass function at all similar to that observed in the Milky Way. It is also unclear whether reionization itself had a major role in regulating subsequent galaxy formation, for example by suppressing star formation in low-mass galaxy halos. As new facilities such as SIRTF and NGST come on line, understanding the causes and effects of reionization will be one of the major goals. With that in mind, we shall briefly review the capabilities of the Space Infrared Telescope Facility (SIRTF), the Hubble Space Telescope Advanced Camera for Surveys (HST ACS), and the Next Generation Space Telescope for studying Lyman break galaxies. We will then look at the stellar population parameters of the $z \sim 3$ samples of Lyman-break galaxies and attempt to turn the clock back to predict their star-formation rates at higher redshift. We find a somewhat surprising result: that the models imply a luminosity density at $z = 4$ significantly below that observed, and fail to produce enough ionizing photons at $z = 6$ to account for reionization. We discuss modifications of the models that might be needed to avoid these problems.

2 Lyman-Break Galaxies in the GOODS Era

When the HDF-N/WFPC2 observations were made, $z \approx 3$ was the frontier for galaxy surveys. Today it stands at $z \approx 6$, where we know very little: only that some galaxies and QSOs already existed, and that their energetic output may have just risen to the point of reionizing the IGM [1,6]. From $z = 6$ to 3, the age of the universe more than doubles, and its density decreases more than five-fold. We expect corresponding changes in the galaxy population, but measuring this evolution will require large large, systematic surveys for galaxies at $z > 4$ to compare with more than 1000 Lyman break galaxies (LBGs) now known at $z \approx 3$ [13]. Color selection at $z \approx 5$ to 6 requires deep imaging at $\lambda \geq 0.9 \mu m$. However, from the ground even the brightest $z > 5$ galaxies are barely detectable. If there is no evolution between $z = 3$ and 6, we expect an L_{UV}^* LBG to have $m_i = 25.7$ at $z = 5$, and $m_z = 26.0$ at $z = 6$. The HST Advanced Camera for Surveys (ACS) will easily reach these limits thanks to the dark sky and sharp image quality, which also provide greatly reduced photometric error, incompleteness, and contamination due to confusion with other (mostly foreground) objects. An L^* LBG detected at $z = 6$ by ACS will also be detectable in ultra-deep integrations with the IRAC detector aboard SIRTF. Depending on the quality of the point-spread function such detections will either be marginal (requiring prior knowledge of source positions and possibly statistical co-addition of multiple galaxies for a secure detection) or clear cut (if the PSF is as good as measured in pre-flight tests). Figure 1 illustrates the expected yield of $z \sim 5.5$ LBGs from the GOODS survey from a simple extrapolation of the $z = 3$ luminosity function. We expect several hundred to of order 1000 candidates.

3 Lyman-Break Galaxies in the NGST Era

The Next Generation Space Telescope, slated for launch before 2010, is expected to be a ~ 6.5 m passively-cooled telescope with exquisite sensitivity from 0.6 to 28 μm. Three focal-plane instruments are planned: an optical/near-IR camera with a $4' \times 4'$ field of view, a low-resolution ($R \sim 100 - 1000$) near-IR spectrograph (to be built by ESA), and a mid-IR camera/spectrograph. At 4.5 μm, an exposure time of 5 seconds will be required to match the 50-hour depth of IRAC GOODS images. Presuming that observations from SIRTF, ACS and WFC3 are successful, by the time NGST launches galaxy populations at $z < 6$ should be reasonably well understood. However, the fact that reionization occurs at $z \gtrsim 6$ suggests that the earliest luminous structures formed at still higher redshift.

By redshift $z = 3$ the massive stars that reionized the universe (if indeed stars were responsible) exist only as undetectable remnants. However, if early

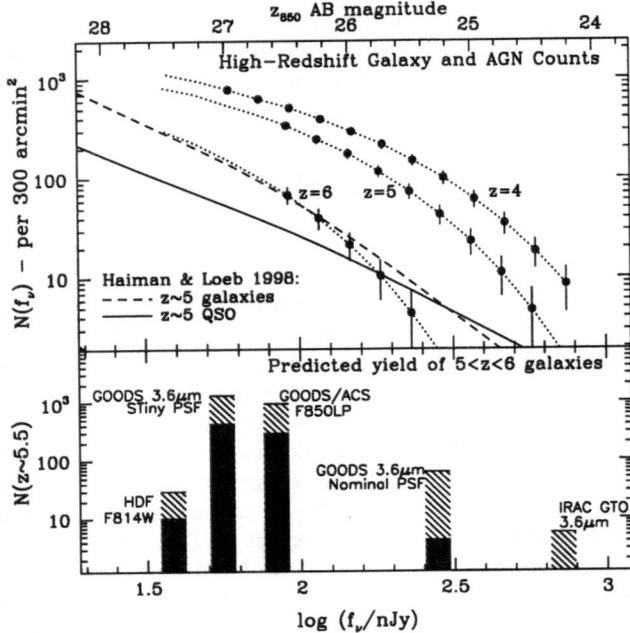

Fig. 1. *Top:* Number of sources per unit redshift per logarithmic flux interval anticipated in the GOODS survey region. Solid and dashed lines show the average number of galaxies and QSOs, respectively, per unit redshift $5 < z < 10$ from Haiman & Loeb (1998). Dotted lines show a model based on the measured luminosity function of LBGs at $z \sim 3$, with L^* scaled with redshift as $(1+z)^{-3/2}$, i.e., following the Press-Schechter relationship for halo mass. Error bars represent Poisson statistics on galaxies recovered at $z \sim 4, 5$, and 6 using simple B, V, and i-dropout criteria. Fluxes have been scaled to equivalent z-band fluxes using a typical LBG SED. *Bottom:* Total galaxy counts at $z = 5.5 \pm 0.5$ vs. the 10σ limiting depths (in nJy) of various surveys, for a non-evolving LBG luminosity function (shaded bars) and for one with the evolution adopted in the upper panel (solid bars). Under these assumptions the HDF should have had 10-30 $z \sim 5.5$ galaxies, consistent with tentative identifications via photometric redshifts. The GOODS ACS survey will yield 300-1000 secure identifications. Estimates of high-z galaxy numbers are also shown for the GOODS *SIRTF* 25-hour depth survey, under two assumptions that bracket the range of expected sensitivities due to uncertainties in the on-orbit IRAC PSF. Predicted numbers (nominal PSF) are shown also for the planned $(0.3\ \mathrm{deg}^2)$ 3-hour IRAC GTO survey of the Groth strip. A cosmology with $\Omega_M, \Omega_\Lambda, h = (0.3, 0.7, 0.65)$ is adopted

generations of stars formed with a Salpeter or Scalo initial mass function (IMF), lower-mass cousins of the ionizing sources would still be on the main sequence at $z = 3$, and would perhaps reside in Lyman-break galaxies. Existing optical and near-IR observations of LBGs yield constraints on stellar populations that can be used to explore this possibility. Papovich, Dickinson & Ferguson [10] studied a sample of spectroscopically-confirmed LBGs from the Hubble Deep Field North (HDF) in the redshift range $2.0 \lesssim z \lesssim 3.5$. The UV-optical data were drawn from WFPC-2 observations, and the infrared from NICMOS J and H-band observations and from K_s-band observations with the infrared imager IRIM at the KPNO 4m Mayall telescope [5]. Stellar-population models from 2000 version of the Bruzual-Charlot [2] code were fit to 31 galaxies, varying metallicity, e-folding timescale τ_{SF}, age, IMF (Salpeter, Miller-Scalo, Scalo), extinction, and extinction law [3,4]. The geometric mean of the best-fit ages for the sample is 0.12 Gyr for the solar metallicity case. Thus a typical galaxy observed at $z = 3.0$ would have "formed" at $z = 3.15$. Shapley et al. [12] analyzed G, $\mathcal{R}$, J, and K_s groundbased photometry for a sample of galaxies with spectroscopic redshifts $2.2 < z < 3.4$. The published paper reports results for the best-fit continuous star-formation models ($\tau_{SF} = \infty$) to the 74 galaxies for which acceptable fits were obtained. The median best-fit age for this sample is 0.32 Gyr, implying a formation redshift $z = 3.4$ for a typical galaxy observed at $z = 3$.

In Figure 2a, we show the star-formation histories derived for each galaxy in the two samples, under various assumptions. The top panel shows exponentially-declining models. The second and third panels show models with a constant star-formation rate (where only the age, extinction, and total stellar mass are free parameters). In the exponentially decaying models only one out of the 31 galaxies would have been present at $z = 6$. In the oldest continuous-star-formation models from [10], six out of 31 or 19% would have been present at $z = 6$. The Shapley et al. [12] models imply that only 17% of the galaxies were present at $z = 6$.

Models with two distinct episodes of star formation allow more star formation at higher redshift. Papovich et al. [10] fit maximally-old models to their LBG sample, deriving constraints on the mass of an old population that formed with a Salpeter IMF in an instantaneous burst at $z = \infty$. This model quantifies how much stellar mass can be hidden "underneath the glare" of the young population that dominates the UV/Optical radiation from each galaxy. However, the star-formation rate predicted at $z = 6$ from such maximally old components is zero, because all star-formation happened at higher redshift. It is more likely that starbursts induced by mergers are spread out over some range of redshift and do not occur in all galaxies simultaneously. If the older burst in the LBGs is put at redshift lower than $z = \infty$, the mass in the burst must be lower. Rather than fit a whole suite of models of different burst redshifts, we can, to a good approximation, scale the allowable mass in the old component by a power-law fading model. By fitting the B-band luminosity vs. time for $10^7 < t < 2 \times 10^9$ yr, we find $L_B \propto t^{-0.8}$ for a Salpeter IMF for an instantaneous burst in the Bruzual & Charlot solar-metallicity models. If each galaxy had an instantaneous probability $P(z)$ of forming stars at redshift z, and a typical burst had a duration Δt the average SFR from an ensemble of such galaxies would be $\xi(z) = M(z)P(z)/\Delta t$, where $M(z)$ is average mass formed in each burst and Δt is the average duration of each burst. For simplicity we adopt a constant $P(z)$ from $z = 10$ to the observed LBG redshift z_{obs}.

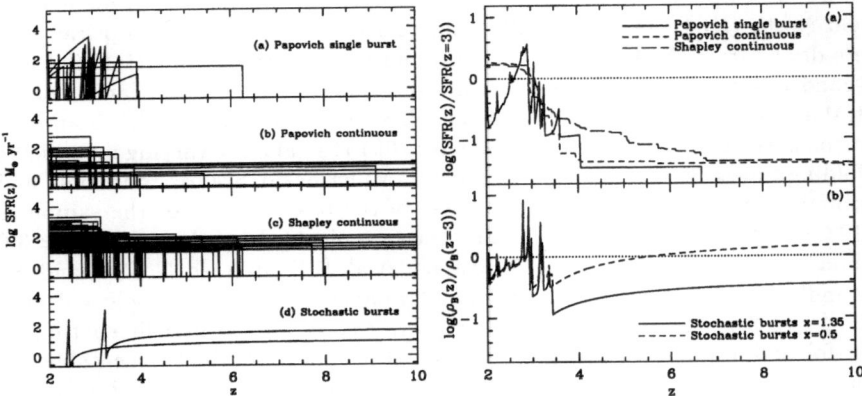

Fig. 2. *Left:* Star-formation rate vs. time for individual galaxies, as inferred from the SED models. The top panel shows the best-fit models with exponentially declining SFR. Panel (b) shows the star-formation histories from continuous star-formation models from [10] characterized by a stellar mass M and an age. Panel (c) shows the same kind of continuous star-formation model for the Shapley et al. [12] sample. Panel (d) shows two examples of the stochastic burst model described in the text applied to galaxies 97 and 1115 in the PDF01 sample. *Right:* Top panel – star-formation density vs. time for the monotonic models, normalized to the value at $z = 3$. The solid curve is for the Papovich et al. [10] single-burst models. The short-dashed curve is for their continuous star-formation models. The long-dashed curve is for the Shapley et al. [12] continuous star-formation models. Bottom panel – rest-frame B-band luminosity density vs. time for the stochastic burst models with a Salpeter IMF (solid) and a top-heavy IMF with $x = 0.5$ (dashed). Luminosity densities are normalized to the mean in the range $2.5 < z < 3.5$. For the Salpeter IMF the ratio of the B-band luminosity densities and the star-formation rate densities are the same. However when the IMF is varied it is more relevant to look at luminosity densities. The B-band is shown because that is what we calculate from the power-law fading model. The UV luminosity density is more relevant for the discussion of reionization and the $z = 4$ luminosity function. We have verified that even for an extreme IMF slope of $x = 0.3$ the 1500Å-B colors and 860Å-B colors are within 0.2 and 0.3 mag, respectively, of the colors for the Salpeter IMF

Figure 2d shows the SFR vs. redshift implied by such a stochastic model for two individual galaxies (numbers 97 and 1115) in the PDF01 sample. The low-redshift spikes in the star-formation rate correspond to the young component that dominates the light at the observed redshift; the star-formation progressing to higher redshift represents the mean for an ensemble of stochastic bursts. Obviously any single galaxy would simply show two spikes of star formation for this kind of model, but if we consider such a galaxy as a proxy for millions of others, the star-formation history shown in the figure represents the maximal rate of star-formation due to stochastic bursts as a function of redshift.

The constraints become clearer if we consider the entire sample of galaxies. Figure 3 shows the evolution of $\dot{\rho}_{SFR}(z)$ with time relative to that $z = 3$ computed by summing up the models shown in the previous figures. The top panel shows the smooth star-formation histories. For these cases the inferred co-moving density of star formation declines dramatically from $z = 3$ to higher redshift. Even if we put the maximum mass allowed in stochastic-starbursts at redshifts $z > z_{observed}$, the star-formation rate at $z = 6$ is still a factor of 3 below that at $z = 3$, as shown by the solid curve in Fig. 3b.

Adopting $f_{esc} = 0.1$, the required density of star-formation for reionization in the Madau, Haiman & Rees [8] model is a factor of 1.3 times higher than the dust-corrected $\dot{\rho}_{SFR}$ at $z \sim 3$ measured by [13]. In contrast, the star-formation rates inferred from the SED fits imply a sharp decrease in $\dot{\rho}_{SFR}$ between $z = 3$ and $z = 6$. The problem becomes even more severe if a significant fraction of the baryons are already collapsed into minihalos at the time of reionization. In this case the required number of ionizing photons increases by a factor of 10-20 [7], and all models fall short even if $f_{esc} = 1$.

There is another, perhaps more serious, problem with the star-formation histories derived so far: all the models imply a dramatic decline in star-formation by $z = 4$. But the observed LBG rest-frame UV luminosity functions are very similar at $z = 3$ and $z = 4$, and the integrated star-formation rates derived therefrom differ only by a factor of 1.1 ± 0.4 [13]. Thus the star-formation *histories* derived from the $z = 3$ LBGs are in direct conflict with the star-formation *rates* derived for the $z = 4$ LBGs.

More star formation can be hidden in bursts if the bursts fade faster. If the fading exponent over an age range $10^7 - 2 \times 10^9$ yr is ζ, the additional increase in stellar mass that can be hidden relative to a Salpeter IMF is roughly $t_7^{-\zeta - 0.8}$, where t_7 is the age in units of 10^7 yr. At an age of 1 Gyr a factor of four more stellar mass can be hidden if $\zeta = -1.1$ than if $\zeta = -0.8$. Indeed, changing the fading exponent to $\zeta = -1.1$ is sufficient to make the allowed star-formation rate at $z = 6$ equal to the star-formation rate at $z = 3$. The global star-formation rate from such a model is shown as a dashed curve in Fig. 2b. If the IMF is a powerlaw $\phi(M)dM \propto M^{-(1+x)}$, a fading exponent $\zeta = -1.1$ requires an IMF slope $x = 0.5$ compared to the Salpeter value $x = 1.35$ (for an instantaneous-burst solar-metallicity stellar population). A steeper fading slope $\zeta = -1.2$ (corresponding to an IMF slope $x = 0.3$) is needed to bring $\dot{\rho}_{SFR}$ at $z = 4$ to within a factor of 1.3 of that at $z = 3$. Lower metallicities require even more top-heavy IMFs. Options other than varying the IMF are of course possible (e.g. evolved stellar populations could be hidden by dust that builds up over timescales of 10^8 to 10^9 yrs). However, the requirement for faster-than-Salpeter fading is robust. Furthermore, the fading must be even faster if galaxies on average have more than two burst episodes.

Constraints on the star-formation histories of LBGs will improve greatly over the next few years with the advent of SIRTF and ACS. Observations with these instruments will further narrow the parameter space available for bursty or episodic star-formation. Such observations will set the stage for the detailed exploration of galaxy formation in the "pre-reionization" era at $z > 6$ with NGST.

We would like to thank our collaborators on the HDF observations for their many contributions to this work. We thank Jennifer Lotz and Mauro Giavalisco for valuable discussions. Support for this work was provided by NASA through grant GO07817.01-96A from the Space Telescope Science Institute, which is operated by the Association of Universities for Research in Astronomy, under NASA contract NAS5-26555.

References

1. Becker, R. H. et al.: astro-ph/0108097 (2001)
2. Bruzual, A. G. and Charlot, S.: ApJ **405**, 538 (1993)
3. Calzetti, D., Armus, L., Bohlin, R. C., Kinney, A. L., Koorneef, J. and Storchi-Bergmann, R.: ApJ **533**, 682 (2000)
4. Cardelli, J. A., Clayton, G. C. and Mathis, J. S.: ApJ **345**, 245 (1989)
5. Dickinson, M.: *In The Hubble Deep Field.* ed by M. Livio, S. M. Fall and P. Madau (Cambridge University Press, Cambridge 1998) p. 219.
6. Djorgovski, S. G., Castro, S., Stern, D. and Mahabal, A. A.: ApJ **560**, L5 (2001).
7. Haiman, Z., Abel, T. and Madau, P.: ApJ **551**, 599 (2001)
8. Madau, P., Haardt, F. and Rees, M. J.: ApJ **514**, 648 (1999)
9. Moorwood, A. F. M.: this conference (2002)
10. Papovich, C., Dickinson, M. E. and Ferguson, H. C.: ApJ **559**, 620 (2001)
11. Pettini, M., Shapley, A. E., Steidel, C. C., Cuby, J.-G., Dickinson, M., Moorwood, A. F. M., Adelberger, K. L. and Giavalisco, M.: ApJ **554**, 981 (2001)
12. Shapley, A. E., Steidel, C. C., Adelberger, K. L., Dickinson, M., Giavalisco, M. and Pettini, M.: ApJ **562**, 95 (2001)
13. Steidel, C. C., Adelberger, K. L., Giavalisco, M., Dickinson, M. and Pettini, M.: ApJ **519**, 1 (1999)
14. Yamada, T.: 2002. this conference.

Measuring the Mass of High-z Galaxies with NGST

Tommaso Treu[1] and Massimo Stiavelli[2]

[1] California Institute of Technology, Astronomy 105-24, Pasadena CA 91125
[2] Space Telescope Science Institute, 3700 San Martin Dr, Baltimore MD 21218

Abstract. We discuss dynamical mass measurements of high-z galaxies with the Next Generation Space Telescope (NGST). In particular, we review some of the observational limits with the current instrument/telescope generation, we discuss the redshift limits and caveats for absorption and emission lines studies with NGST, and the existence of suitable targets at high redshift. We also briefly summarize strengths and weaknesses of proposed NGST instruments for dynamical studies.

1 Introduction

During this meeting we have heard of mass measurements at redshifts that were out of the realm of possibilities just a few years ago. However, several obstacles stand between us and the future when dynamical mass measurements will be routinely feasible at $z > 1$. Among them, the lack of spatial resolution and the atmospheric emission/absorption at infrared (IR) wavelengths (see e.g. the contribution by M. Franx). The Next Generation Space Telescope (NGST) – with its unique combination of large collecting area, superb spatial resolution, and low background – will provide a major contribution to the extension of mass measurements at $z > 1$.

Here we discuss mass measurements with NGST, focusing on dynamical measurements (lensing and stellar mass measurements are discussed elsewhere in these proceedings, see e.g. the contribution by H. Ferguson). As for any dynamical mass measurement, a spatial scale and a velocity scale are needed. In the following, we will assume that spatial scales are easily measurable even though this is not the case for sub-galactic clumps at very high redshift. Our focus in this contribution will be on emission and absorption line measurements (Sec. 2 and 3) of velocity scales. In particular, we will briefly review the limits of what is feasible with the current technology and present some detailed simulations on the capabilities of NGST for mass measurements. Finally, we will discuss the implications of the choice of NGST-instrumentation for mass measurements.

2 Emission Line Measurements

Kinematical measurements based on emission lines are easy to carry out but hard to interpret. This is because the gas may be far from equilibrium and thus gas kinematics may not be telling us anything about the mass of the host galaxy. One

way to overcome this difficulty is to have access to two-dimensional kinematical
information and this is indeed what was needed to solve the same problem in the
context of mass measurement in nearby spiral galaxies using HI [19] and black
hole mass measurements from the kinematics of nuclear gas disks [7]. NGST
presents several advantages in this context because of its high angular resolution.
Its uninterrupted wavelength coverage guarantees that strong emission lines will
be available over a large range of redshifts. The low background redwards of 2.5
μm will guarantee high sensitivity for measurements using Hα at z$\geq$ 3. On the
basis of the visibility of Hα one can argue that for objects at z$\leq$ 3 a detailed 2D
kinematic mapping can be carried out from the ground using an adaptive optics
fed integral field spectrograph on an 8-meter class telescope. At z$\geq$3 oxygen lines
are still accessible from the ground but may be suppressed by the expected low
metallicity of most objects at that redshift. Thus, the availability Hα and the
sensitivity of NGST make it the ideal instrument at z$\geq$3.

In addition to the availability of lines another issue is the resolving power
needed to carry out a measurement. In a nutshell, faint galaxies will on average
have lower mass and smaller internal velocities, requiring higher resolving power
to be studied. Since on average we expect galaxy mass to decrease with redshift,
one will need progressively higher resolving power to study galaxies at increasingly high redshift. This intuitive result is illustrated in Figure 1 where we show,
in the left panel, simulated NGST spectra for a typical Milky Way progenitor
(OBJ=13) at z=3, 5, and 7 and for a bright Milky Way progenitor (OBJ=9)
at z=7. These models have been obtained with a merging tree code [4,12]. The
resolving power needed to measure internal kinematics is in the range 6,000 to

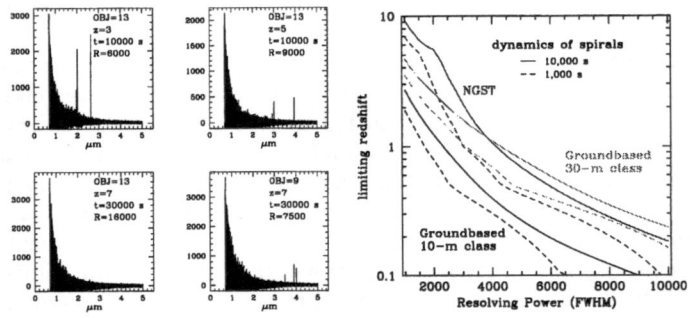

Fig. 1. Left panels: simulated spectra for the typical (OBJ=13) Milky Way progenitor
and for the brightest (OBJ=9) Milky Way progenitor. Resolving powers in excess of
6,000 are needed for z>3 even for the most massive progenitors of L$\star$ galaxies. Right
panel: correlation between available resolving power and maximum redshift for which
a the internal kinematics of a spiral galaxy can be resolved. For relatively large objects
at high redshift NGST is superior to even a 30m ground based telescope

16,000. In the right panel we show a relation connecting the available resolving power to the highest redshift that can be probed. This plot has been obtained by assuming the validity of the Tully-Fisher relation and by requiring that for the given exposure time and resolving power sufficient signal-to-noise is achieved to carry out the measurement. The optimal emission lines is used for each redshift [13]. It is clear that for low mass objects NGST is not competitive with large ground based telescopes. However, NGST is superior to 30m class ground based telescopes for massive objects at $z > 2$.

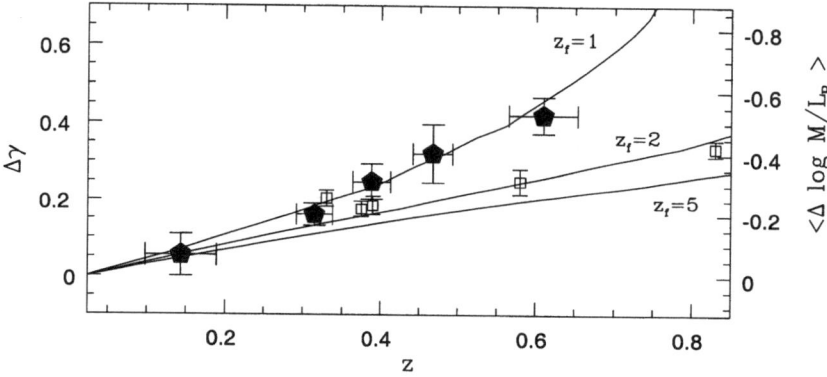

Fig. 2. Absorption line dynamics at $0.1 < z < 1$, ground-based limits: evolution of the Fundamental Plane of E/S0 galaxies from $z = 0.8$ to $z = 0$ (from [17]). The average offset of the intercept of field galaxies from the local FP relation as a function of redshift (large filled pentagons), is compared to the offset observed in clusters (open squares). See [17] for references, description, and details

3   Absorption Line Measurements

Stellar absorption lines are a very good probe of velocity scales for several reasons: *i)* a large fraction of galaxies do not have emission lines, and therefore absorption lines are the only way to go; *ii)* stars are a good tracer of the velocity distribution of the system, as opposed to emission lines coming from HII regions which might not be in dynamical equilibrium with the galaxy; *iii)* stellar absorption lines kinematics tend to suffer less from the effects of interstellar absorption, local motions, and winds then emission lines kinematics; *iv)* if streaming motions are not significant as in massive E/S0 galaxies, interesting dynamical constraints can be gathered without the need for spatially resolved information.

However, stellar absorption lines have to be present, and stellar populations need to be in dynamical equilibrium, in order to obtain meaningful information.

Typical optical absorption lines (e. g. the Mg triplet) need of order ~ 1 Gyr to develop in stellar populations, while the time scales can be relatively shorter for other frequently used lines such as the near-IR Ca triplet. On the one hand this is good, because it guarantees that if such lines are present then the stellar populations are almost certainly old enough that the system is in equilibrium. On the other hand, it is worth asking the question of whether systems with stellar absorption lines exist at high redshift. In the following we will illustrate with a few examples the current limits of ground based measurements, we will show why we believe there are interesting targets beyond these limits, and we will explore the feasibility of such measurements with NGST.

Recent studies of the evolution of the Fundamental Plane (FP) with redshift show that E/S0 galaxies do not undergo major structural changes from $z \sim 0.8$ to the present (e. g. [20,16,17]). Such measurements, based on a combination of spectroscopy (velocity dispersion) and imaging (photometric structural parameters) provide important information on the evolution of the internal structure of E/S0 galaxies and their stellar populations (see also the contribution by G. Illingworth). For example, [20] inferred from the evolution of the FP that cluster E/S0 have old stellar populations, while [17] used this technique to show that secondary episodes of star formation are common in massive field E/S0 at $z \sim 0.5$. It would be interesting to extend such measurements to higher redshifts, where we know that some old E/S0 exist, at least in the range $z = 1 - 2$ ([3,18,11]; see Figure 3). However, long integrations on large telescopes are needed to push such studies significantly beyond $z \sim 1$ because E/S0 become very faint in the optical and sky emission lines severely limit what can be done in the IR. For example, [8] measured the central velocity dispersion of the lens galaxy in the gravitational lens system MG2016+112 ($z = 1.004$) by integrating 8.5hrs with the Echellette Spectrograph Imager (ESI) at the Keck II telescope. Given the excellent sensitivity and resolution of ESI (that allows for a good removal of sky lines in the red/IR), it seems likely that with the current generation of 8-10m class telescopes such measurements will be limited to bright E/S0 at $1 < z < 1.5$.

As shown in the right panel of Fig. 3, the redshift range $z = 1 - 2$ would be easily within reach with NGST (see also [14]), and even objects beyond $z \sim 2$ could be targeted, if they exist. Based on calculations similar to the one used in Fig. 1 (right panel), adopting the FP instead of the Tully-Fisher relation as luminosity-velocity scaling law, at $z > 1$ NGST will be significantly better than a 30-m telescope, especially for large masses. Recent studies suggest that old stellar populations indeed exist at $z > 2$. For example [10] found that the colors of Lyman-α selected galaxies at $z \sim 2.4$ are consistent with those of an old stellar population ($\sim$ Gyr of age) – responsible for the observed red colors – with a sprinkle of starbursting population – responsible for the observed emission lines – (see also [5]).

3.1   The Internal Mass Distribution of High Redshift Galaxies

So far, we only discussed absorption line kinematics in the absence of spatially resolved information. However, spatially resolved kinematics have been very important in our study of the local Universe. For example it has provided crucial constraints on the internal structure of E/S0 (e. g. [2,6]), and Hubble Space Telescope (HST) subarcsecond resolution spectroscopy has been used to prove the existence of black holes at the center of massive E/S0. Unfortunately, the small aperture of HST and the seeing-limited spatial resolution of ground based telescopes have so far basically confined spatially resolved kinematics to the local Universe. As an example of the current observational limits, [15] have recently started a kinematic survey of gravitational lenses. Under the best observing conditions (seeing 0.''6) – with long integration times at the Keck-II Telescope – they were able to measure velocity dispersion profiles extended beyond the effective radius for lens galaxies at $z < 0.5$. Although the extended profiles provide valuable and unique information on the mass distribution, a higher spatial resolution is needed to study in detail the mass distribution within the central kiloparsecs. NGST – with its combination of large collecting area and superb spatial resolution – offers the tremendous opportunity to explore spatially resolved kinematics at cosmological distances. With a typical resolution $\leq$ 0.''1 NGST will be able to study in detail the mass distribution of E/S0 and its evolution with redshift.

4   Mass Measurements with NGST: Instrumentation

Both emission and absorption line studies of the kinematics of high redshift galaxies require a resolving power of at least 3,000, i.e. somewhat higher than what is currently in the baseline for NGST. The need for high resolving power is stronger for emission line measurements since they can be carried out on intrinsically fainter objects. Because of this reason there is a strong rationale for a two-dimensional spectroscopic capability at R$\simeq$3,000. This capability could be

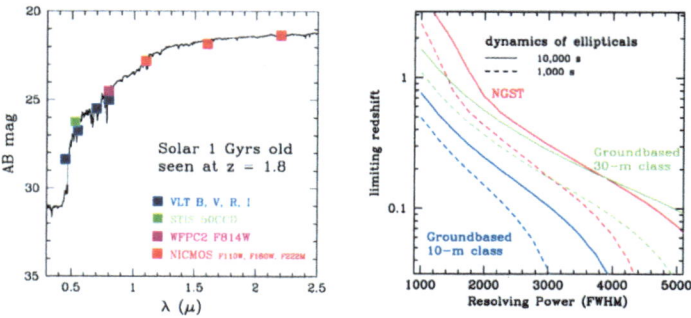

Fig. 3. Absorption lines dynamics at $z > 1$: targets for NGST. Left: spectral energy distribution of and extremely red object with E/S0 morphology at $z \sim 1.8$ (from [18,11]). Right: comparison of the performance of NGST, a 10m ground based telescope and a 30m ground based telescope

in the form of an integral field spectrograph [1] if budget, volume, and weight constraints allow it. Alternatively, it could be provided by a multi-object spectrograph based on a configurable slit array provided the configuration time is short enough. Such a spectrograph would allow one to step the slit over resolved objects to be mapped in two dimensions while integrating on fainter sources [9].

Acknowledgments

We acknowledge financial support by NASA through a grant from STScI (HST-AR-09222), which is operated by AURA, under NASA contract NAS5-26555.

References

1. R. Bacon: 'The Power of Integral Field Spectroscopy for NGST'. In: *NGST Science and Technology Exposition*, eds. E. P. Smith and K. S. Long, ASP, vol. 207, pp. 333 (2000)
2. G. Bertin et al.: A&A, 292, 381 (1984)
3. A. Cimatti, et al.: submitted (2001) astro-ph/0111527
4. M. Fall, S. Movshev, M. Stiavelli: unpublished
5. P. J. Francis, et al.: ApJ, 554, 1001 (2001)
6. O. Gerhard, A. Kronawitter, R. P. Saglia, R. Bender: AJ, 121, 1936 (2001)
7. R. K. Harms, et al.: ApJ, 435, L35 (1994)
8. L. Koopmans & T. Treu: submitted (2002)
9. J. W. MacKenty, M. Stiavelli: 'A Multi-Object Spectrometer using Micro Mirror Arrays'. In: *Imaging the Universe in Three Dimensions: Astrophysics with Advanced Multi-Wavelength Imaging Devices*, ed. W. van Bruegel and J. Bland-Hawthorn, ASP, vol. 195, pp. 443 (2000)
10. M. Stiavelli, C. Scarlata, N. Panagia, T. Treu, G. Bertin, F. Bertola: ApJ, 561, L37 (2001)
11. M. Stiavelli, et al.: A&A, 343, L25 (1999)
12. M. Stiavelli: 'Internal Structure of high-z Galaxies with NGST'. In: *The Next Generation Space Telescope: Science Drivers and Technological Challenges, 34th Liège Astrophysics Colloquium, Belgium, 15-18 June 1998*, ed. by P. Benvenuti, P. Madau (ESA Publications 1998) pp. 71
13. M. Stiavelli: 'NGST Performance for Studying the Internal Dynamics of High-z Galaxies'. In: *Science with the NGST*, ed. by E. P. Smith and A. Koratkar, ASP, vol. 133, pp. 279 (1998)
14. T. Treu: 'The Fundamental Plane of Elliptical Galaxies at Low and High Redshift'. In: *The Next Generation Space Telescope: Science Drivers and Technological Challenges, 34th Liège Astrophysics Colloquium, Belgium, 15-18 June 1998*, ed. by P. Benvenuti, P. Madau (ESA Publications 1998) pp. 255
15. T. Treu & L. Koopmans, in preparation (2002)
16. T. Treu, M. Stiavelli, Bertin G., S. Casertano, P. Møller: MNRAS, 326, 237 (2001)
17. T. Treu, M. Stiavelli, S. Casertano, P. Møller, G. Bertin: ApJ, 564, L00 (2002)
18. T. Treu, et al.: A&A, 340, L10 (1998)
19. T. S. van Albada, R. Sancisi: Phil. Trans., A, 320, 447 (1986)
20. P. G. van Dokkum, M. Franx, D. D. Kelson, G. D. Illingworth: ApJ, 504, L17 (1998)

Prospects with ALMA

P.A. Shaver

European Southern Observatory, Karl-Schwarzschild-Str. 2,
D-85748 Garching bei München, Germany

Abstract. The Atacama Large Millimeter Array will greatly enhance our ability to study the dynamics and masses of galaxies and their components at both low and high redshift. It will open an entire range of the electromagnetic spectrum, one containing a wealth of molecular and atomic spectral lines, at wavelengths unaffected by dust obscuration. It will be sensitive to objects at the highest redshifts, making it possible to follow the processes of mass buildup and interactions over the history of the Universe. A brief summary of some of these possibilities is given in this paper.

1 The Atacama Large Millimeter Array (ALMA)

The Atacama Large Millimeter Array is a merger of the major millimeter arrays projects – the European Large Southern Array (LSA) and the U.S. Millimeter Array (MMA), and possibly the Japanese Large Millimeter and Submillimeter Array (LMSA), into one global project. This will be the largest ground-based astronomy project of the decade after VLT/VLTI, and, together with the Next Generation Space Telescope (NGST), one of the two major new new facilities for world astronomy coming into operation by the end of the decade.

ALMA will be a unique facility, having powerful capabilities for making measurements relevant to the masses of galaxies at both low and high redshift. It will be comprised of 64 12-meter antennas, giving a total collecting area of 7238 m². The antennas surfaces will be accurate to 25μ rms or better, to provide high sensitivity at the highest frequencies. The antennas will be movable, so that the configuration can be changed and provide a "zoom lens" capability. The most compact array will be concentrated within 150m, and the largest array will have baselines up to 14 km, giving an angular resolution of 10 milli-arcsec at the highest frequencies. The receivers will ultimately cover all atmospheric bands in the frequency range 30-900 GHz. With 2016 baselines, ALMA will have excellent "snap-shot" capability, making images of high fidelity in just seconds. The site chosen is the Llano de Chajnantor, near San Pedro de Atacama in northern Chile. At an altitude of 5000m it has exceptionally low atmospheric opacity, and is certainly one of the very best sites in the world for such an array. The combination of large collecting area, high antenna precision, low receiver noise and low atmospheric opacity will result in an rms sensitivity of 0.01 mJy ($1\,\sigma$) at 1.3mm in one hour of integration.

The construction phase of ALMA is scheduled to begin in 2002. The first production antennas should arrive on the site in 2005, and limited scientific operations will be possible by the community in 2007. The complete array should be available by 2011 – a powerful new addition to the world's astronomical facilities.

Much more detail can be found in the proceedings of the 1995 ESO-IRAM-NFRA-Onsala workshop on Science with Large Millimetre Arrays (Shaver, 1996) and the proceedings of the 1999 meeting on Science with the Atacama Large Millimeter Array held in Washington (Wootten, 2000). Other references can be found at the ALMA websites http://www.eso.org/projects/alma/, http://www.alma.nrao.edu, and http://www.nro.nao.ac.jp/~lmsa/index.html.

1.1 ALMA Capabilities Related to High Redshift Galaxies and Mass Measurements

High redshift galaxies are particularly accessible to observations in the millimeter and submillimeter wavebands. The copious dust emission from star-forming galaxies dust produces a large peak in the rest-frame far-infrared, which, when redshifted, greatly enhances the millimeter and submillimeter emission from these objects. This negative K-correction is so strong that galaxies at $z \sim 5-10$ can appear brighter than those at $z \sim 1$. Thus, ALMA will be a very sensitive instrument for the study of galaxies at high redshifts. Surveys with current instruments already reach the milliJansky level, and sources at redshifts of a few. However, they are already becoming confusion limited at this level, so ALMA's angular resolution as well as its great sensitivity will be required to go much beyond. Dust emission at 1.3mm has been detected out to the highest redshifts - in one recent case from a $z = 5.5$ QSO using the MAMBO array on the IRAM 30-meter telescope (Bertoldi & Cox, 2001). This is the deepest mm integration so far, and the emission comes from the optically-faintest QSO known at high redshift. The implied star formation rate is 600 $M_\odot$ yr^{-1}. This QSO would be detected by ALMA in just a few seconds.

The same dust that enhances the millimeter emission also diminishes the UV and optical emission from these galaxies. Thus, ALMA will be highly complementary to the HST or VLT in finding the high-redshift, dust-obscured star-forming galaxies that would otherwise be hard to detect. In fact, the total observed emission from the high-redshift Universe is dominated by the millimeter and far infrared wavebands. Hidden sources imply hidden mass. The millimeter/submillimeter observations can provide the total dust and gas content, and the complete line profiles. They make possible the determination of rotation curves in otherwise obscured galactic nuclei at high spectral resolution.

ALMA's high angular resolution, up to 0.01 arcsec, will of course be essential to identify the sources detected, remove confusion, separate AGN from extended emission, determine whether the sources are lensed or mergers, map line emission for dynamical studies and make measurements of rotation curves, and resolve individual clouds in nearby galaxies, essential for determining virial masses. Lower angular resolution is, however, needed in some cases, to give adequate surface brightness sensitivity and detect the total emission from extended regions – for example, to derive accurate Tully-Fisher measurements at low redshifts.

The dust continuum emission itself gives important information on components of galaxy masses. First, it is important to determine whether the millimeter-wave continuum is indeed dominated by dust. The dust mass depends on the temperature and dust properties, and a large wavelength coverage is required to obtain both. There could be several temperature components; cold components could contain much mass, and be missed from the observations. A possible case in point is APM08279+5255. This is a luminous lensed source at $z = 3.91$. The CO(4-3) and CO(9-8) lines indicate warm molecular gas, about 200-300 pc in size, with mass $\sim (1\text{-}6) \times 10^9$ $M_\odot$. Recent CO(1-0) and CO(2-1) observations however suggest a much more massive reservoir of low-excitation gas, extended over 30 kpc with total mass $\sim (0.6\text{-}3) \times 10^{11}$ $M_\odot$ (Papadopoulos et al. (2001). This illustrates that observations of both low and high excitation gas are required.

Accurate dynamical studies are made possible by the myriad molecular and atomic spectral lines in the millimeter and submillimeter wavebands. The "ladder" of molecular transitions makes it likely that a redshifted line will appear in one of ALMA's observing bands. The detection of two will determine the redshift. Already CO lines have been detected from some of the highest-redshift sources known (eg. Omont et al., 1996), and with ALMA this will become routine. CO lines from our own Galaxy could be detected by ALMA out to $z \sim 4$, and the 158 μm [CII] fine structure line will be detectable from star-forming galaxies across the Universe.

Molecular lines from high-redshift objects are particularly important, because at very high redshifts most of a galaxy's mass could be molecular, quickly enriched by starbursts. With ALMA, a wide range of lines will be available, covering all redshifts. Again, a wide frequency range is required to cover both low and high excitation gas. The molecular lines will make possible determinations of both molecular and dynamical masses, and studies of the high-redshift Tully-Fisher relation. Over ten high-redshift CO detections have so far been made, at redshifts ranging from 1.4 to 4.7, about half of them boosted by gravitational lensing. The implied masses of molecular hydrogen are in the range 10^9 - 10^{11} $M_\odot$.

ALMA will provide a wide menu for spectral resolution. For example, it will be able to cover a velocity range of over 21,000 km/s at 20 km/s resolution, and it will reach to spectral resolutions of $\sim$ a meter per second and better. The flexibility is important. On the one hand, adequate spectral resolution is needed to see structure and sub-components for dynamical mass estimates. On the other hand, broad bandwidths are required to search for and determine the systemic redshifts of quasars, to obtain adequate baselines for the study of broad spectral lines, and to detect components that are far apart in velocity. And optimal spectral resolution is needed for deep searches for objects of various masses.

Spectroscopy with ALMA will contribute considerably to the studies of gravitationally lensed sources. The fraction of lensed sources may be much greater in the mm/submm than in other wavebands because of the steep sources count. In the case of strong lensing, high resolution mapping of both the continuum and lines will be possible free from complications due to obscuration in the lensing object. Cluster arcs will be mapped at high angular resolution, greatly facilitating source reconstructions.

ALMA will revolutionize the study of molecular absorption lines in quasar spectra, making thousands of quasars accessible. An example is the $z = 0.9$ absorption system in the spectrum of PKS 1830-211, which was found purely by spectral scans at millimeter wavelengths (Wiklind & Combes, 1997). Such observations will make possible the study of the kinematics of intervening masses at very high spectral resolution, and probes of the obscuring region close to the AGN.

Quasars and AGN can also be studied "in depth", because of the low synchrotron and dust opacity and the unprecedented angular resolution of millimeter VLBI – tens of microarcsec, corresponding to spatial scales of $10\text{-}10^4$ Schwarzschild radii, the size of the expected accretion disk. ALMA would provide milli-Jansky VLBI sensitivity, corresponding to brightness temperatures as low as $10^2 - 10^4$ K, making the environs of AGN and extragalactic mega-masers accessible. Millimeter VLBI might even detect a gravitational shadow from the black hole in the centre of our Galaxy (Falcke et al., 2000)

In nearby AGN, the optically-obscured molecular tori and rings, and the circumnuclear starburst activity, can be studied in detail. The circumnuclear starbursts in most of the known Seyferts will be resolved – NGC 1068 will be observed with parsec resolution. The presence of central black holes can be studied kinematically in a large number of galaxies. Centaurus A will be a wonderful target for ALMA, providing a unique laboratory to study molecular gas in emission and absorption in a nearby edge-on active nucleus. All of these studies will have a direct bearing on unified schemes for AGN, fueling mechanisms, and the relation between AGN and circumnuclear starbursts.

Thus, in many ways, ALMA will provide a powerful new tool in the study of the dynamics and masses of galaxies and their components, at both low and high redshift.

Acknowledgements: I am grateful to F. Combes, A. Omont and A. Franceschini for helpful discussions. A particularly useful paper on the subject has been published by Combes et al. (1999).

References

1. Bertoldi, F., Cox, P. (2002), astro-ph/0201330
2. Combes, F., Maoli, R., Omont, A. 1999, A&A 345, 369
3. Falke, H., Melia, F., Algol, E. (2000) ApJ 528, L13
4. Omont, A. et al. (1996) Nature 382, 428
5. Papadopoulos, P. et al. (2001) Nature 409, 58
6. Shaver, P.A. (ed.) (1996) Science with Large Millimetre Arrays (ESO Astrophysics Symposia, Springer, Berlin)
7. Wiklind, T., Combes, F. (1998) ApJ 500, L129
8. Wootten, A. (ed.) (2000) Science with the Large Atacama Array (Publ. Astron. Soc. Pacific)

Author Index

Printing: Strauss GmbH, Mörlenbach
Binding: Schäffer, Grünstadt

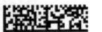